This is the first modern ethnography of the Murik, a relatively small but important community settled on the Sepik River estuary in Papua New Guinea. It is also the first full-length account of a non-Western culture to make comprehensive use of the conceptual framework of the Russian literary theorist, Mikhail Bakhtin. Among the Murik, as in other Pacific societies, women are conceptualized as the source of nurture, generosity and love. However, this conceptualization creates a kind of existential problem for men, who have political power, and their claim to sustain and reproduce society requires them to appropriate the nurturant qualities of women. So they must, in some sense, feminize themselves or model certain aspects of themselves after women. A "maternal schema" or "poetics of the female body" therefore underlines the sociocultural patterns of these societies. Lipset shows how this schema or poetics expresses itself in a number of different domains of Murik life: in kinship relations, life-cycle rituals, the men's cults, and in disputes and processes of conflict resolution. These issues are important for Melanesian anthropology more broadly and tie in with some of the major contemporary debates in the social sciences: the relationship between ideas of male and female power.

Cambridge Studies in Social and Cultural Anthropology

106

MANGROVE MAN

The monograph series Cambridge Studies in Social and Cultural Anthropology publishes analytical ethnographies, comparative works and contributions to theory. All combine an expert and critical command of ethnography and a sophisticated engagement with current theoretical debates.

A list of books in the series will be found at the end of the volume.

Founding editor:
Jack Goody: University of Cambridge

Frontispiece: A firstborn child, decorated with heraldry, is carried through Darapap by her mother's brother

MANGROVE MAN

Dialogics of culture in the Sepik estuary

DAVID LIPSET
University of Minnesota

CAMBRIDGE
UNIVERSITY PRESS

PUBLISHED BY THE PRESS SYNDICATE OF THE UNIVERSITY OF CAMBRIDGE
The Pitt Building, Trumpington Street, Cambridge CB2 1RP, United Kingdom

CAMBRIDGE UNIVERSITY PRESS
The Edinburgh Building, Cambridge, CB2 2RU, United Kingdom
40 West 20th Street, New York, NY 10011-4211, USA
10 Stamford Road, Oakleigh, Melbourne 3166, Australia

First published 1997

Printed in United Kingdom at the University Press, Cambridge

Typeset in 10/13 Monotype Times [SE]

A catalogue record for this book is available from the British Library

Library of Congress Cataloguing in Publication data

Lipset, David, 1951–
    Mangrove man: dialogics of culture in the Sepik estuary / David
Lipset.
        p.   cm. – (Cambridge studies in social and cultural
anthropology; 106)
    Includes bibliographical references and index.
    ISBN 0 521 56434 4 (hc). – ISBN (invalid) 0 521 56435 2 (pb)
    1. Murik (Papua New Guinea people)   I. Title.   II. Series.
DU740.42.L565   1997
306′.089′9912–dc21   96–50213 CIP

ISBN 0 521 56434 4 hardback
ISBN 0 521 56435 2 paperback

To
two brothers
Max and Michael
and
their mother
Kathy

Contents

x   *Contents*

Illustrations

Plates

Figures

Maps

Tables

Acknowledgments

This account of men's dialogues with images of womanhood in Murik culture has received many gifts. It has benefited, in the first place, from multiple periods of conjoint fieldwork undertaken by Kathleen Barlow and myself. Funding for the initial sixteen months of research (1981–2) came from the Graduate School of the University of California, San Diego, the Wenner-Gren Foundation and Sigma Xi. In 1985, Barlow and I organized "The Sepik Documentation Project" and got the opportunity to return to spend two months on the Murik coast with funds provided by the Australian Museum in Sydney and the Graduate School of the University of Minnesota the following year and then again in 1988. I made a brief visit by myself in 1993 under the auspices of the Fowler Museum of Cultural History at UCLA and the National Endowment for the Humanities. My tenure as a McKnight-Land Grant Professor at the University of Minnesota and the receipt of a Bush Foundation sabbatical grant there also contributed resources and release time which allowed me to prepare this book.

While the ethnographic present which I present is the 1980s and early 1990s, secondary sources upon which I have been able to draw go back to about 1913. When they are listed, indeed, these sources condense much of the twentieth century of Papua New Guinea into a paragraph. I have used Barlow's translations of three rich, ethnographic essays (1922/3; 1926; 1933) written by an Austrian Catholic missionary, P. Joseph Schmidt (SVD) who lived among the Murik for thirty years beginning just prior to World War I. At Margaret Mead's suggestion, Louis Pierre Ledoux, who was then a young Harvard college graduate, spent five months in a Murik village in 1936 and took about 600 pages of fieldnotes as well as an equal number of splendid photographs, both of which he made available to

Barlow and me in 1982. In the late 1960s, Frank Tiesler, the ethnology curator at the Staatliches Museum für Volkerkunde in Dresden, did an exhaustive review of missionary and anthropological literature on maritime and regional trade along the North Coast which cast the Murik as major players (K. Barlow, trans. 1969/70). During the 1970s, as the pace toward Papua New Guinea statehood began to quicken, indigenous Murik voices were heard for the first time. Sir Michael Somare's autobiography (1975) discussed the relationship of his statecraft to his membership in Murik culture. And Matthew Tamoane, a linguistics graduate of the University of Papua New Guinea under the sway of Seventh Day Adventism, wrote a nostalgic piece (1977) about his father's sister, who was one of the last household shamans.

Upon our arrival in Darapap village, Barlow and I immediately encountered the complexities of the Murik system of adoption, a system in which custody claims in children are negotiated and disputed on through adolescence. One particularly salient claim in this process is the personal names by which a child is called, the name being the property of the donor, which is a gift in an ongoing series of exchanges which define identity and therefore loyalty. As we were married, Barlow and I were adopted by two separate families, assigned genealogical positions in them and given local names (as were our two boys after they were born). Through these adoptive identities, and the Murik insistence that we behave "properly,' we attempted to learn what was entailed in maintaining them and slowly entered into an increasingly elaborate field of obligation, entitlement, courtesy, disrespect and avoidance, the very stuff, in short, of Murik personhood.

We did the best we could with the demands of Murik kinship. Perhaps our most effective tactic in this regard was learning to eat the Murik diet, which largely consists of seafood and sago jelly, concluded by areca nuts and tobacco. In so doing, we presented ourselves as culturally human, that is to say, as people who were ready to risk subordinating ourselves to the many implicit sociopolitical claims that are made through foodgiving among the Murik. We went along to work occasionally, but being unskilled canoeists, we could not but interfere with the day's catch. We assiduously presented ourselves at public events, rituals, village meetings, outrigger canoe departures and so forth. We spoke Melanesian Pidgin English (Tokpisin), the lingua franca which is widely used throughout the Murik Lakes and is becoming a first language to some children. We used the grammar of the Murik vernacular worked out by Stan and Jodeine Abbott, formerly of the Summer Institute of Linguistics (1978). While the level of our fluency in Murik may have satisfied the fancy of many Murik people, by the end of our third fieldtrip, we were continuing to learn to understand and speak the language, and still had to rely on translators, when the pace of discourse quickened.

Our debt to the Murik, who looked after us, bore with our cultural stupidity, occasionally took advantage of our clumsiness (see Chapter 6) and, above all, helped us to understand their lives, is irredeemable. Sir Michael Somare, "the father" of his country, and "native son," favored our interests from the very first week we arrived in Papua New Guinea, but his pride of place and hospitality were of a piece with that which we received from people in each of the five Murik villages in which we eventually worked. In Darapap, to which Somare sent us for logistical reasons, Minjamok and Murakau adopted me as Abu, Murakau's younger brother. Dakuk, Kiso and Sauma adopted Barlow as their daughter and me as their son-in-law. By no means did this work become key informant ethnography, however. I was in constant dialogue with many people. In Darapap, I want to mention the names of Lydia, Frankie, Tabanus and Daniel Wambu, Kangai, Pame, Sivik, Ginau, Marabo and Tamau, Bate and Kanjo, Tamoane, Yarong and Saimbu, Mwaima and Jamero, James Kaparo, Joel Gobare, Kaibong, Gidion Mwagun, Tekla, John Sauma and Luke Manambot. In Jangaimot, Pita Kanari and Komsing helped me. In Wokumot, Bai and Mwandekama were generous with their time. In Aramot, Wangi and Manag and Patrick Komba'u were my main informants. In Karau, Bujir and Jaja Kanung, Darai, Maiwa and Sarakena, Raphael Maiwa, Saub Sana and Anna, Akam and Maia were always willing to discuss their views of ongoing events with me. In Mendam, Boita and Kubisa, Aris and Nangumwa, Yamuna, Bate and Gobare were of particular help. Finally in Kaup, I spoke mainly with Sakara and Vincent Tanep. Several urban Murik were extremely forthcoming: Clara and Sailus, Paul Kangai, Ginau Mowe, Aupai, Sarewa and Matthew Tamoane.

While we were living in Darapap in 1981–2, the Summer Institute of Linguistics gave us permission to live in the house which the Abbotts had abandoned in the village. Jock Campbell, then the Coastal Fisheries Officer for the Department of Primary Industry in the East Sepik Province, provided information about the history of the Murik fishery. Several members of the Catholic clergy resident in Marienberg, Father Louis Kovacs, Sister Marianne Peer and Father Joseph Pierskall, gave us data about religious change and illness as well as lodging for several months in 1988. And, Wallace "Mac" Ruff, a landscape architect, has given me permission to reproduce the remarkable line drawings of the Murik buildings and village plans he did in 1981.

Although the moral status of anthropology in Papua New Guinea was in flux during the 1980s, Barlow and I had nothing but the most gracious dealings with our sponsors in the country: the East Sepik Provincial government, the Departments of Anthropology and Politics at the University of Papua New Guinea, the Institute for Papua New Guinea Studies and the National Museum and Art Gallery. The hospitality of Barry Craig, then the anthropology curator at the National Museum, and Pamela Swadling, its prehistorian, introduced us to the art and geomorphological background of the region. Siroi Marepo Eoe, its head, gave us unqualified support and seconded John Salau, then an assistant curator, to work with us in 1986 on a regional survey of exchange.

Since this book remains a palimpsest of the doctoral dissertation it once was, I would be remiss to omit mention of my teachers from those distant days: F.G. Bailey, Gregory Bateson, Raymond Fogelson, Anthony Forge, Gananath Obeyesekere, Fitz John Porter Poole, Donald Tuzin, Ted Schwartz, Mel Spiro, but, most of all, Michael Meeker. Each of these men helped me, whether wittingly or not, to understand how I did and did not want to do anthropology.

My colleagues in the Department of Anthropology at the University of Minnesota, Stephen Gudeman, John Ingham, Mischa Penn, Gloria Raheja, and especially Eugene Ogan encouraged me with dialogue and more material forms of support. My students, Eric K. Silverman, Paul Spicer, Lori Jervis, Susan Schalge, Antonia Schluter and Jolene Stritecky listened to, and also argued with, me about things Murik, as did Jamon Halvaksz, who also prepared the index.

The clarity of the argument I have achieved in this book, such as it is, has benefited from criticism by Simon Harrison, Paul Roscoe, Jessica Kuper and two anonymous readers at Cambridge University Press.

I simply cannot reduce the complicated contribution that Kathleen Barlow has made herein to a clause. All I can say is what I always say in one way or another, which is that this minimal acknowledgment and expression of gratitude is wholly inadequate, for which I am sorry.

1

Introduction

In March 1981, having just arrived in the Murik Lakes, I went to a meeting of men from two feuding villages. They had assembled in the male cult house of a neutral community to begin to resolve a recent sequence of brawls which had broken out between their respective youth. The men of Darapap, numbering ten or twelve, arrived first and engaged their neutral hosts in a riotous comedy, the content of which was far beyond my comprehension. However, new as I was then to the culture of Murik men, and uninformed about its meanings, what they did made a lasting impression.

One elderly man hobbled into the male cult house with the aid of a staff, stood to rest for a moment a few steps inside the doorway, surveyed the scene until he spotted a junior man lying motionless on the floor, supine. Darting across the hall, the man launched himself crotch first onto the face of the prone figure. Squatting over him, he began to bob up and down, as he shouted something about the younger man's penis. Another man rubbed his buttocks on the leg of a youth. I saw a man lift his leg over someone's head. Others groped each other in the genitals. At the same time, the men engaged in a mock repartee which I gathered (because it was partly spoken in Tokpisin) was about adultery. I did not know what to make of the gleeful yet casual effrontery with which adult men called each other "dickheads," "pricks" and "lechers." I nonetheless felt a distinct sense of *déjà vu* and asked Marabo, a relatively senior man sitting next to me, if mothers' brothers were joking with sisters' sons. I could hardly contain my astonishment when he told me that they were.

Despite this man's verification of my guess, I knew nothing about these abuses, profanities and improprieties (or even the extent to which the Murik considered them as such). I understood little about the acceptability of "this peculiar combination of friendliness and antagonism"

1

(Radcliffe-Brown 1965b: 91) in that, or any other, context. My knowledge of the sacred male space in which the meeting took place, or about the organization of kinship or marriage among the Murik, was nil. Except for a vague awareness that some sort of peacemaking ritual was underway, I knew nothing about the background which had necessitated these negotiations. Amid the anxiety created by such ignorance (Devereux 1967), my initial encounter with Murik slapstick was uncanny, at once familiar yet unnatural.

Uncanny? Why uncanny? Because I had read about something like it in *Naven*, Gregory Bateson's (1936) classic monograph of ritual and society among the Iatmul, a people living upriver from the estuary-dwelling Murik. At the culmination of the rite for which Bateson named that book, a transvestite man paid tribute to his sister's son by rubbing the cleft of his buttocks along his nephew's shin. Put simply, Bateson's complicated, functional analysis of this tributary gesture was that its performance inhibited the outbreak of structural tensions in Iatmul society through the force of its condensed poetics. The alternative to the clefting gesture and the rites it climaxed was an unchecked escalation of conflict into violence – "symmetrical schismogenesis," Bateson called it – among individuals or groups of kinsmen who, by virtue of their bravado and weak affinal ties, rejected the performance of any form of tribute as culturally demeaning to their gender identity.

Of the many responses to Bateson's difficult, but prescient, analysis of *naven* rites,[1] rethinking the clefting gesture *per se* – locally viewed as a male pantomime of the maternal body during sexual intercourse and birth – has not been among them. I myself had given a careful reading to Bateson's Iatmul ethnography while preparing his biography (Lipset 1980) but did not expect to discover this contradictory, androgynous image of reproduction when Barlow and I arrived in the Sepik estuary. The uncanny experience of seeing *naven*-like behavior in a context of social control excited me and gradually I turned my attention[2] to the ethnographic puzzle Bateson and others (Mead 1935; Hogbin 1970; Forge 1966) had found in the Sepik so many years earlier: namely, the equivocal relationship between men and the maternal body.

Masculinity, the maternal body and hidden dialogue

In this book, I argue that in Murik culture the dynamic in this relationship is a particular kind of dialogue. Underlying the ethics, satire, crises and expiatory experiments by which men understand their part and participation in cultural relationships is a set of values in personified form. A prototype of the body, which I call a maternal schema,[3] is taken for granted as Murik men think about and make order out of lived experience. As they create or renew moral relationships, for example as elder brothers, or in positions of ceremonial leadership, they see themselves as possessing qualities or attributes they associate with a chaste mother who is surrounded by hungry, dependent children in need of "her" nurture, hygiene, protection and instruction. This schema may take on different, but identifiable, forms depending on the context. Missing values may be filled in completely or partially, or inverted, for example, in the kind of obscene joking relations depicted above. Actors may explicitly acknowledge being guided by the maternal schema in certain situations, such as with their inland trading partners, while in others they presume it and do not admit of "her" prototypical guidance, as when reinventing a rite of reconciliation. What is culturally distinctive about "the mother" is that, while the exterior and visible conduct of her mothering offers to them a model for moral behavior and order, qualities and substances inside of her body, particularly her powers of sexuality and fertility, the men stigmatize.

The most important ethnographic argument I shall make is that this inner/outer split in the maternal body is part of a "hidden dialogue" in terms of which men think through and negotiate the reproduction of Murik culture. In a hidden dialogue, "the statements of the second speaker are omitted, but in such a way that the general sense is not at all violated. The second speaker is present invisibly, his words are not there, but deep traces left by these words have a determining influence on all the present and visible words of the first speaker" (Bakhtin 1984b: 197). This book analyzes contexts in Murik culture in which men are that present speaker. No matter how reified or authoritative they would purport themselves and their voices to be, I argue that "their masculinity" remains a conditional and stylized disposition in the shadow of a barely disguised figure. The mysteries definitive of the Mangrove Men are marked by paradoxical metaphors, rather than assertions of unilateral control. Ideas of domination, autonomy and so forth fail to apprehend the complexity of their engagement with and responsiveness to what I am calling the maternal schema.

If dialogue is a more appropriate characterization, I need to specify and clarify the kind of dialogue to which I am referring: it is not a literal one. I shall not be concerned to analyze the conversation of men actually speaking with women or vice versa, to analyze women talking to them (but see Barlow and Lipset 1997). Nor will I be concerned with whether or how the maternal schema may be differentially presumed or answered by men and women. Nor do I mean to suggest that only men respond to it while Murik women do and say little more than present an image of their bodies to men. Empirically, nothing could be farther from the case (see, e.g., Barlow 1985a; 1992; 1995). I am relying upon the idea of hidden dialogue as a metaphor and a methodology to explain why the meanings of men's practices and performances coalesce to form an ambivalent whole. The men's answers are not to women *per se*, but to and about a culturally particular image of womanhood.

The view of dialogue to which I subscribe here is part of a more general theory of discourse and culture developed by Mikhail Bakhtin, the anti-Stalinist, Russian semiotician. As its critical framework is neither intuitively nor conceptually obvious, it is necessary to present a brief exposition of what he meant by "dialogism." According to Bakhtin, culture is saturated with multiple, equally authoritative discourses whose "essence . . . lies precisely in the fact that . . . [they] remain independent and, as such, are combined in a unity of a higher order" (Bakhtin 1965/1984a: 21). Multiple points of view, which are not kept in check by self-regulating mechanisms or terror, make up a contrapuntal unity. At the same time, dissent is neither random nor revolutionary. Dialogized thought and speech are molded by objective and stable forms of expression – genres – which are a "means of collective orientation in reality" (Bakhtin 1978: 131–5; 1981: 249, n. 17). Culture therefore consists of grand, indissoluble ambivalences. Centripetal and centrifugal voices contest each other *within generic forms* to make up open-ended, rather than canonical, wholes. Personalistic imagery of the threshold between freedom and dogmatism, or between irreverence and piety, that appeared in the human states of becoming, particularly absorbed Bakhtin. Laughter, eros, birth and death preoccupied him. The mockery of fools, novels whose heroes were independent of their authors' intentions, the fantastic temporality of carnival, fascinated him. Characteristic of the fiction of Rabelais and Dostoevsky, the subjects of two of his main works, was not just dialogicality but equivocality. What Bakhtin said of Dostoevsky applies to himself: "In every voice, he could hear two contending voices, in every expression a crack, and the readiness to go over immediately to another contradictory expression; in every gesture he detected confidence and lack of confidence simultaneously; he perceived the profound ambiguity, even multiple ambiguity, of every phenomenon" (Bakhtin 1984b: 30). For Bakhtin, the unity of culture is and must remain paradoxical: the moral diversity in the nineteenth-century novel was irreducible.

Anthropologists have tended to neglect two related dimensions of his position (see, e.g., Tedlock 1983; Bauman and Briggs 1990; Bandlamudi 1994). One is that dialogicality in discourse, or in discursive performances – the view that the meaning or design of expressive action is always formulated in response to the other – does not require the physical presence of an interlocutor. It rather becomes all the more powerful in the literal absence of the other, who nonetheless goes on exerting unseen material and historical effects on language and meaning. The second is that Bakhtin not only loathed reigning political (or theoretical) centers which he condemned as monologic, but also viewed dialectical progressions of order and conflict from which new, totalizing syntheses emerge as no less objectionable. His concept of culture was not teleological (Kristeva 1967: 58; Todorov 1984: 104). What he extolled in discourse were antithetical "two-in-one image[s] . . . after the manner of the figures on playing cards" (Bakhtin 1984b: 176) which admit to no finalizability.

But what might such a nonliteral, nonteleological view of dialogue and culture, a view which, in a sense, is "really" a liberal attack upon repressive state power via exegeses of Western poetics and a critique of structural linguistics, possibly have to do with rural Papua New Guinean men? I see several kinds of relevance. Instead of the orderly metaphor of function, or the moral metaphor of text, it offers a more prosaic, human image of culture. Institutions, discourse cannot even be conceived outside of semiotic relations between self and other, relations that are presumed to make up a rupture-prone, unfinished environment. Instead of privileging a local voice (e.g., one gender), or a foreign one (e.g., a postcolonial regime of value), as either noble or demonical, dialogism would seem to emphasize ambivalent systems of meaning that accept no resolution. An objection which could and ought to be raised here is whether a plurality of rejoinders all made within a single schema of images and values does not contradict the Bakhtinian notion of unfinalizability in culture. This is not, of course, a problem that would bother Murik actors. The relationship between the plurality of lived reality and generic, taken-for-granted, assumptions about discursive form, which do indeed limit fields of representation and do aim for completion, raises a thorny methodological problem for a dialogical view of Murik culture. Bakhtin of course did distinguish between more and less dialogical genres; but even ones that were most dissonant and "free," such as carnival, were still rejoinders to an official world.

I should nominate one further advantage of the dialogical approach to ethnography. Instead of a methodology of the reader unilaterally discovering patterns of meaning in symbols and action as if culture was "poetry"

(Geertz 1973: 443), understanding "burn[s] from the borrowed light of alterity" (Bakhtin quoted in Todorov 1984: 100). Knowledge is and can only be known in relations from without, through the irrevocable position adopted by all ethnographers, that of the outsider. Thus, in its vision of being as always in discursive relation to the other and in its rejection of unilateral voices, Bakhtinian dialogism offers a constructive, inexhaustible methodology that is appropriate for the study of changing Melanesian cultures, where, as Leenhardt (1979: 153) once remarked, the person was understood as an "empty" intersection at which the others to whom he or she is related meet.

The monologic body

My ethnographic project here is to analyze the terms in which men think through, live, create and negotiate culture, as if they were responding to a particular schema of the maternal body. Neither in Melanesia, nor anywhere else in the world, however, has this relationship been understood as a dialogue, hidden or otherwise.[4] Rather, the body has been unilaterally defined by "fundamental principles of . . . culture" (Bourdieu 1977: 94).[5] According to Emile Durkheim, a founder of this unilateral view, persons have two bodies: a high moral one, and a low physical one. The latter, possessed by the individual, is profane, if not simply evil. The former, possessed by society, is a *tabula rasa* for sacred representations of collective, moral authority (Durkheim 1915: 307). Beliefs about comportment – say, about courtesy, or the appropriateness of laughter – convey the degree of control exerted upon the individual, or a lack thereof, by collective institutions (Radcliffe-Brown 1965b;1965c; M. Douglas 1975: 87). Alimentary and gestative processes conceptualize moral process itself (van Gennep 1914). The duality of good and evil is projected onto duplicated limbs and bilateral symmetry, the body's left and right sides, as well as onto its vertical posture, whether low or high (Hertz 1960). Short hair symbolizes and publicly communicates sexual restraint or repression (Leach 1958; cf. Hallpike 1969). Substances entering or exiting through the thresholds of the body – foods, sexual fluids and so forth – define the moral status of a person (Frazer 1922) in accordance with the legitimate norms of society. On the one hand, "the uncontrolled orifices of childhood and senility and the unmediated flow of menstrual blood widely signify infrasocial states of being, and a less than optimal containment of the person within his/her bodily margins. On the other hand, bodily closure [often] signals a clearly distinct, centered identity and the capacity to engage in stable . . . relations with others" (Comaroff and Comaroff 1992: 73). Anthropologists, it seems, have assumed that categories of moral order impose themselves on the body. Bakhtin would have called this approach monological.

In Melanesia, the terms and idioms of this monologue inevitably arise from symbolic relationships between men and the feminine body. I want to cite two brief examples of what I mean. Among the Hua of the eastern highlands, certain foods were deemed to have toxic effects on the health and agency of initiated men. The men used to eat no possum, because its hair resembled women's pubic hair, and its odor was like that of menstruating women. Possums, moreover, lived in holes that looked like vaginas and bore children like human women. Eating possum, men believed, made them pregnant (Meigs 1976; 1984). My second instance comes from Wogeo Island, just a few miles offshore of the Sepik River, where men and women were understood to pollute each other. Women menstruated, men knew, to cleanse themselves. Men, for their part, used to isolate themselves from women and bleed their genitals, an ablution which they likened to male menstruation (Hogbin 1970). There is no question that ritual practices of men could be rethought as if in dialogue with culturally variable concepts of the feminine body for the whole of Melanesia, if not the entire insular Pacific. In addition to its analytical and empirical value, such a survey would vindicate the contructivist theory Margaret Mead made famous in her Sepik tryptich, *Sex and Temperament in Three Primitive Societies* (1935); namely, that gender is neither a natural nor an inevitable category of bodily meaning. More theoretically, it would demonstrate how understanding "the feminine" functionally, namely, as serving reigning masculine authority (Herdt and Poole 1982; M. Strathern 1988a), obscures the astonishingly complex androgynous stylizations in which the latter realize their agency and positions in society.

Feminists and "the feminine" in simple societies

The functional relationship of the feminine body to male authority in technologically simple societies was the central problem that preoccupied feminist anthropologists during the 1970s and 1980s. Relying upon an analytical framework adopted from Meyer Fortes, they assumed that society could be divided into "domestic" and "politico-jural" domains in which the latter encompassed or dominated the former (Fortes 1969: 72; cf. p. 100). Michelle Rosaldo observed that, typically, "the gender" of the subordinate domain was feminine, although in differing degrees, while the superior politico-jural, or public, domain was masculine (1974: 23; see also Lamphere 1974; Ortner 1974). Debate about Rosaldo's dichotomy ranged from outright empirical rejection of it to identifying variations, whether balanced, unisex, segregated, or hierarchical.[6] Ortner and Whitehead (1981) then refined it by contending that kinship, gender and political authority ought to be viewed as parts of a single, variable field of meaning, rather than as separate domains. Nevertheless, the politico-jural domain, which they now called "prestige structures," or how men established and legitimated conjugal rights (see also Evans-Pritchard 1965), was still said to determine or explain the feminine. Collier and Rosaldo (1981) devised a typology of prestige structures. The *bridewealth* type included large sedentary economies of food producers among whom social standing was won and validated through affinal exchanges of valuables between kin groups. The *brideservice* type referred to small, nomadic, hunter-gatherers among whom men initiated and validated conjugal relations through a groom's reciprocal provision of a woman to his spouse's kin in addition to working for them in the course of daily life.[7]

Marilyn Strathern acclaimed the Collier/Rosaldo typology as having distinguished nothing less than a "profound symbolic shift" in modes of collective representation of the feminine (1985a: 198; see also 1988a). When, for example, a groom gave meat to his affines as an act of brideservice, his gift would not square a debt. It was rather part of a continuous claim a man had to make in his wife. Neither labor nor any transfer of wealth would stand in any categorical or aggregate sense for this, or any other, relationship in such a prestige structure. Only the exchange of another person, usually a sister, and the other services provided by the groom, could create and maintain a marriage (cf. Kelly 1993). In the bridewealth type, by contrast, exchanged valuables do create new relationships, or mediate old ones. The transaction of wealth-objects for women, in other words, allows for systematic cultural differences in how the feminine might be understood. Woman is no longer an embodied cluster of indivisible capacities, as she is construed in brideservice systems, but, being more thing-like, "she" can indeed be represented as a "gift" given from one group to another (see also Lévi-Strauss 1969). Her "labor may be conceptualized as detachable from the person and, like [her] fertility or sexuality . . . a disposable asset" (M. Strathern 1985a: 198). That is to say, as well as the wherewithal to achieve prestige, Strathern concluded that bridewealth-based prestige systems yield metaphors for a different kind of feminine body.

Sherry Ortner (1981; cf. Howard and Kirkpatrick 1989) distinguished a third type of feminine status and body (see also Collier 1984; 1988; and Collier and Yanagisako 1987). Western Polynesian women, she observed, held elite status in society along with men. They maintained politico-jural status, either as sisters or fathers' sisters, after they married; and did not lose ascribed rank. Ortner was not disavowing her earlier argument developed with Whitehead (1981) which had explained the feminine in terms of how men legitimated conjugal rights. Marriage might otherwise account for womanhood in western Polynesia but did not because a more important social value – being high-born – superseded it.

Some years later, however, Ortner (1990) did indeed argue that women's status might be determined by multiple, possibly contradictory, historically specific "gender hegemonies" rather than a single, timeless prestige structure. This refinement was part of a general movement among feminist anthropologists of the late 1980s and early 1990s away from the Fortesian assumption that politico-jural authority was "a dominant" (Jakobson 1971: 82) that defined or silenced women. Some feminists had now begun to focus on the intricate lives of self-possessed, sometimes insubordinate, women living at the margins of local authority and the world system (Geiger 1986; Abu-Lughod 1986; 1993; Tsing 1993; Behar 1993). Their studies sought to extol and commemorate "voices claiming no final authority" (Ong 1987: xv; Serematakis 1991; see also Abu-Lughod 1990) not only as resistant but as "creatively constructing a complete social world" (Behar 1990: 229).

Perhaps they had begun to develop a more dialogical view of the feminine. But the legacies of adopting the domain model, and contending with oppressive, androcentric forces, left them with a somewhat limited appreciation of contradictory images of the feminine. "Women's . . . commentary . . . shows a range of response – acceptance, resistance, subversion and opposition – to dominant, often male discourse" (Gal 1991: 193). Certainly, the feminine was no longer being viewed as a Pygmalian subject in the 1990s, created and defined by masculine centers of authority, but as an independent voice in the great dialogue (see also M. Strathern 1988a: 57, 127). But is theorizing the feminine exhausted by substituting a metaphor of defiance or autonomy for one of muteness? If, as Bakhtin argued, culture is a contrapuntal and unfinalized environment, then other, more ambivalent images of the feminine than those handed down either by male prestige structures or created by women resisting them ought to exist. Such refractory imagery is the gift that Pacific island men have presented to Western social thought. For in them, the reference of the feminine is neither to wealth nor to the politico-jural authorities who transact it. This reference "cast[s] a sideward glance to an absent interlocutor" (Bakhtin 1984b: 206) living in and presiding over another context. Rather than women viewed in terms of men, or as if living apart from men, here we find the mysteries of culture expressed in androgynous, double-voiced, images of agency that do not lie beyond dialogue but exude polyphony (see, e.g., Mead 1935: 256).

In an otherwise perceptive reanalysis of exchange theory based upon an interpretation of Trobriand mortuary rites, Annette Wiener (1978; 1979; 1980; 1983) overlooked this point. Ritual yam exchanges were there understood as a phase of "reproductive process" (Wiener 1979: 330). Because Wiener made the then orthodox Fortesian assumption that politico-jural institutions organize domestic relationships, she did not conclude that this image might be part of a larger cultural dialogue. Among the matrilineal Trobriand Islanders, ritual exchanges explicitly disperse to men, or to one descent group instead of another, procreative forces that regenerate society. Wiener even went on to discover the image of reproduction in Melanesian societies practicing every conceivable lineal principle, but never recognized the salience of the maternal body in which it is located. However masculine in voice, the hidden dialogue Melanesian men carry on with embodiments of "woman," particularly in "her" variable role as mother, provides them with artifices through which authority over and agency in culture may be asserted (Meeker, Barlow and Lipset 1986).

The Mangrove Men

My empirical goal is to illustrate the relationship between Murik men and the maternal body as a Bakhtinian dialogue. There are several reasons why they provide an excellent case in point. To begin, the Sepik River spills out onto a great cultural frontier. Here institutions that recall non-Austronesian-speaking inland New Guinea, and its deeply misogynist male cults, meet the stratified, more gender-equal cultures of the Austronesian-speaking Pacific. Among the groups living in the estuary of the great river, without doubt, the Murik are the leading intertribal actors. Although Murik culture is distinctively "Sepik," that is, distinctively non-Austronesian, the gaze of the people is nevertheless fixed on the Schouten Islands offshore as well as along the coast, rather than upriver. Their fishery is supplemented by extensive overseas trade. Subsistence, and the attribution of political status, depend upon access to intertribal relationships and resources. Today, the Murik engage this cultural frontier in motorized outrigger canoes, voyaging to hereditary trading partners or to market in the provincial capital.[8] Rather than living self-sufficiently, the pivotal role Murik men have played in this region has underscored representations of interdependency, representations that admit rather than resist their prototypical imagery.

Further, prestige in Murik is distinctive for combining concepts of personhood associated with both the brideservice and bridewealth types differentiated by Collier and Rosaldo. Marriage among them is ritually unmarked. Perhaps the people once practiced sister-exchange. Today, they rarely bother with it, although they do go on expecting brideservice. Politico-jural rank is vested in titles and ornamental regalia that are held by a gerontocracy. Senior firstborn brothers and sisters legitimize ceremonial status by organizing exchanges of ornamental wealth. In the male cult, however, exchanges of objects did not win rank. Rank was contingent upon services provided by married couples to its reigning authorities, services which were understood to square ritual debts. The logic of the brideservice/bridewealth typology is thus confounded. But so is Ortner's western Polynesian model in which women's status is explained by aristocratic birth rather than marriage. In Murik, men and women – as firstborn siblings – seek and deploy authority *together*. But though their prestige is ascribed by birth order, it is eventually won or lost via the transaction of *both* goods and services. In terms of gender, Murik is not a highly stratified culture, much less a misogynist one. Relations between male and female are rather understood to be deeply interdependent. What do the Mangrove Men admit? "'All power,' said the ancestors, 'comes from women.'" Their revelation motivated no strategy of cooptation or denial. It acknowledged, indeed venerated, a notion of agency – which reflects the influence of a culturally particular maternal schema – that men contrive, albeit ambivalently, to answer. What is exemplary is their candor about their relationship to this image.

A feminine image of interdependency – of a self that forever remains part of a maternal other – is elaborated by Murik men (Meeker, Barlow and Lipset 1986). Within the boundaries of collective space, ideals of nurture, care and help for the other coalesce in a schema of motherhood. Outside of these boundaries – on the beach or during overseas trading expeditions – the force of maternal norms withers. Representations of the self as opposed to the other in strategies of self-assertion become possible for men and youth, less so for women and elders. Circumstances of normative boundedness, in other words, would seem to shape how the self apart from others may be imagined. Among the Murik, the potential for extrasocial or antisocial opportunities at or beyond the margins of collective space yield very few images of a self opposed to or separated from the other.[9] Neither warfare nor work offer unambiguous metaphors for such representations. Instead, these realms of experience only further confirm the salience of the self as part of the maternal other.

Mangrove Men struggle to marshal resources which constitute mother-hood. The artifices of agency for which they contend often appear to be tactics that amount to an ethical claim that they – in addition to women – reproduce, sustain, and make the community prosper. Their control is an "artifice" because its symbols foreground what is not the case while back-grounding what is true. One of the signature tactics by which Murik men claim to represent agency in themselves is anthropomorphized gift exchange. However, the moral significance of their transactions is not as universal as the romantic Mauss wished they were (1925). Exchange, as a system of cultural meaning, only reflects the relative poverty of extrasocial or antisocial opportunities in their normative circumstances. Apparently, sensitized by ethical deficiencies, men are led to try to figure capacities of women as masculine. Gift-giving, I argue, is part of a broader strategy in answer to an image of interdependency whose quintessential exemplar is found, not in them, but in mothers.

Recall that Mauss (1925: 9) derived his three "laws" of prestation – to give, to receive and to return a gift – from the Maori *hau* spirit contained in forest game, the spirit which obliges a recipient to return a countergift to a benefactor. Rethinking the *hau* concept, Sahlins pointed out that the latter gift was subsequently used to compensate the forest spirits for that which the warrior/hunters had taken from them; their sacrificial offering worked to preserve the abundance and fertility of the forest (1972: 157–8). In addi-tion to obligating reciprocity, the *hau* implicated the presence of a

nurturing feminine other. If . . . there is a metaphor of maternal nurturing behind the *hau* of the forest, it is interesting that we discover this metaphor in the domain of hunting, where men's role as food suppliers is set apart from their relationships with women. For it is precisely in this context, where the powers of men as food suppliers are unchallenged by women, that their claims over feminine powers can be most confidently figured.   *(Meeker et al. 1986: 66)*

Like the Maori gift, which repays a generous, maternal other living in a masculine space, the forest, Murik men "fight" to outgive other men during feasts. They deploy resources to "counter the disturbing fact . . . [that] feminine powers . . . [which] exemplify a [moral] relationship between . . . self and other" are foreign to them (Meeker et al. 1986: 24). Men thus appear deficient with respect to other men, rather than women. But their ritual displays of abundant nurture only expose this ethical defi-ciency all the more, since the feasts elicit and confirm the very ethos of interdependency men lack. Notably in Murik, such events bristle with masculine frustration and are often subject to outbreaks of violence.

In other strategies, Murik men once sought to assert control over women's bodies through marriage rules. But these claims also failed. They inevitably remained triangulated: wives went on being daughters and sisters who were divided by their sexuality but not by sentiment from kin. Relations between father/brother and husband resisted institutionalized attempts to deny sexual tensions through incest prohibitions, sister-exchange and affinal courtesies. Marriage did and does not integrate the community but disrupts it. Women lost through marriage can never be replaced. Murik men go on feeling a chronic deficit of love and commit themselves to achieving the impossible goal of controlling women's sexuality (see Chapters 8 and 9). They seek to monopolize this "scarce resource" through the use of covert devices, seductive constructs which magically compel women's love, or subvert the amorous advances of rival men. And through the male cult, they tried to detach themselves emotionally from women's sexuality (see Chapter 7). Mangrove Men once sought "freedom" from their sexual dependency upon women. But the achievement of this very limited sort of autonomy was impossible. The qualities with which women are culturally endowed went on challenging their masculinity. Masculine assertions of agency and authority remained compromised by women's bodies. Processes and attributes ethically associated with women – e.g., nurture, instruction and, most of all, peace – are evaluated as honorable and prestigious. Processes and attributes ethically associated with men – e.g., collective aggression – recede into secrecy. The result is a rupture-prone, unfinalized dialogue about the reproduction of moral order.

I have divided this book into three parts, each of which addresses different positions in this dialogue. Part I has four chapters. The first assesses contrary interior and exterior images of womanhood in Murik legends, ethnohistory, village organization and work. The second analyzes reproduction beliefs, in which the uterine body is stigmatized, and juxtaposes them with sibling norms, in which the role of elder brother is defined as indulgent and as emotionally inhibited. The third chapter then argues that this juxtaposition is not just heuristically convenient but reflects a culturally dialogical relationship; the values of elder brother are epitomized by ceremonial leaders, the senior men and women who respond to the uterine body by ritually reproducing the groups they lead. They do so by decorating children in ornamental regalia. A case study of succession, which I present in the last chapter of Part I, goes on to document a last, crucial attribute of the maternal schema: while the sexuality and procreative forces in "her" interior body are conflict-ridden, "her" exterior form is all facade, a hollow, androgynous "canoe-body."

Part II consists of two chapters in which I analyze the male cult as an ambivalent rejoinder to the maternal schema. Chapter 6 contrasts a masculine, grotesque body that appears in its satirical discourse with the unconditionally maternal role men assume during ritual. Chapter 7 discusses how the men's most powerful war spirits, whose aggression used to require the ritual prostitution of women, were a training against sexual dependency. The history of this institution, I argue, is a dialogue between local sociopolitical organization and early childhood experience, on the one hand, and colonial/postcolonial repression, on the other.

Lastly, Part III concludes with two case studies of conflict resolution. The first one is an extended account of how, following the brawls mentioned at the outset of this Introduction, Murik men recreated their maternal schema in the course of negotiating a peacemaking rite. The second case reprises this process, except that courts and constables of the state were brought in to arbitrate, which remedy failed to resolve the conflict. The relationship between local processes of social control and those of the state, I suggest in conclusion, was no less dialogized, no less ambivalent, in the 1980s and early 1990s, than the relationship of Murik men to the schema of the maternal body.

PART I

DIALOGICS OF THE MATERNAL SCHEMA AND THE UTERINE BODY

Plate 1: Jamero of Darapap: central in the schema through which Murik order is lived, contested and created is a culturally particular image of a nurturant, maternal body

2

A predicament in space

As seen from a motorized outrigger canoe making its way along the coast from the west, the estuary region of the Sepik River (Map 1) initially looks like glimmering specks on the horizon rather than a single body of land. Gradually, these images take on the appearance of substance and depth. Chugging along the Murik coast, one sees a bleak, 25-mile stretch of grey beachsand, littered with occasional shrubbery, vines, coconut palms, and stands of casuarina trees. Just behind it are vast inland lagoons. Consisting

Map 1: The North Coast region of Papua New Guinea

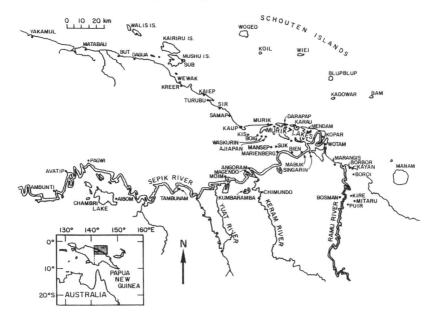

of approximately 135 square miles of shallow, brackish waters, the Murik Lakes are inundated daily by marine tides as well as the seasonal flooding of the Sepik (Plate 2). Several strong currents simultaneously flow through two large channels which make navigating into the lakes a somewhat hazardous, although not particularly daunting, challenge for Murik sailors. Once through the channel, the sight of an open, and relatively calm, translucent body of water offers relief. However, to individuals, or small groups of people, who paddle about these waters to fish, collect firewood, travel to buy sago from inland trading partners, or to visit kin in other Murik villages, that is to say, to people not returning home from foreign ports, but leaving communal space, the lakes provide the setting for a variety of discursive experience and sentiment. During the dry season (May–October), they are simply viewed as a resource which tests the mettle of the fishwives and fishermen who ply them. But at certain times of the day, even during this season, they can be more frightening than utilitarian or beautiful. Paddling alone, some people fear the dusk, when ghosts, or malevolent, capricious spirits, who lurk in whirlpools, within trees, or in the bodies of sharks and saltwater crocodiles, may attack. Particularly during the dry season, the lakes can also offer an opportunity for individual expression. The verdant mangrove forests which line them serve as a text for masculine

Plate 2: The Murik Lakes

bravado. Young men carve the faces of their war spirits (or today their initials) in bark and climb up the taller trees to strip off the branches beneath the treetops to signify their courage. But the lakes may also grant repose from the tensions and pressures of collective life and prompt nostalgia for love lost, or longing for the departed. For here there is privacy; here individual men and women compose verses of *woyon*, the "songs of the mangroves."

The Murik Lakes are divided into broad lagoons and narrow channels by thick forests of *rhizophora* mangroves (see Plate 2), the trees which are adapted to the intertidal zone between mean sea level and the highest tides (Recher and Hutchings 1980). The slimy floor of the lakes is permanently waterlogged, rich in organic matter but low in oxygen. In order to reproduce and grow, the mangrove tree has had to solve the problems posed by this world of water (see Map 2). Its propagules float in order to lodge in shallow water before being swept away by winds and the tide. And its root system forms an intricate network of plank buttresses, the surface and aerial extensions of which resemble a tangled thicket of fingers, legs and knees. Rising up as much as six feet above the brownish mud, these roots tend to jut out from the trunk almost horizontally and then arch

Map 2: Ecology of the Murik region

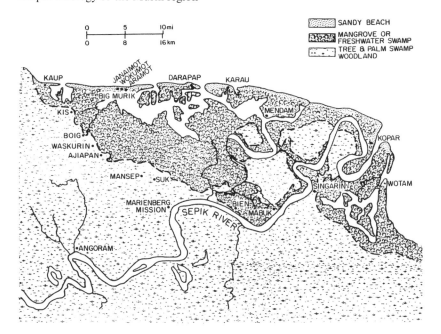

downwards toward the muck, branching repeatedly in different directions away from the trunk, forming loops of diminishing sizes. Occupying a niche between land and sea, the adaptation of the mangrove forest to this intermediate riverine-lacustrine-marine ecology is notable for its exquisite permeability. Birds, mammals, reptiles and insects enter these waters from land, and with high tide. Saltwater fish swarm through them feeding upon crustaceans, mudcrabs, worms and more than two dozen species of mollusc which also inhabit the slimy, glutinous mud. Suffice it to say, the people who live here acknowledge their identity with the mangrove swamps, calling themselves *Bar Nor*, which means "Mangrove Man."

After a brief survey of the prehistory of this region, I will begin to sort out the contrary images of womanhood in terms of which the people who preside over this world of water understand their history and exploit its resources. I shall introduce Murik social structure, as well as the sexual division of labor, in which moral elements in the maternal schema appear to stigmatize interior aspects of "her body," namely feminine sexuality and male aggression. The chapter ends with a summary of the effects of twentieth-century economic change on this dialogue.

An intertidal frontier

Any study of a Lower Sepik culture should begin with mention of the massive geomorphological transformations in the Sepik–Ramu basin that were recently discovered by Swadling (1989; 1990; see also Roscoe 1994). According to her chronology, after the last glacial period reached its peak about 14,000 years ago, the earth warmed up and ice melted. Sea levels rose as much as 140 meters, flooding the basin. Until about 6,000 years ago, this was a saltwater estuary, an enormous inland sea lined with mangroves, supporting coral reefs and associated marine life. During the next 4,000 years, the sea receded. Silt washed down from the mountains north and south of the basin and slowly infilled the estuary until the extensive floodplains and backswamps which today compose the river basin began to form. By 2,000 years ago, the inland coastline was probably located at what is now the confluence of the Sepik and Keram Rivers (see Map 1). In other words, the lower river stations at "Angoram and Marienberg would have been [located] on the coast" (Swadling et al. 1988: 14–15). The same slow process of infill continued to go on until the current geography of the region would have been in place by about 1,000 years ago. The Murik Lakes on the western side and the Watam Lakes on the eastern side of the river are all that remain, in Swadling's view, of this vast, vanished, inland sea. Given such a dramatic, albeit gradual, process, and the many disruptions of indigenous populations it would have entailed, perhaps the incredibly diverse linguistic situation in the contemporary Sepik estuary begins to come into focus. "The challenge of adjusting to this changing landscape and circumstances," Swadling and Hope conclude, "probably made change and diversity a way of life in this region" (1992: 36). If one adds to this environment the further complication that, as the inland sea receded, Austronesian speakers started to reach the region, the linguist William Foley's conclusion that "nowhere in New Guinea is linguistic fragmenta-tion greater than in the drainage system of the Sepik and Ramu Rivers" becomes comprehensible (1986: 279). Along the Sepik River, two language groups, the Lower Sepik and the Ndu families, belong to a single phylum. In the estuary, this phylum meets two others (Laycock 1973). Buna, which is related to Arapesh, enters the picture and is spoken by inland communi-ties, while an Austronesian dialect, related to languages throughout Seaboard Melanesia and the insular Pacific, is spoken on most of the off-shore islands as well as in certain coastal enclaves (see Map 3).

Now the Lower Sepik languages extend in a contiguous series along the river up from the coast as far as the Moim Lakes, the area in which contemporary Murik believe their ancestors lived before migrating to

Map 3: Sepik River and North Coast language groups

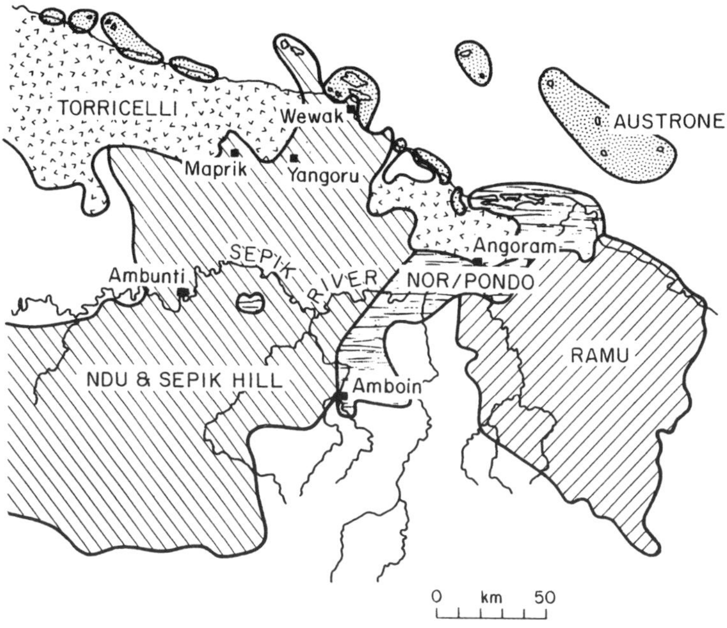

present-day coastal settlements (see Map 1).[1] According to Foley, these languages are grammatically distinctive for possessing, as they do, an intricate pattern of agreement for number and noun class among nouns, possessive pronouns, adjectives and verbs in a system of prefixes, infixes and suffixes (1986: 222–9). Although contemporary Murik lacks noun classes, it does have the complex pattern of concord that Lower Sepik languages share and Foley worked out in his version of proto Lower Sepik (1986: 228). Why Murik grammar should have become simplified, he cannot explain. A solution to his problem can be found, I think, in adaptive strategy. Prior to the introduction and spread of Tokpisin as a lingua franca, a pidgin Murik was used by trading partners in the lower river region. While the Murik vernacular still remains full of loan words, they are not "borrowed" from inland languages but from the grammatically simpler Austronesian languages spoken on the islands. The simplification of Murik can be understood in terms of the extent to which its speakers were (and remain) intertribal actors who assimilated technology, language and social forms from the island Austronesians. While the Mountain Arapesh were once dubbed an "importing culture" by Margaret Mead (1938), the Murik were, and continue to serve, as a cultural nexus of far greater magnitude. They are an entrepot people, who constantly engage in import and export, a people whose dependency on intertribal trade is far more complete (see Tiesler 1969/70; Lipset 1985; Lipset and Barlow 1987).

Early twentieth-century, contact-era trade relations along the North Coast have been reconstructed, catalogued and mapped in a monumental work by Frank Tiesler (1969/70; see Map 4). Relying exclusively upon reports by pre-World War II explorers, ethnographers and particularly the reports of the SVD missionaries, Tiesler sought to account for the development of a series of overlapping trade networks in the region by reconstructing a prehistory in which Austronesian-speaking sailors play a central role. Approximately 2,500 years ago, after having settled uninhabited offshore islands, the Austronesians began to be pressured by demographic forces to initiate trade relations with indigenous non-Austronesian-speaking groups living along the North Coast and then to establish compact, densely populated, mainland "colonies." Trade for foodstuffs was eventually supplemented by the export of such crafts as pottery, adornments, weapons and, most importantly, outrigger canoes. Gradually, the regional value placed upon these latter goods gave rise to the creation of a common material culture that was independent of linguistic affiliation but was an "accomplishment of all the groups living within it" (Tiesler 1969/70: 118; cf. Mead 1938: 165). Thus, of the two leading outrigger canoe-building peoples in the region, the Ali and the Murik, one speaks an Austronesian language, while the other does not.

The most comprehensive contact between indigenous coastal groups and the maritime Austronesians arose in "the lagoon region of the Sepik estuary" (Tiesler 1969/70: 122). The people living along this intertidal frontier became the leading intermediaries in the region, living as they did at a transfer point where Austronesian and non-Austronesian peoples, ideas and institutions met and mixed. The non-Austronesian cultures located on the lower and middle Sepik River became elaborated (see e.g., Bateson 1932; 1936/1958), the dugout canoe having permitted an intense, although hostile, pattern of intertribal relationship. By contrast, the more isolated and smaller coastal groups lacked the means of travel and communication for maintaining extensive regional relationships. With the arrival of the Austronesians, however, the outrigger canoe came into use and seaboard societies, in particular the Murik and the Ali, became culturally differentiated from those of the Sepik hinterlands, where materially simpler peoples lived who were unable to engage in canoe-based, regional trade. The Murik continue to recognize this, devaluing inland peoples for their poverty and valuing the dry season between May and October (*awar*), when the delicacies and specialized crafts of the coastal-dwelling and island peoples can be imported, as the "good time" (*akun ariito*, literally the "good sun"). The producers of such manufactures, Tiesler went on, were typically the more active players in the region because they tended to live in relatively impoverished environments. In spite of subsistence deficits, the paradox of the outrigger canoe peoples was that they acquired prestige in the region "as against the groups upon whom they . . . [were] dependent . . . They . . . [were] dealers of many desired goods and thus attain[ed] a key position" (1969/70: 120, 68). Their regional status, in short, did not arise from economic, political or military hegemony but from culturally valorized transactions.[2]

The Murik view of the intertidal frontier

The Murik acknowledge their entrepot identity in legends and ethnohistory they call "ancestor-spirit talk" (*pot ai'iin*). These tales do not tell of the divine creation of a social universe. They detail the wanderings of ancestor-spirits, magical refugees and defeated immigrants who bind the region together willy-nilly. The "ancestor-spirit talk" is made up of romances, tragedies and comedies. Its heroes appear not as gods who are physically and morally superior both to other men and nature but as "ancestor-spirit men" and "ancestor-spirit women" (*pot nor* and *pot merogo*) who are only superior in degree to other men and their environment. The powers they possess are less than divine. They are ancestors whose actions are human, albeit enchanted. The laws of nature are only slightly suspended for them. Their deeds of courage and endurance, their passions, are prodigious to be sure. And the world they inhabit is occupied by talking animals, animated ogres, invisible spirits and such. But it is a world through which these spirits, refugees and immigrants wander as exiles, wayfarers and rejected lovers, rather than as omnipotent beings. They are only slightly more than, or else precisely equal to, the agency and moral qualities of human beings. Conflict in the world of the Murik

Map 4: The Murik trade network

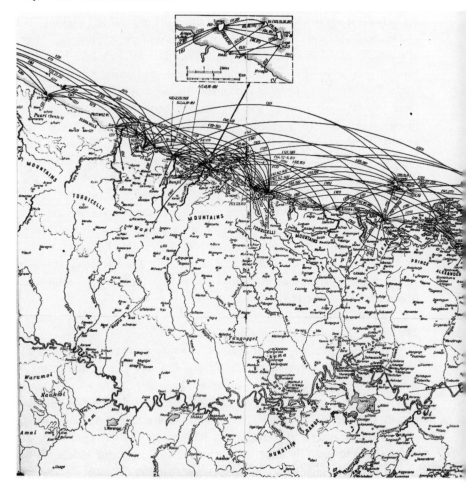

ancestral narratives is not resolved through divine intervention but through more egalitarian, shall we say, more human, actions. In other words, if I can apply Northrup Frye's Aristotelian classification of modes of fiction here (1957: 33–5), the stories by which the Murik account for themselves are legends, not myths. Spirits and men are not depicted as possessing a radically different set of powers or moral standards in them but rather appear to be similar (see also Sahlins 1981: 15).

Up and down the North Coast, the origin and urgency of the regional economy is explained in terms of a story about a pair of ancestor spirit-men, said to be brothers (see also Kulick 1992; Pomponio 1992; Meeker,

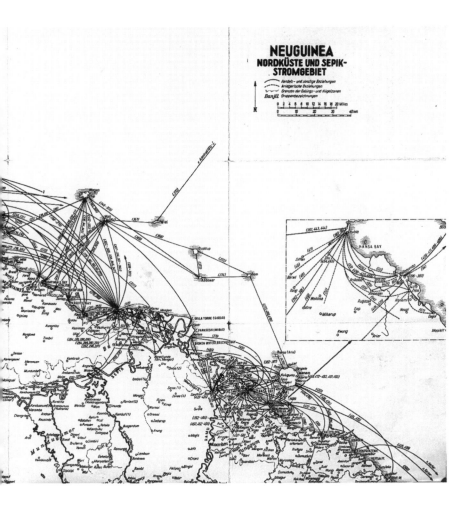

Barlow and Lipset 1986; and Lipset n.d.b). The pair go by many aliases. Lower Sepik peoples tend to call the elder brother Andena, and the younger brother Arena, whereas the elder is known as Wankau and the younger Dibariba among coastal peoples and islanders. The Murik, of course, know them by both pairs of names. While some island groups start the story sociocentrically (e.g., Hogbin 1935: 378–9, 407), from the Murik perspective, the brothers come from a tributary of the lower Sepik called the Porapora River. The two had come to blows over elder brother's wife, whom younger brother had seduced. In his rage, elder brother attempted but failed to kill his junior rival. Taking flight, younger brother sailed off in his outrigger canoe toward the east, creating "the region" and its various adaptive strategies as he went. In search of his lost sibling, elder brother went west. There will be subsequent opportunity to discuss the images of siblingship, masculinity and Oedipal rage represented in the story; just now I shall draw attention to elder brother's role in the creation of "the region." Although I did not collect the following episodes in Murik, they are known there. I recorded them in Tokpisin in the lower Sepik River village of Singarin and in the coastal village of Turubu in 1986 and 1988 respectively (see Map 1). A Lower Sepik language is spoken in the former community, while the latter group speak an Austronesian dialect.

Andena, the Elder Brother

Andena, the elder brother, travelled in the *Sea Eagle*, his outrigger canoe, magically creating the stands of wild sago along the banks of the lower Sepik with his spittle and founding the male cult. He went from village to village. He gave yam digging sticks to some gardening villages, sago mallets to others. Other communities received from him magical stone axes for carving slit-drums, pottery-making paddles or drills to make shell rings.

Only the coastal and island communities got his outrigger canoe and the tools, prow and sideboard designs, the mast decorations as well as the names, magic and ritual for building and consecrating it. Andena identified more closely with his outrigger canoe, his ornaments and hourglass hand-drum, rather than with gardening, sago or pottery-making tools. So he only gave the coastal and island peoples these things because, as he told them, "you are more like me."

Andena came ashore in each village [he visited] decorated from head to toe in his regalia. His skin was painted flaming red. A net bag with cassowary feathers hung on his chest. He wore a bark cummerbund around his waist and plaited wicker bands on his wrists and ankles. He also wore a dogs' teeth headband and necklace, *Nassa* shell bandoliers and a wicker hairpiece. His loincloth was beautiful. The basket he carried bore his own name.[3]

Upon arrival, he would show villagers how to trade. A host should lay down canoe rollers and then offer only the most bounteous generosity to his guest, carry his basket, protect him during his stay and finally treat him to a lavish feast, the night before his departure. Andena also set the rates of exchange between trading partners. In return for his hosts' hospitality and gifts, Andena gave all of his ornaments, except for his dogs' teeth headband which he kept for himself.

This part of the story presents an image of a male body brilliantly arrayed in jewelry. The body of Andena is, as Marilyn Strathern might say, eminently partible (1988a). Nearly everything the elder brother spirit owns, his ornaments, magical knowledge and technology, are detachable and transactable. But observe that, with the exception of the wild sago stands along the lower Sepik River, which are associated with the elder brother's magical spittle, the rest of his paraphernalia are simply cast as gifts rather than tokens of consubstantiality. The relationship between adornments and the body will take on increasing import below. Here, I would just conjecture that this image of the partible hero is an indigenous representation of the Austronesian "colonization" of the region. For the supralocal order created by this senior, foreign guest during his equivocal search for his "lost brother," does not merely explain the origin of the male cult, valuables, the means of production, exchange and the etiquette of visiting trade. During his voyages, the elder brother is not only credited with constituting the North Coast as a region, but with creating its mosaic of distinct local ethnicities, based upon a narcissism of adaptive strategies.

The outrigger peoples, the pottery peoples, the gardening peoples all distinguish themselves by reference to an emblematic adaptation in the regional context, which they attribute to this personified image of exogenous, masculine power, the "elder brother" spirit-man. If, as Swadling has argued (1977), the distribution of this story seems to correspond to the cultural boundary of Austronesian influence, then the exogenous power exerted upon the indigenes would seem to represent these mariners. Therefore, it is of no little moment, given Tiesler's view of the pivotal position of Murik in North Coast prehistory, that while some Murik explained the origin of their male cult and the wild sago growing along the lower Sepik as "gifts" of the culture hero, none of them explained either the origin of their canoe technology, or their role as the leading overseas traders in the region, solely in terms of "his" actions in the tale. To do this, they tended to supplement several other stories, the first of which concerned two groups of their refugee-ancestors (*nogam*), who quit the Moim

Lakes, along the middle Sepik River, eventually to take up residence on the coast.

The migrations of the Murik ancestors

Following a dispute over the distribution of crocodile meat during a feast, Murik ancestors "fled" downriver and out to sea aboard crude bamboo rafts. They were blown [west] along the coast toward Walis Island.

They settled among several peoples further and further [to the west of the mouth of the Sepik]. Each time, conflicts arose and they went on in search of a new place to live. On Walis Island, drinking water was scarce. Markets were disrupted by sexual jealousies on Muschu. In Matabau village, an ogre with huge testicles ate their sons. A pair of twins killed the ogre by severing his testicles, which turned into two boulders, still visible today. The Murik ancestors eventually drifted back to the coast by the mouth of the Sepik. Their failed settlements began hereditary trade relations throughout the region. The petrified testicles of the ogre used marked the [premodern] boundaries of trade relations to the west.

Although Swadling has "little doubt that major cultural traditions such as . . . the departure of the Muriks from the Moim Lakes and their settlement sometime later at the the Murik Lakes actually coincides with significant . . . events" (Swadling et al. 1988: 14–15), I find the *nogam* migration impossible to evaluate from the point of view of prehistory. The point I take from both it and the "Two Brothers" legend is that the stories do not explain or comprehend the regional world, and the Murik position within it, in terms of either the Murik or their culture heroes. Nor do they locate a Murik homeland (the Moim Lakes to which they also arrived from elsewhere). Instead of some kind of an image of uterine fertility – a cave, or a hole in the ground – they depict a world of social process, conflict, exchange and migration. The Murik see themselves as an exogenous people, not an authocthonous one. They are a people whose culture lacks a center in space: both their *nogam* ancestors and their culture hero Andena originate elsewhere. The center of their world is therefore empty. Its essence or substance is thus a fluid matrix, or a social field, rather than a finite body. A nonlocus has displaced a fixed locus: the center is not the center. A gap, in need of augmentation, is depicted in these stories. The polysemy of this gap, and the many rejoinders to it, constitutes the ongoing predicament in Murik culture, which is, in my view, most acutely felt by men.

Womanhood in Murik ethnohistory

Headhunting ended around the mouth of the Sepik in 1918. Prior to this time, the disposition and right to ambush, kill and decapitate were retained by secret men's spirit cults in defense of the autonomy and welfare of their communities. No supralocal authority existed to repress them. The male cults, as well as the occasional rogue warrior, possessed "free recourse to force" (Sahlins 1972: 172). The Murik were culturally, but not militarily, central in the estuary region during this era. That is, the region was politically acephalous: no village or group of villages was militarily dominant, as Gewertz has claimed to have been the position of Western Iatmul over adjacent, inland peoples, a position which she has gone so far as to label "hegemonic" (1983: 100). Rather, Murik military power in the Lower Sepik had limited goals. Its pursuits were not primarily for material or political gain but the fulfillment of ritual protocol for the consecration of cult houses, outrigger canoes and the initiation of novices. Murik men were a peripatetic force in those days. Semi-annual war parties, armed with multiple weapons – long spears triggered by spearthrowers, as well as bows and arrows – traveled about the region in river canoes. The zone of their warfare extended inland, upriver and along the coast. The offshore islanders, however, were never Murik enemies. Military alliances were made between trading partners, dispersed, classificatory siblings, or between male cult groups. As Harrison has pointed out (1989; 1993b), these relationships were *not* strictly coterminous with village or "tribal" boundaries. The Murik who attacked inland sago suppliers did not have trade relations with their victims, although some of their co-villagers might have done. Moreover, internecine warfare between Murik villages broke out sporadically. Most commonly, local violence escalated from a purported rape, seduction or infidelity which had been touched off by jealous, Oedipal rages between coeval men. As in the "Two Brothers" legend, disputed rights to the sexuality of a woman was an impassioned call to arms (see Chapters 7–9). In the following episode, which accounts for the subdivision of water rights in the Murik Lakes, not only are relations within villages subverted by adultery disputes between men, but so are relations between the Murik and their sago suppliers.

Weg and Wira; or, the founding of Karau village

The ancestors split up into two groups during their migration. Each one settled separate villages – called Wokumot and Iji'gob – which were on opposite sides of a big channel into the western lakes. Individuals from both groups entered into trading partnerships with inland sago suppliers who pitied them as being ignorant, poor "children." The landed people became known as "trade mothers" (*asamot ngain*) to the Murik. They taught them how to build canoes, fish and smoke their catch as well as how to trade.

But during a masked dance of the water spirits (*ig brag*) mounted in Iji'gob one of the sago suppliers was murdered. A lone Murik woman had been raped while trading for plumage for the headdresses of the costumes. The rapist was invited to attend the celebration so the woman's husband could retaliate when he took his turn to dance inside the costume of the water spirit.

At the same time, a second intrigue was going on within Iji'gob. A young man, named Wira, was having a love affair with the wife of Weg, his age-mate. The husband discovered the liaison and paid a sorcerer to kill his wife's lover who died during this same feast. In the recriminations that followed, the husband falsely charged a man from another group living else-where in the lakes for having "hired" the sorcerer. Believing his accusation, the male cult of Iji'gob began to attack the village of the accused. Their attacks continued until the cuckold's own father, who knew about his daughter-in-law's liaison, finally exposed the truth: the guilty party was his own son, a man within Iji'gob itself. The Iji'gob attack had been for naught. Fearing counterattack, they fled down the coast to found the village of Karau on beaches of the eastern lakes.

Two images of women appear in this complicated story. A vulnerable, sexual one: a Murik woman is raped and a man's wife is seduced. These wrongs lead to dishonesty, retaliation, collective violence among men and the collapse of community. And the maternal one: the sago suppliers look after their dependent children as "trade mothers," providing them with not only food but the technology for food production and trade. The moral contrast between the two kinds of women could hardly be rendered more plainly.

In the story of the founding of Karau village, Murik ancestors are said to consist of two subgroups. There is also a third group of Murik-speak-ers, today called Kaup. Located on the shores of a smaller lagoon which is isolated from the main system, Kaup is the westernmost Murik-speaking community (see Map 2). Kaup is distinctive for its large land holdings. In 1988, when I finally managed to visit it for the first time, I learned from Vincent Tanep how the village became affluent.

How Kaup got land

The "beach people" [of Kaup] used to barter fish, shellfish and women for sago. We had no land. [The bush people] . . . owned land . . . We marketed [with them] by exchanging food and "paying skin." Their way was to give us sago, but they would first demand that we allow them to have sexual inter-course with women. We fought over this. Kaup got angry and agreed to kill the "landed men" . . . Now we still talk about this warfare. Some say that [our ancestors] were fighting over landrights, over ground. They did not fight over land. It was women. Their war was caused by women, just women. It has always been this way. It has no beginning.

This stunning image of moral exchange ruptured by feminine sexuality speaks for itself with impunity. No less extraordinary is the Oedipal view of the causes of warfare my informant generalized from it: warfare is the work of desire. Indeed, the whole mode of the "ancestor-spirit talk" could be seen as low mimetic comedy whose theme is the violence which sexual violations of women continually precipitate, violence which ruptures moral order.

Other moments in Murik ethnohistory shed light on the divisive relationship between the sexuality of women and society. Tales of conflict account for the residence of descent groups in particular Murik villages. In the denouement of these stories, defeated, landless sibling pairs marry Murik women, take up matrilocal residence and bring their own, distinctive material culture with them: ornamental regalia, named dwellings, the designs and names of outrigger canoes, male and female cult sacra as well as weapons, canoe paddles, headrests, dishes, utensils, tools, and so forth. Most important, the refugees persist in maintaining relations with dispersed kinsmen, by claiming bilateral genealogical ties and ethnohistorical knowledge as well as by staging ceremonial activity. These relationships are not any kind of complementary filiation attached to some principle of unilineal descent. The Murik reckon descent cognatically.

The stories imbue the Murik entrepot adaptation with a thoroughly exogenous identity. As Margaret Mead once declared about the upriver Tchambuli, the Murik take no "absolute attitude toward their culture" as one which always existed (cited by Errington and Gewertz 1986: 99). This is evident in the meaning of Murik village names. "Wokumot" refers to a legendary serpent-spirit from Walis Island which is held to have created the Murik Lakes with a snap of its tail. The derivation of "Karau" village is particularly telling. This name is attributed to island trading partners on Muschu Island. The refugee-ancestors were guests of Sub villagers during a phase of their migration. "Karau" is a contraction of the Sub name of the point – Utim'kara – where they are believed to have camped.[4] The putative origin of the Murik people is upriver. Their village names derive from island trading partners. The origin of the Murik Lakes they also associate with the islands. The name of the lakes, by the way, was taken from the name of a mountain where the Wokumot group of *nogam* ancestors bivouacked during a phase of their migration. Their sago and knowledge of canoe construction and fishing were imported from inland trading partners. The origin of their ancestor spirits and much of their magico-religious paraphernalia are similarly viewed as exogenous. Only overseas trade relationships are believed to be authocthonous. They were not established by the "Two Brothers" but by Murik ancestor-refugees as they migrated along the North Coast in search of a new village site for themselves. They took up several residences, each of which failed for various reasons, including sexual jealousies. Their failures initiated the trading partnerships.

Tribe, village and house

Each of the three Murik-speaking groups – Kaup, Wokumot and Karau – eventually divided into three ritually autonomous hamlets, or villages, called *nemot* in the Murik vernacular (see Figure 1).[5] The three Karau villages, the three Wokumot hamlets, today called "Big Murik," and the Kaup hamlets share no moral obligation to ally in warfare or settle feuds (see Chapters 8–9). They accept the authority of no superordinate institution (save the state) and above all, share little common sentiment (cf. Evans-Pritchard 1940: 122). Whatever solidarity these groups possess is based upon dispersed kinship, affinity, the secret spirit cults, a corpus of ancestral narratives, a common adaptive strategy, shared off-shore fishing rights, and a mutually intelligible language. The Murik both expand and contract the inclusiveness of 'tribal' attributes, adding to or dispensing with any of them when the need arises. The maximal unit may thus include seven non-Murik-speaking, inland sago-supplying communities, today known as the "Bush Murik," as well as the "Number Two Murik," who are the non-Murik-speaking villagers living around the mouth of the Sepik (see Map 3). Both the "Bush Murik" and the "Number Two Murik" identify themselves as "Murik" because of affinal ties, common descent and perceived cultural similarities. Only rarely, however, do the "Beach Murik" reflect upon their "tribal" identity. Instead, they are much more inclined to stress the cultural and political divisions among their constituent groups:

Figure 1: Murik subgroups

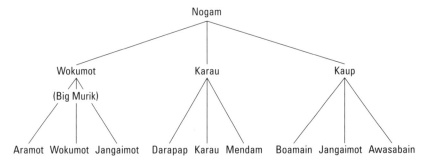

Table 1. *1982 Murik population*

	Big Murik	Karau	Kaup	Wewak	Total
Population	376	576	350	176	1,478
Households	58	103		25	186
Mean household size	6.48	5.2		6.1	5.9

Note: In 1928, Father Schmidt, the resident Catholic priest, estimated that about 1,000 people were living along the Murik coast. According to the census Barlow and I conducted in 1981–2, the population had risen to 1,304. In addition, another 176 were living in Wewak, the provincial capital and market town, and 100 or so people resided elsewhere in the country. The village of Kaup was inaccessible to us in 1981–2.

dialectical and ethnohistorical differences, the politico-ritual autonomy of each descent group, locality and village. Indeed, the local government councilors, of which each Murik village elects one, do not cooperate with each other very willingly.

The Founding of Darapap village

Karau gave rise to Mendam and Darapap villages. The beginnings of Mendam seem to have been uneventful. Darapap (the literal meaning of which is "that spot") was founded following an armed brawl in Karau. In the 1880s, a jealous husband attacked his wife's lover with a stone axe. The immediate cause of the violence was said to have been the birth of a son which the husband believed to have not been his own. In order to protect the lover, who was considered a capable, strong youth, about ten to fifteen families quit Karau, and resettled on an unoccupied stretch of beach a few miles to the west (see Map 2).

The inevitable cause of suspicion and conflict among them, as in bride-service types of prestige structure, is personal honor challenged by seduction, abduction or rape: Mangrove Men fight over women.

Murik villages vary in terms of available subsistence resources, isolation from town and state, and use of space. I have stated that Kaup owns considerable land, sago stands, as well as a small lagoon. It is a large, well-kept community which was formerly subdivided into three named hamlets but is now made up of just two. The one road which links any Murik village to the district center at Angoram terminates here. In 1982, the most deficit-ridden village was Big Murik. Being then set on a sandbar less than 300 yards wide, the "Three Hamlets" (*Nemot Kerongo*), as the Murik tend to call this community, own fishing rights to the western lakes but have no

Table 2. *1982 village population*

Big Murik=Aramot	Wokumot	Jangaimot
132	123	121
Karau=Mendam	Karau	Darapap
233	96	249

Note: The population of Darapap had risen to 453 in 1993.

convenient garden land (except for the coconut groves and betel nut trees growing at the site of an abandoned village on the inland side of the lakes, many hours away by lagoon canoe). In 1981, a storm completely gutted an entire row of beachfront houses in Big Murik and "littered" the beach with shells. But by 1986, the coastline was expanding again and people were planting coconut palms in front of their houses. Although water is always in sight in this locale, effectively squeezed as it is between the lakes and the sea, there is a chronic shortage of drinking water, not to mention shortages of garden produce and sago. Big Murik presents the most extreme version of the Murik condition, the world of water lacking in water. Rather than closed and exclusive, here the boundaries of space are permeable, fluid and open. Indeed, they are catastrophically inundated from time to time, and regularly traversed for purposes of trade. In Murik, land does not present itself as a metaphor of perpetuity, fertility or identity.

In the 1980s, only Kaup, and the hamlet of Aramot in Big Murik, conformed to what was once an ideal use of space. Being located on the beach, these two communities were laid out in orderly rectangular fashion around a sandy square, which served as a dancing ground or the site of village meetings. By contrast, the three Karau villages were located on the lakefront, rather than seaward. They had been rebuilt facing the lakes in the aftermath of severe weather, without regard for this or any plan. Darapap, Karau and Mendam villages were close clusters of dilapidated, weathered-down, thatch houses, standing over mangrove stumps and muck, or over tide flats (see Figure 2). Darapap and Karau villages, which were separated by a treacherous channel that was about 600 yards wide in 1982, shared joint fishing rights to the eastern lakes. They also shared rights to a small bit of highly saline land adjacent to the communities in which the people planted coconut palms and made small swidden gardens. Being built in a long line along an edge of the lake without a foreshore, the village of

Mendam, located closest to the Sepik River, had the look of a riverfront community (see Figure 3). It also resembled a riverine village insofar as it was self-sufficient in sago, owning rights to the rich stands which lie between the lakes and the river, along Majop, an artificial canal which connects the two. Although the village owns an abundant sago resource, Mendam controlled the smallest section of the lakes.[6]

Whether or not they diverge from the ideal use of space, the sago thatch dwellings and male cult houses of which each Murik village is composed all stand up on spindly six-foot mangrove stilts (see Plate 3). The dwelling (*iran*) is an undivided rectangular space inside, lacking in many partitions between private and collective space (except at night when mosquito nets are hung). The house is said to be centered around its "mother's" hearth (*maig*) and firepit, located in the back third of the floor (see Figure 4). On the rectangular rack built over it, one can usually find a few fish left to smoke over a smoldering fire, or a piece of sago bread for the children's snack. Rear walls are lined with cooking utensils, heaps of wooden and ceramic plates, and huge clay pots for storing sago flour. Baskets of smoked fish hang from the ceiling on wooden hooks to keep from rats. On side walls, half-finished baskets are pinned. In the rafters, piles of dried basket reeds and fishing spears are stored. By day, mosquito nets and sleeping mats are rolled up, so that the space by the entrance is left

Figure 2: Bird's-eye view, Darapap village

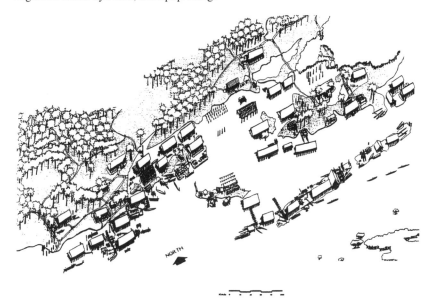

Figure 3: Site plan, Mendam village

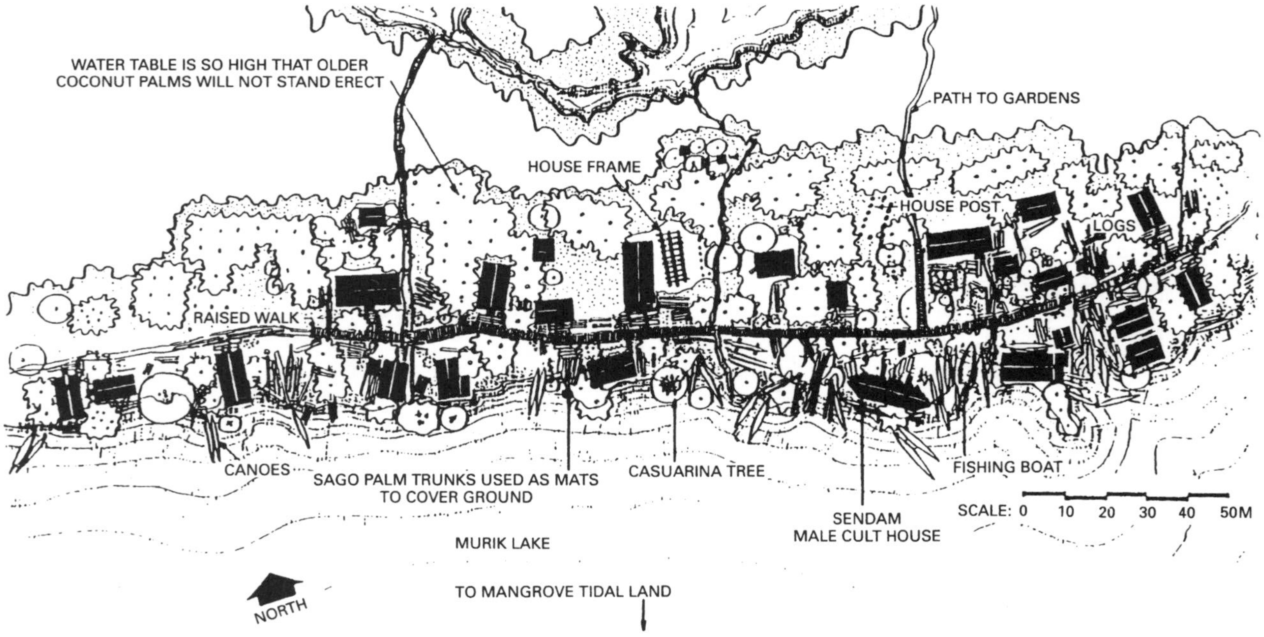

WATER TABLE IS SO HIGH THAT OLDER
COCONUT PALMS WILL NOT STAND ERECT

PATH TO GARDENS

HOUSE POST

LOGS

HOUSE FRAME

CASUARINA TREE

SENDAM
MALE CULT HOUSE

FISHING BOAT

SAGO PALM TRUNKS USED AS MATS
TO COVER GROUND

CANOES

RAISED WALK

MURIK LAKE

TO MANGROVE TIDAL LAND

NORTH

SCALE: 0   10   20   30   40   50M

relatively open. Nevertheless, a Murik house is filled by the clutter, industry and presence of women. Indeed, the "soul" or "ghost" of the house (*iran nabran*) is said to be "a woman," who oversees the activities of the "mother of the house." This female spirit is the shadow who peers over a woman's shoulder as she sits cooking by her hearthfire. Interior, domestic space, in other words, is classed as feminine (see Plates 1 and 4).

The domestic group lives out its cycle of development here. The dwelling is a context of intimate care and protection, nurture, commensality, conversation and the moral transformation of children into autonomous persons. It is a space in which the elderly die and their bodies are prepared for burial in family plots located either within, or on the periphery of, the community. But the house is not a space in which mothers give birth (cf. Munn 1986: 34–5). Because childbirth is thought to be toxic to male elders, it is excluded not merely from the house but from the community as a whole. Each village has a birth house (Murik: *be iran*; Tokpisin: *haus karim*) located on its periphery into which men may not enter. This *ne plus ultra* feminine space is a miniature version of the dwelling, except that it may be partitioned into two small rooms, the rear of which serves as the delivery room. In ethnohistorical tales, the moral integrity of the community was depicted as vulnerable to contested feminine sexuality. Now a second source of its vulnerability has come into view: the reproductive force of the youthful, maternal body.

Plate 3: Dwelling in Karau village

By contrast to the birth house, from which men are banned, women and children are forbidden inside the male cult houses. But these buildings often stand in the center of villages rather than along their edges. The male cult houses were once used as staging areas for war parties as well as ritual performances. They went on serving as halls for many kinds of assemblies through the 1980s and on into the 1990s. I will discuss the dialogics of the male cult in relation to the maternal schema in Chapters 6–7; now it is enough to say that the discourse of the former is in large measure a barely hidden polemic about the inhibitions of the latter.

To sum up: as a tribal unit, the Murik Lakes peoples consist of a tripartite congeries of cognatic descent groups that are domiciled in five villages. Each village is divided into heterosexual but largely feminine spaces, and exclusively masculine domains, which are the dwellings and the male cult houses, but encompasses them both within its collective space. Only birthing women are marginalized into peripheral buildings.

Figure 4: Dwelling house

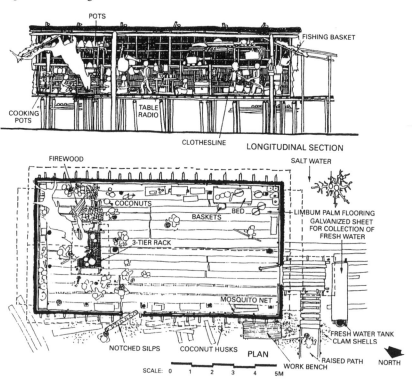

The sexual division of labor

Studies of horticultural Sepik societies have made much of the consub-stantial ties between men and ceremonial yams. In particular, the health, virility and vitality of both have been said to be compromised by women's sexual and reproductive bodies (Kaberry 1941–2: 356; see also Forge 1966; Scaglion 1981; cf. Harrison 1982). As men suffer from contact with femi-nine substances, so their yam gardens are put in jeopardy by women who are therefore tabooed from them during the growing season. "Arapesh yams . . . abhor the smell of female sexual secretions present in a man's body after intercourse" (Tuzin 1972: 237). Arapesh men also consider yam production to be a form of paternity. Men establish "filial" relations with yams in the ritual process of gardening. Their yams are "children" to be tended and then decorated like adolescent cult initiates, after which they become tokens in competitive ceremonial exchange. To attack a yam is also construed as an attack against a man's "son," as well as his jural status, since the yam represents his future position in society. In Murik, where gardening is culturally insignificant, the male body and person are not projected onto substances grown in land. But the implicit concern inspiring Sepik yam religion – the attempt to sustain an autonomous cosmic masculinity in response to women's uterine and sexual bodies – is no less at issue. In Murik, the male body and person are also defined as

Plate 4: The interior of the dwelling is said to center around its "mother's hearth"

contingent upon ceremonial exchanges of food, but the food is either "gathered" from the lakes or imported from trading partners living elsewhere in the region. The system of status attribution depends upon magico-religious work no less than it does in horticultural Sepik societies. But rather than gardening tubers or raising pigs, it calls for the building of and voyaging in canoes, as well as the trade relations which provision the ceremonial economy that underwrites it. Thus it is that the growth of foodstuffs is not viewed as vulnerable to feminine impurities, but canoes and trade are.

An aquatic hunter-gatherer division of labor organizes daily relations of production between Murik men and women. But lacking a routine domain of work, the Mangrove Men are minimal hunters (although there are one or two exceptions in each community who occasionally hunt saltwater crocodiles or wild pigs). Young men spearfish for pleasure. Individually and in small groups, they regularly cast store-bought, nylon

Plate 5: A huge trevally, such as this one, could be cooked and served in male cult competition

drift nets at night (nets which they maintain by day) and become especially active when a surplus of fish is necessary for feasting, trading, or marketing. They do not harvest mudcrabs, but do dive for other shellfish, such as clams, mussels and oysters. The various other fishing techniques employed by men include fencing fish in channels, chasing them into nets, and nighttime spearfishing by torchlight. In addition, men carve artifacts to sell to tourists and make ornaments for overseas trade. They also spend a great deal of time working on the construction and repair of buildings: dwellings and cult houses. However, of greatest salience to their political identities are the vehicles they build, maintain and operate. For the most part, two types of dugout canoe are used in Murik, a motorized oceangoing outrigger (*sev gai'iin*) and a riverine vessel (*gai'iin* or *bor*) which has no outrigger.[7] While food production and ornament manufacture are not single-sex activities, the construction of vehicles which move through regional space is an exclusively masculine enterprise. "Even though canoe making and canoe travelling represent a purely masculine power . . . [they] also reveal that the masculine self is marked by a feminine other" (Meeker, Barlow and Lipset 1986: 52). Rendered audible in outrigger canoe construction is a dialogic relationship between the fertility and sexuality of the maternal body and collective male agency (see Barlow and Lipset 1997; cf. Hogbin 1935 and Munn 1973).

Plate 6: Men building a lagoon canoe

After a canoe log is floated back to the Murik Lakes from an inland or coastal trading partner, men and women haul it to a peripheral part of the village, where men erect a temporary "canoe-shed" (*gai'suumon*). As an outpost of their cult house, this structure is an exclusively masculine space from which women are debarred, for, inside, the building of the vessel is seen as the result not only of male skill but of cosmic, male agency. The *gai'suumon* is a symbolic equivalent of a birth house, the space from which men are debarred and the space within which birth is understood to result from women's skills and collective, magico-religious agency. Isolated as they are from the domestic community, men may work naked there, "naked," say the women, "like children." The hull must be protected, men believe, from the sexuality of fertile young women while it is being fashioned. Male youths must remain chaste as they carve so as not to cause the hull to split open "like a woman's genitals" (Barlow 1985a: 116). Cracks will break, it is thought, should a boatbuilder have made love the night before he comes to work. Spells must be murmured and ginger leaves (*gorongol*) may be dropped at the doorway of the canoe-shed in order to cleanse the hull of the "smell" of sexual intercourse. As the log is carved into a canoe hull, men also view it as being metamorphosed from a feminine entity into a powerful, masculine vehicle by means of a process of cultic initiation. To be made into a strong warrior, it must be secluded from contact with the debilitating effects of women's blood.

The vessel is eventually brought out of the canoe-shed as a finished product. When it is ready for launching, a clay pot and a small fire are placed on the ground before it. The hull is shoved over these images of domesticity, shattering the pot and extinguishing the fire, before becoming independently mobile in the water. As symbols of female nurture, these items represent the acts of cooking which the canoe leaves behind. But they are simultaneously symbols of birthing. During the final moments of labor, buckets of seawater are poured over the woman's belly and genitals both to relieve the burning sensation women are said to suffer and to cleanse the mother. As a representation of masculine identity, the outrigger canoe is literally constructed by men working together as a cosmic body. But its manufacture is also imagined in terms of feminine creativity, which is otherwise banished from the community. In the canoe-shed, the men symbolically "give birth" to the canoe, which is protected by their chastity and collectivity. Canoe construction challenges, even as it is challenged by, the sexual and procreative forces inside of women's bodies.

The destructive potential of feminine sexuality upon collective order appears no less clearly in beliefs about the safety of the premodern canoe during its voyage. Women are acknowledged by men to possess magical power "to send" the canoe and "return" it to port. "It is women," goes the Murik proverb, "who take the canoe to the islands. It is women who bring it back. The canoe travels on the strength of women." The hull of the outrigger canoe is, as we have seen, iconically identified with women's genitals. But while at sea, the steersman's wife is held to possess magical influence over the outrigger canoe as it travels to and returns from the islands (see Plate 7). This woman is therefore subjected to a series of taboos (*numaruk garu'a*) which have an immobilizing effect upon her, putting her in the care of others while her husband, the steersman, is at sea (see Barlow 1985a: 118). She may not, for example, cut grass. If she does, the outrigger lashings will break. She may not chop firewood. If she does, the canoe will split open and sink. She should not speak to a lover, who is one of her husband's rivals. If she does, the canoe will not return. Above all, she must

Plate 7: A premodern, sailing outrigger

not have intercourse with him. If she does, the canoe will mimic ﹅ undulations she makes while making love, and then sink into the sea. By imagining the safety of the outrigger canoe – in one sense, a collective male body – as mimicking the movements a woman makes during intercourse, men would seem to recognize not only the problem feminine sexuality poses to their cosmic agency, but the dialogicality of this relationship. Moreover, by symbolizing the construction of these vehicles as a gestative and birthing process through which a female body becomes masculine, men acknowledge their ambivalence about the originary powers of "her" uterine body, for these are the very powers they would otherwise stigmatize.

Literal women, for their part in the Murik adaptive strategy, are largely responsible for subsistence fishing, shellfish gathering and cleaning the catch (for use in daily, ceremonial and market contexts). Women use store-bought hooks and nylon drop-lines and prefer to fish sitting alone in small canoes. Women also attend to childcare, household maintenance, and collect firewood. Women go to market in town to sell seafood and baskets. While both sexes make shell and teeth ornaments for trade purposes, it is women, and only women, who weave the Murik baskets which are so highly prized throughout the region. This is all to say that the majority of women are constantly at work throughout the day.

Absences due to trade, education, employment, or hospitalization constantly create a manpower problem in Murik. In the expediency of the moment, the sexual division of labor is perhaps less strictly observed here than elsewhere in the Sepik. As the proverb goes about dividing a cooked fish: "the man should eat the tail, the woman the head, unless she is more hungry than he." Men have the duty of arranging and coordinating both the inland and overseas trade. But both men and women trade and inherit their own separate trading partners (*asamot najen*). Both genders have distinct use and transfer rights in the goods they produce and import. Women go by themselves to trade with inland sago suppliers, leaving a husband by himself to look after their children. As a woman will climb up and repair holes in a roof in the absence of her husband, so a man will also cook for himself on the occasion when absolutely no women are around. But about this duty both sexes are unanimous: cooking and serving food in domestic contexts are uniquely feminine activities. Even a wife's brother will not defend his sister when she fails to feed her husband properly and will do nothing to protect her from the latter's rage.

It is difficult to apply Munn's spatial contrast in the Gawan sexual division of labor (1986: 31–2) – that the women are sedentary, while the men are mobile – to the Murik situation. Here women's work is not more immobile than men's. Although women do sit down as they paddle their canoes, sit down when they shuck shellfish, sit down when they weave baskets and, above all, sit down to cook, both genders climb coconut palms and both routinely leave the community for various reasons. Men are therefore neither less sedentary nor more mobile than women. Both travel to and beyond the moral edges of Murik space, both sit down to work.[8] In Murik, if there is a contrast between the way the genders work in space it might be better phrased as a matter of degree rather than of kind. Women "gather" and trade both in and out of the community. While men's voyages sometimes take them further afield, their carving, house and canoe building projects are no less sedentary than the women's cooking or weaving activities. And this pattern reflects the minimal differentiation of male and female in this part of the culture.

The regional world

Murik do not characterize food in terms of gender categories (cf. Meigs 1984), but cooking and foodgiving are unambiguously feminine activities. Cooking is associated with the highly charged, but asexual, context of mothering and infantile dependency. The mother who feeds sago pudding to all her "children" whenever they are hungry is said to be a "big woman" (*numaruk apo*). Of course, when she sometimes fails, or is unable, to do so, the frustration she creates arouses a legitimate expression of anger not only in children but also in husbands. The equation between foodgiving and mothering is clearly expressed in the inland relations with the sago suppliers. As in other parts of the Sepik River, the relationship between inland and waterfront peoples is styled as one between "mothers," who are the sago suppliers, and "children," the fish suppliers. The sago suppliers are termed, by both the Murik and by themselves, as "trade mothers" (*asamot ngain*) because, at least from the Murik point of view, a good child should bring fish to his or her mother from an early age, to be rewarded with sago, which the mother cooks to make a meal. The Murik thus model themselves as the dependent "children" of their sago suppliers, offering fish to them, expecting to be rewarded with sago. But sago is a labor-intensive product available in limited quantities. The sago exchanged is carefully limited at fixed rates and is never abundant. The Murik complain of the small amounts of sago they receive and the inhospitable demeanor of their inland trade partners. While the Murik see themselves as "good children," the sago suppliers become "bad (stingy) mothers," and a legitimate object of Murik disgust (see Barlow 1985a). Even in the 1980s, men still disparaged and mocked the poverty of the "people of the bush" (*sibu, gata*) for their lack of ornamental regalia. However, from what I could gather from talking to inland people, their attitudes to the "Beach Murik" were and are not particularly deferential or envious. They scoffed at their coastal neighbors, whom they once regarded as "rich" in shell valuables, baskets and the like, as little more than "poor children" to be pitied for their food shortages, lack of fresh water, landlessness, and whom they even had to teach to fish and carve river canoes.

In precapitalist, premodern times prior to 1918, Murik men not only used to look upon their sago suppliers with a distaste that bordered on contempt, but they would ambush inland peoples, the "Bush Murik" with whom they did *not* have trade relations. The exchange of fish for sago took place between pairs of hereditary trade "brothers" whose relationship was otherwise kept to a minimum.[9] Traders arranged "silent" exchanges so as not to meet face-to-face (Grierson 1980; Heider 1969). Every few weeks, according to Behrmann (1922: 313), Murik men mounted armed expeditions across the lakes to deposit smoked fish and shellfish on a platform built in neutral territory. Upon calling the sago suppliers by sounding a conch shell or beating a drum, they would withdraw overnight and return next day to pick up their goods. Exchange rates were fixed and the trade promoted little solidarity. Given the state of war, the silent fish-for-sago trade was circumspect and strictly limited to foodstuffs. With this tense background in mind, perhaps it is not surprising that fish-for-sago trade was and remains conceived as a custodial relationship of infantile dependency and maternal nurture, imagery which excludes the more subversive aspects attributed to the female body, namely "her" sexuality and fertility. Indeed, Murik women were forbidden to marry the sago-suppliers, and even in the 1980s affinal relationships between them were virtually nonexistent.

By contrast to this ambivalent relationship, in the overseas world Murik men and women were and remain known as flamboyant purveyors of exotic, beautiful, delicious and practical goods. They used to transship inland and overseas material in both directions: inland foodstuffs they took to islanders and vice versa. In the precontact era, Murik baskets were the most valued goods in the overseas world. In the 1930s, Margaret Mead observed that

Murik canoes voyage all along the northeast coast, vending their material and ceremonial wares to the Beach people, who, in the course of time, sell them to the more inland villages. While this Murik development is a particularly self-conscious exploitation, it may be regarded as a logical extension of the continual buying and selling of . . . [ritual] objects and dances which goes on all over the entire region.   *(1938: 176)*

Murik men were the leading impresarios in the region. In return for extending rights to a folk opera, for example, they would receive large quantities of foodstuffs, pigs, garden produce, *canarium* almonds, tobacco, as well as such other items as weapons and canoe logs. Or, as Mead once put it to Clark Wissler, "Murik lives on diffusion; selling job lots of culture to the poor Papuans" (quoted in Forge, n.d.; see also Tiesler 1969/70 and Gell 1992).

In the imagery of overseas trade, the awkward admixture of maternal nurture and continuous male aggression is replaced by seductive designs and the powers of feminine sexuality. Both men and women make and contribute goods to this sector of the economy, and it is not unusual for women to travel with men to overseas destinations. However, it is the duty of men, and men only, to sail the outrigger canoes and organize the expeditions, particularly out to the Schouten Islands, which the Murik visit during the "dry season." Making use of several different exchange rates (unlimited, fixed, negotiated and credit), the Murik export ceremonial wealth as well as seafood, sago and plaited baskets, the hallmark of their regional identity, to these hereditary trading partners. For overseas expeditions, women collect shellfish, cook sago breads and plait baskets for their husbands to trade. Differentiated by size, texture, color, use and name, these baskets are traded on a one-for-one basis in different places in the region for sheaths of tobacco, buckets of *canarium* almonds, carved wooden plates, clay pots, and even pigs. Being central tokens in their ceremonial economy, importing pigs and almonds was and remains the *raison d'être* of overseas trade for Murik men.

"The islanders," a man once remarked to me when we were discussing overseas trade relationships, "treat us like white people. We go and bring them a couple of crab, a small packet of sago and a few baskets. For this pittance, they work very hard for us." That is, in exchange for an unquantified "gift" of seafood and imported sago, the Murik receive relatively unlimited access to the gardens of the trading partners, access to which is only bounded by the limited space in the holds of their outrigger canoes. Overseas exchange is conducted as "visiting trade" (Heider 1969). Its moral compass (Helms 1988) is decidedly Maussian. The islanders and the Murik exchange personal names and overwhelm each other with hospitality. Guests are "infantilized." When the Murik visit the islands they are expected to do little more than sleep and eat bowl after bowl of banana and taro soup while their goods are being prepared. In return, when the

islanders come to visit Murik, they are treated identically. A young man
from Koil Island recalled visiting Murik for the first time in 1973:

As soon as you come, if they don't know you they ask you the name of your father.
Then they direct you to the house of the person there with that name and they feed
you. People ask you to several houses according to the names of your relatives and
you get so full from all the food that you can't eat any more. Finally at someone's
house, when they offer you yet another bowl of sago pudding you put out your
hand and say "no more." They are not cross because they know you've eaten at
other people's houses too.   *(Huber n.d.)*

Although overseas relations are construed as relations between fictive
siblings rather than mothers and children, the social context of trade
clearly accords with an image of abundant nurture and childlike dependency. But there is an additional dimension of island trade which departs
from this image. Overseas trade takes place amid a solidary ethos and
raises the tantalizing prospect of acquiring precious and delightful goods,
the goods through which not only social standing but love is won. These
relationships provision the grand, nightlong feasts, which create opportunities for amorous intrigue. The "silent" sago trade takes place between
potentially hostile people; both its prosaic goods and social relationships
are restricted. In the overseas relations, in addition to the wider variety of
exotic goods, the inhibited, chaste norms of maternal nurture and dependency are eased rather than asserted, and subversive dimensions of the
body, enchanted designs and sexual powers surface. The overseas trading
partner is likened by Murik men to a fickle woman who must be magically
seduced to give up "her" special treasures. And both genders may, and are
expected to, have sexual liaisons during their island sojourns.

All along the North Coast, from east of the mouth of the Sepik as far as
the islands around Aitape, the arrival of visiting traders by sea – but only
by sea – occasions sex-role reversals (Barlow 1985a). When men come
ashore, women do not approach them seductively, but aggressively; while
men do not approach women forcefully, but seductively, even coyly. In the
cultures around the mouth of the Sepik River, hostesses are known (by
men) as "those who itch," itch for sexual intercourse. When men climb
down from their outrigger canoes, the women may attack them with sharp
little clam shells (a common Murik euphemism for female genitalia). In
compensation for the blood the women may draw, they offer the men their
sexual services, and, when their guests are ready to return home, they
present the men with small gifts of betel nuts and tobacco. When women
arrive at these villages, their male hosts may approach them by slyly tick-
ling their legs. If the women are not attracted, their escorts defend them

against further advances. The reversals of male and female roles convey once again an image of women's sexuality as powerful and dangerous to men, who must parry it carefully and defensively.

The two different images of womanhood in inland and overseas trade confirm the existence of a contentious, ambivalent relationship between representations of moral, custodial motherhood and feminine sexuality, uterine force and violence that also appeared in Murik legends, ethnohistory and outrigger canoe construction. In a rupture-prone situation (inland trade), the more conservative side of the schema, that of a nurturant mother and her dependent children, prevails, and sexuality is excluded. In the more solidary situation (overseas trade), erotic images of womanhood prevail. Feminine sexual powers challenge men, who then feminize their trading partners whom they seek to seduce through their magical charms. The two faces of womanhood, I argue, comprise a single referential object which engages Murik men in an unceasingly quarrelsome dialogue.

The twentieth-century economy
Sporadic exchange with the West took place during the latter nineteenth century. However, ongoing colonial relationships with white men, Christianity and capitalism began when a permanent Catholic Mission outstation was established in Big Murik in 1913. This post was manned for thirty years by Father Joseph Schmidt (SVD). While Schmidt taught German to many Murik men and women, he also learned to speak fluent Murik, translated a number of prayers and wrote several rich ethnographic essays. He held mass on a weekly basis and trained catechists to do so in the rest of the Murik and Lower Sepik villages. Schmidt died during World War II and was not replaced. The Catholics began to retrench after the war and a number of Sepik outstations were closed down (see Huber 1988). The European priests in the Lower Sepik withdrew to live in the Marienberg Mission, from which they went on making rounds in the Lower Sepik together with an Austrian sister who ran several health clinics each year. In the early 1950s, Seventh Day Adventists founded a mission outstation in Darapap which remained an ongoing concern in this one village through the 1980s. Murik catechists, meanwhile, went on conducting services in the other four.

Murik men and women worked outside villages as early as the 1920s, first as service personnel and domestics for the Catholic missions. In the 1930s, they were among the first natives trained as medical orderlies and schoolteachers in the Sepik River region. They also worked as manual laborers in plantations and in the gold fields in Wau. The Murik adaptive strategy and moral ethos remained largely unchanged, however. Murik outriggers, still powered by coconut bark sails and sympathetic magic, continued to journey along the Sepik coast and out to the Schouten Islands in search of ceremonial wealth to redistribute in local prestations.

The year-long Japanese occupation of the Murik coast during 1944, the last year of World War II, left an eight-year-old Michael Somare with not unpleasant memories. He went to the primary school which the soldiers established and learned to count and sing a few songs in Japanese. He watched youths learn to use rifles, while adults harvested oysters and collected firewood for the soldiers. When they first arrived, Somare was given to understand that they were "white" ghosts of deceased ancestors who were coming to rescue them. This is why they were well received from the outset: "Our people found the Japanese extremely friendly . . . [They] did not plunder our gardens and they respected our women, I suppose they had to respect the women because they were among their chief suppliers of food . . . We often danced for them and they performed their sword dances for us" (Somare 1975: 5–6).

After the war, Father Louis Kovacs made a visit to the lakes in a canoe driven by a 6-horsepower Seagull outboard motor. Kovacs remembered that the year was 1954 and the instantaneous uproar the sight of the machine caused. "The Big Murik saw the machine and said, 'Father, that is the real thing. We want to buy it.'" By 1970, Murik outrigger canoes had become completely motorized, and Wewak, the provincial capital, the principal source of gasoline, not to mention outboard motor maintenance in the region, had become the principal market for smoked and fresh seafood, the only Murik "cash crop." While attending the urban market, Murik families overnighted with employed kin who built thatched pile-houses at three beachfront camps, on land which belonged to their hereditary trading partners. These urban settlements became small replicas of village society.

Since the early 1960s, when the first Murik fishing cooperative was founded, village fishing groups have vacillated between whole or multi-village business organizations and smaller, kinship-based work groups, the latter usually being made up of a collateral sibling group, its affines and offspring. The so-called "family business group" typically owns nylon nets, an inboard boat, dinghy, or outrigger motorcanoe, which hauls baskets full of smoked or iced fish to sell in town markets, the profit usually being targeted for ceremonial exchange, the purchase of Western goods, the payment of taxes or children's school fees.[10] In the 1970s, several fishing cooperatives started and collapsed. The visible detritus of this cycle, boats sitting up on blocks, their engines rusting away in the salty air, supposedly waiting for the arrival of spare parts, was left behind as a monument to the difficult trial-and-error process of fisheries development (see Plate 8). Two rival cooperatives were competing in Karau and Darapap during the 1980s. Both were selling smoked and fresh fish in town and were following the original gill-net, trade-store, boat-centered pattern. The Karau group owned a 30-foot plank copra boat, the *Mangrove Man*, while the Darapap group was trying to pool enough money to buy their own boat. The three other villages, Kaup, Big Murik and Mendam, were exclusively engaged in "family business."

The village economy was also supported by remittances from urban kin, the most conspicuous member of whom was obviously Sir Michael Somare. In the Sepik region, however, the notoriety of Murik culture did not originate with Somare, but from the role played by Murik men as impresarios of the dance and as purveyors of ceremonial wealth. Apart from a modest celebrity around the country, no doubt a boost to Murik pride, village economy had still not been transformed by either the famous "native son" or other members of its urban elite. Through his actions and inactions, Somare simultaneously stood as a model of and rival to the rest of this emerging elite. While he fulfilled ritual obligations as they came due, Somare observed that, for reasons of statecraft, he had to maintain a strict neutrality in village business ventures. By contrast, other members of

Plate 8: A new outrigger motor-canoe and an old, broken-down fishing boat

the urban elite participated less in ritual but more in the fishery. In every village, several families owned and operated 25- or 40-horsepower outboard motors which employed kin had remitted to them. In Darapap, Matthew Tamoane, a graduate of the University of Papua New Guinea, founded a fishing cooperative and trade store in 1980 which sought, but ultimately failed, to overcome sublocal rivalries. The overall impact of this urban elite upon the economy, in short, was slight (cf. Smith 1985).

Under the influence of centrifugal forces exerted by capitalism, the provincial market and Western goods, the value of Murik goods decreased. In the 1980s, some, but not all, inland trade partners were selling sago for cash while refusing to pay money for Murik fish (Lipset 1985; see also Barlow, Bolton and Lipset 1988). Because of their relative isolation from town and their ongoing preference for sago, however, the Murik had to put up with this inequity. In overseas trade as well, pigs had largely been put on a cash-only basis (see Chapter 8). With the regional decline of jural institutions, the value of Murik ceremonial wealth had diminished. The demand for Murik baskets and dances remained strong, but the value of ornaments, shells, shellrings, dogs' teeth headbands, and so forth had atrophied in favor of the currency of cash (Lipset and Barlow 1987). Overseas travel, moreover, had simply become more expensive than it had been in the era of the sailing outrigger (see Chapter 5). The advent of petty capitalism truncated the frequency and scope of regional exchange, making the Murik poorer. Their source of cash was located six to eight hours away. The fishery not only required a relatively large capital outlay for outboard motors or boats, but a stable sociopolitical organization to exploit it. In addition, the costs of gas, urban subsistence and mechanical maintenance reduce the profit margin for family businesses. Nonetheless, long before the economic decline of recent years, the chronic deficits imposed by their landless environment had given rise to an adaptive strategy supplemented by masculine vehicles passing through a multiplicity of moral thresholds for the purposes of trade. A vehicular metaphor of and for work and gender identity was already in place. The festooned image of Andena, the "elder brother" spirit-man, travelling through the region in his "canoe-body," carefully professing a supralocal ethos of "the gift," befitted the equivocal position of the Mangrove Men of the 1980s no less than it had in 1900. And, as we shall go on to see, the vulnerability of the splendid hero, his jealous, Oedipal desire for women and his ambivalence about their fertility, continued to engage him in an intense dialogue about the reproduction of moral society.

3

The maternal schema and
the uterine body

It should be apparent that the Murik adaptive strategy suspends the society in an elaborate web of regional trade relationships. Less apparent may be that, except for the provincial market, the right to implement these relationships was and remains a hereditary privilege. Expanding the range of kinship therefore expands an actor's potential access to imported resources (see also Pomponio 1992: 100). Less apparent is also that Murik assumptions about kinship constitute an argument for and against the role of the uterine, maternal body in the reproduction of society. If we attend to how kinship is understood, further dimensions of this argument can be heard. Rather than an attack by a collective, masculine body against women, as is often the case in Melanesian cultures, this is a struggle to honor the custodial, or pre-Oedipal, mother, "who" is personified by both women and men, and to contain an image of an interiorized, uterine body, "who" is also personified by both genders. This chapter examines evidence that such an androgynous dialogue is going on in the contexts of reproduction beliefs, siblingship and the norms of affinal relationships.

The "bat mother"
Of all the views of reproduction that have been reported in the Sepik region,[1] perhaps those of the Arapesh present the most informative contrast to the Murik case. According to both, conception and birth are held to be saturated in feminine blood, which fluid is feared to contaminate the masculine body and agency. According to both, pregnancy results from multiple heterosexual encounters. In Arapesh, the male role in intercourse is thought to feed and help the fetus grow. In Murik, however, the phallus is not afforded a capacity to nurture; a man's semen does not induce fetal growth. Intercourse simply causes the mixing of a man's "semen" (*sabiin*)

or "genital liquid" (*minjir*) with a woman's "egg" (*da'ug*), a process of
fertilization the Murik explicitly liken to the reproductive physiology of
fish. As pregnancy becomes apparent, a Murik couple should then abstain
from sexual intercourse for two or more years, until, as it is said, the child
begins to walk and talk. One explicit reason for this taboo is to space
births. But these pre- and post-partum taboos are also meant to protect
the husband's health from contact with the impurities of pregnancy and
birth as well as to benefit the infant by immunizing the mother's milk, as it
were, from contact with the sexual impurities with which the father's philandering may mystically infect it (see, e.g., Godelier 1986, among many
others).

According to the Arapesh gestation theory, an unborn child "sleeps"
like a chick in an egg in the mother's womb where "first there is just blood
and semen, then the arms and legs emerge and finally the head" (Mead
1935: 32). In Murik, gestation is no self-sufficient, uterine process. Instead,
the mother's reproductive organs are dislodged from her body. The mother
is not held to be the "true mother" (*nogo ngain*) of the fetus, only its
"canoe" (*gai'iin*). The "true mother" is a womb spirit called the "bat
mother" or the "flying fox mother" (*nabwog-ngain*) "who" is located in the
placenta. This somewhat capricious spirit does not cause conception. She
only determines the sex, the facial resemblance of the baby to one of its
parents, as well as its personality (cf. Meggitt 1965: 110). Although she is,
in this sense, separated from the process of gestation going on inside of her
body, and made into a passive vehicle or host, an expectant mother (particularly of a firstborn child) still remains the identified lifegiver. As soon
as she becomes visibly pregnant, she will begin to refer to herself by the
first person dual pronoun and will begin to observe food taboos to protect
the body and personality of the fetus (Barlow 1985b). During any pregnancy, a woman should eat as much or as little of virtually anything she
wants to, so long as it is fresh, unblemished and well-formed. Lest her
infant take on their distinctive characteristics during her first pregnancy,
she should avoid eating various animals from the inland bush. Partaking
of flying fox may cause her child to have shriveled buttocks or become a
person who steals by night. Eating small nocturnal marsupials may cause
a child to have bulging, greedy eyes that long for other people's food.
Eating the Malay apple, a seasonal bell-shaped fruit, is thought to cause
big ears and a tendency to eavesdrop on other people's affairs. All these
foods are, in any case, seldom available. Two consequences of these
admonitions and injunctions are to concentrate the pregnant woman's diet
on regularly available foods – seafood, sago, coconuts and rice – and to

disperse responsibility for a healthy outcome of her pregnancy from inside a mother's body to the actions of the father and the rest of the kinsmen and kinswomen who nurture her. In addition to these and other dietary restrictions, during the initial months of a first pregnancy, a woman should keep her condition secret from everyone other than her mother and elder sister so as not to expose herself to dangerous spirits or enemies of her husband, father, or bilateral grandparents, who might seek supernaturally to deform the fetus or cause her to miscarry. There is no sympathetic magic concerning the prevention of premature labor as among Arapesh. Rather, the opposite problem is seen to be more likely in Murik. While pregnant, a woman should avoid attending burials, planting, tying knots or lashing canoes. These activities may cause the fetus to suffocate or become fastened within the womb. Setting aside the issue of why the first pregnancy is more vulnerable than subsequent ones, and the moral failings condemned by the food taboos, I want to draw attention to two related ideas that are absent from the Murik view of gestation. One is that a successful pregnancy depends in some sense upon masculine nurture and the other is that pregnancy involves a process of fetal "growth."

In Murik, the body of the firstborn fetus is imperiled by the shape, rather than the substance, of the food ingested by its mother. Avoiding certain foods is meant to prevent deformity, rather than create a positive value, such as sex, or lineal affiliation, in a newborn. The fetus is not symbolically fed by the phallus or its fluids but is rather fed by food from both sexes. The fetus does not "grow" inside its mother so much as it avoids dietary impurities. Unlike the Arapesh, the Murik are not Voltaireans, who retreat to their gardens and perceive the world in terms of cultivation metaphors. They are preoccupied by a particular, aquatic kind of hunting and gathering in which significant food production processes take place *outside* of the body politic by means of social relations and vehicles that criss-cross exterior space. Consequently, the Murik view of intrauterine processes takes a visible, extrauterine, social form.

Pregnancy is susceptible, it would seem, to shapes of food, to conflict and collective activities. Women should therefore withdraw from society during it and confine themselves inside the dwelling houses of parents and siblings. In turn, the most important thing a man can do, apart from provisioning his wife with the best foods, is remain chaste. Since his sexual conduct can cause miscarriage, stillbirth, or infant mortality, during pregnancy and the post-partum period, a husband should try to be faithful to his wife, and suspend his love affairs. Should he tarry, he "sends a message" to the mother and her fetus that they are "unloved and unwanted," a message which may cause an unborn soul to desert the "canoe-body" of its mother. The baby may then die prior to, during, or soon after birth. An expectant father should also subscribe to other taboos. He should not "tie knots" in ropes, "chop" wood with an axe, or "erect" houseposts. Each of these actions is believed to harm the birth and the infant in particular ways which recall the restrictions to which a steersman's wife should submit while her husband is aboard an outrigger canoe sailing on the open sea.[2] In Arapesh culture, the husband is enjoined not to stray from the birth scene. For like his wife, he is also said to be "having a baby" during her labor. This is why the new Arapesh father must restore his masculinity a few days later by "capturing" a shell ring – called an "eel" – which he gives to one of his affinal sponsors (see also Tuzin 1972). The Murik father also gives birth; he should repair to his mosquito net and loosen the clothing around his waist to ease his wife's labor. Along with the image of the "bat mother" spirit, the dietary restrictions, the pre-partum sex taboos and the artifice of the couvade fictions disperse the agency of the uterine body by insisting that, in addition to the birth mother, men also exert some influence over the birth and life of the newborn (Meeker, Barlow and Lipset 1986: 40).[3]

A woman in labor must repair to the birth house (an outbuilding located on the periphery of the community). Although the health of both senior men and women is vulnerable to supernatural residues of impure blood expelled during birth (*be manumb*), birth pollution is feared by Murik men as the most virulent form of female impurity (Hogbin 1970; see also Kulick 1992: 92–9). Men who come into contact with women who carry it develop shortness of breath and painful weakness in their joints in old age. Now "breath" is thought to come from the mother. During the final contractions, women are therefore urged to breathe deeply and long, in order to "give their breath to the baby." A mother also used to hurry a slow birth by smoking a cigarette, then holding her breath to force the child out of the womb to breathe (Schmidt 1926: 45). If the fluids expelled during birth eventually cause something akin to emphysema (or perhaps asthma) and arthritis in male elders, then birth, which is branded in this context as largely a maternal capacity, both gives and takes vitality from the community. But it is important to be clear that this is no "limited good" theory of reproductive fluids (see, e.g., Meigs 1984; Godelier 1986). Murik procreation beliefs do not avow that sperm "lost" during inter-course directly diminishes a man's strength. They rather tie that impairment to ongoing contact with the "mystical residue" of the blood of birth; to spirits in it (*be menumb*) that take the form of invisible maggots, which well up in and deplete the male body. The maternal body is dispersed but not divorced from its own agency.

The defilement of birth is only the worst form of the perpetual affliction created for men by their relationships with fertile young women. Contact with menstrual blood (*usiinog*), or its residues during sexual intercourse (*son menumb*), also debilitates the agency of men as they work outside the community. Men may therefore "worry" (*gara'u*) about such pollution, particularly when they eat dishes of food they suspect to have been prepared or touched by menstruating women. When young, men are advised not "to worry" too much about making contact with feminine blood because to do so is also to invite illness. But as they age, they become increasingly fastidious and will try to avoid treading upon the shadows or footprints of young women (many senior men, one notes, wear thongs for this reason). Nor will they walk beneath skirts which have been hung up in the rafters of a house. At the same time, and herein lies what is distinctive about the ambivalence Murik men feel about the interior of women's bodies, even though it causes illness, both young and old men are held to be absolutely powerless to resist the allure of feminine sexuality. "Black blood" (*yaron nungungu*) will inevitably collect at the base of a man's spine. And this fluid contains the supernatural traces of feminine, sexual impurities. Thus the dilemma: to be a Mangrove Man is to be a lover, but to be a lover is to defile the body. There is a secret, cosmic antidote. But it is not a vaccine, only an ablution (see Chapters 6–7). Now we can see that Murik reproduction beliefs fail to dislodge procreative force from the uterine body; all they succeed in doing is to create more defilement for men than women.

Maternal nurture

The Murik soul comes to a fetus from an "ancestor-spirit space" (*pot kaban*). But the soul is no reincarnation. It has no zero-sum relationship to the recently dead. Being attributed no genealogical substance, the soul plays no role in conception. It is a life-force, to be sure, which animates the "canoe-body" (*gai'iin*) of the unborn child at some unspecified point during pregnancy.

When a child is born in Murik, one of the women present immediately takes a belt and *ties a knot* in it in front of the child's eyes, ears, nose and mouth. *This is to secure the child's . . . [soul] in his or her body.* The Murik believe that the spirit can slip out of the body when a person is upset and cries or when he or she sleeps and dreams. Because babies sleep and cry a lot, they are believed to be in danger of slipping out of their bodies and going back to the spirit world from which they came.   (*Barlow 1985b: 33; italics mine*)

Because of the vulnerability of the soul, both mother and newborn baby ought to remain secluded in the birth house for several months until the infant becomes strong. During this time, the two should be provisioned and attended by the mother's mother and her younger sisters. A new mother should nurse her baby on demand, always assuming more is better. However, as soon as what remains of the umbilical cord falls off, after about ten days, she should also begin to feed the child sago jelly (*nimbon*) dipped in water or fish broth. Learning to like this "social" food, which is imported from inland "trade mothers," is considered essential to becoming "Murik." A mother should feed the sago to her baby prior to nursing, pinching off tiny bites to give it until the child refuses to swallow any more. Adding sago jelly to the infant's diet is not meant to begin a weaning process. On the contrary, breastfeeding is prolonged and admired in Murik. Indeed, rather than claiming that mother's milk (*ninge'arum*, literally "breast liquid") is really male semen transformed as did Godelier's Baruya (1986), Murik men and women are touched by a nostalgia for the bliss of infantile satiation at the mother's breast and gaze longingly at a baby nursing itself to sleep. For the Murik, the peace, pleasure and intoxicated satisfaction which they value in this image is one of their quintessential images of morality. As a powerful ideal of indulgent nurture, this visible, maternal image, which is for them an image of reproduction that stresses dependency, security and solidarity, epitomizes the schema whose elements are recreated and contested in different contexts throughout the culture. I shall return to, and expand upon, this maternal schema below. For the moment, I just want to point out that weaning can come to be something of a problem, given such a hallowed view of nursing. A ceremonial procedure may be staged, usually for a last-born child who, at an age of five or even older, still refuses to accept being weaned. A senior woman, who knows the magic, may scratch a child's belly with a sharp shell, or a bit of glass, and squeeze the skin until a few drops of blood begin to appear. Her intention is to let out the "bad blood" from the mother's womb which is making the child crave the mother's milk and is preventing it from being nourished by other "social" foods.

In Murik, the soul is imbued with no biogenetic fluid contributed by one or the other parent. Indeed, the resulting fetus both is and is not tied to the mother's body. On the one hand, the fetus receives blood and breath from its mother and may be mystically affected by her diet. On the other, the "true mother," the mother who determines the sex, appearance and personality of the unborn child, is not the birth mother at all, but a womb spirit, the "bat mother," who is independent of both birth parents. In her heightened state of vulnerability, enemies of a mother's kin can harm or even kill the fetus. A woman is thus withdrawn from the community while pregnant and during parturition. This is done for her own protection, but also because, being of unclean body, she poses a hazard to the health of senior men (and women). During birth itself, the husband may assist at a distance by loosening his belt or untying his waistcloth and immediately after the birth, one of the midwives ties a knot in front of the baby's face to secure its vulnerable soul in its body. The mother's breast is almost immediately supplemented, not with some kind of masculine substance but with sago pudding, the "bones of the ancestors," a food that is provisioned by "trade mothers." Although an attempt is made to decenter the mother's body as the primary reproductive "vehicle," taken as a whole, Murik conception, gestation, birthing and nursing beliefs nevertheless define the person primarily as a mother's child. The kind of dialogue taking place in this imagery is not between male and female. Processes taking place within the mother are cast out of the community. To these are appended honorable fictions that men influence the health of the fetus and its successful birth. Similarly, nursing, while laden with significance, is ritually discriminated from the ideal of a mother who abundantly and generously feeds her children "social" foods, indulging them whenever they want to eat. These two aspects of a mother's body, her internal physiology and her exernal nurturant and other custodial capacities, are not morally equivalent. The first one, which is an image of her uterine fertility, is defiled. The second, which is an exclusively social, or visible, image of the maternal body, and has to do with androgynous practices of foodgiving (see Plate 1), is celebrated and serves as the schema which organizes the canonical meanings of jural influence and authority. If the exterior image of a custodial mother is a contrary rejoinder to "her" uterine force, perhaps it is not surprising to report the existence of a parallel and entirely "androgynous body," through which reproduction seeks to become contingent upon foodgiving, exchange, and explicitly disengages itself from any and all consubstantial ties to the birth mother.

Cutting the breast; or, taken children

Adoptees are literally called "taken children" (*sangait najen*).[4] Well-defined norms guide claims prospective parents should make "to take" a newborn. In each step, gifts of food must be offered. A couple who wish to adopt a child (usually from a sibling) should work for the birth mother during her pregnancy and post-partum seclusion by supplying her with fish, sago, coconuts and firewood. As well as providing relief for the husband and mother's family from having to support a child they will not rear, these gifts demonstrate the interest in and ability of the adoptive parents to do the necessary work. In order to be able to claim that she has fed the child herself from its earliest days, soon after an infant is born, the adoptive mother may take it and encourage it to suckle. A secondary claim is to give the baby a personal name. A newborn may receive names from both parents as well as from as wide a range of bilateral kin as may want to make a partial jural claim in him or her. But it is notable that, out of the many names an individual might receive, the name given a child by its mother (*ngain ya'ut*), whether biological or adoptive, is said to be the "true name" (*nogo ya'ut*).

Several months after birth, when mother and child emerge from seclusion, the adoptive parents should make a feast for them and deliver a "gift" in exchange for the child (about US$25 might be given in 1981, but formerly the gift was made up of pigs' tusks or dogs' teeth ornaments). This gift, called "cutting the breast," is meant to debar any further jural claims by the biological parents who, upon accepting it, are supposed completely to surrender rights in the newborn. No one should subsequently advise an adopted child of the identity of his or her birth parents until the child is grown and has children of his or her own. An adult who is told about his or her adoption should flatly deny knowing anything about their identities by saying: "I have no other parents than the ones who raised me."[5] Reproduction is taken another step from the uterine body; it is moved into a realm of activities (foodgiving and the ceremonial exchange of valuables) and transactable parental rights and duties that are asserted and fulfilled by both genders together or separately (Goodenough 1970). This latter realm, notably, is assisted by masculine forces. Thus the opening paragraphs of the autobiography of Sir Michael Somare:

It was during . . . [a] feast that I was adopted by my Uncle Saub. This adoption ceremony was the first really dramatic event in my life . . . During the height of the festivities, when the *brag* masks were dancing, my father picked me up and placed me between the two masks. The masks lifted me up and, between them, carried me over to my Uncle Saub. Then Mog, the daughter of [my father's] deceased brother . . . was handed to the *brags*. The *brags* carried her to my father.   *(1975: 2)*

In the imagery of adoption, in other words, the last remnant of uterine dependency is divorced by "cutting the breast." Motherhood – as discourse – is retrieved from the moral periphery and restored to the center of collective life where two sacralized "bodies" assert themselves: a jural one which involves both sexes, and a cosmic one which is exclusively masculine. Of course, both of these bodies remain entangled in a complex, albeit implicit, dialogical relationship with the body they shun and fear, the uterine body (see also Lipset and Stritecky 1994).

Creeper vines and canoes

Murik kinship terminology follows a generational pattern. As in Hawaii (see Handy and Pukui 1958), kin are stratified horizontally into layers. However, by contrast to the Hawaiian system, a Murik actor will place himself in one of six, not five, generations. A second difference is that, while same-sex siblings of each parent are terminologically classed as "mother" or "father," the cross-sex sibling of each parent is terminologically discriminated (bifurcate merging). All children of uterine and collateral same-sex siblings are one's "own" children. All children of cross-sex siblings are one's nephews or nieces. The reciprocal grandparent and grandchild terms do not discriminate sex. Lastly, a single great-grandparent/great-grandchild term is used reciprocally. That is to say, as in Bali, the generational layers are "bent" so that youngest becomes identified with eldest (Geertz 1973: 372).

For any given ego, the horizontal axis of Hawaiian systems is an extended sibling set.[6] Siblings, half-siblings and collaterals, that are called "cousins" in English-speaking cultures, the Murik aggregate as elder or younger, male or female, "siblings" (*nog*).[7] Now the most common meaning of *nog* is "creeper vine," a thin, dried reed used to bind ornaments, draw women's skirts and lash outrigger canoes (see Plate 9). As a vine, *nog* is thinner than "rope" (*makon*), from which it is linguistically and botanically differentiated. Unlike exchange, leadership or the male cult, "tether" imagery must be one of the most commonplace but least theorized symbols of association in Melanesian cultures (see. e.g., A. Strathern 1971). In the Sepik, I can cite two examples in which the significance of this metaphor appears unequivocally. Among Iatmul, each patriclan has its own "knotted cord," which is a rarely seen, highly mysterious mnemonic device. These cords are used to help men remember nothing less than the creation of the cosmos, "the primal migration and . . . the name of the crocodile whom the founder . . . followed from the place of origin to the region now settled" (Wassman 1991: 51). In other words, the cords represent "the mystery of Iatmul culture itself" (Harrison 1993a: 406). Along the lower Sepik, of course, Margaret Mead (1935) discovered the use of another tether metaphor, the Mundugumor "rope." Literally, the rope in question was an interwoven twine used to make fishing line and nets. Metaphorically, the Mundugumor rope denoted an interrelated, double pattern of inheritance of sacra, ritual services, personal names and the like that linked mothers and sons on the one hand to fathers and daughters on the other, "as if such lines were the plies of a two-ply cord or rope" (Fortune quoted in McDowell 1991: 245). The Mundugumor ropes "began and ended" with bilateral cross-cousin marriages, according to McDowell's reanalysis, and were not descent groups, as Mead believed. Indeed, they were not groups at all but were rather temporary systems of reciprocal ceremonial exchange. Nonetheless, Mead was right to emphasize them, for even forty years later they still seemed to constitute "the core of . . . [Mundugumor] social organization" (McDowell 1991: 263).

In Murik, the *nog* is no two-ply tether, no pair of intertwined relationships. Nor does it represent the origin of the cosmos. Here a "creeper vine" is any sibling group, composed of the uterine, patrifilial, adopted and step-children of a common parent, who eat together regularly, or occasionally. What symbolically is "lashed together" is a group of domestic kin who speak of themselves as commensals and as a bilateral intragenerational network (much less commonly, siblings refer to each other as sharing "one blood"). In everyday life, the term *nog* is also applied as a supplemental prefix to distinguish such commensal kin from collateral kin. According to this usage, the *nog* is differentiated from classificatory relatives who

Plate 9: Murakau (left) and Ginau (right) re-lash an outrigger to its hull with creeper vine (*nog*)

represent the male and female cults. A sister's son will distinguish his *nogo* mother's brother from his classificatory mother's brother who belongs to the male cult. In both the Murik vernacular and Tokpisin, the prefix *nogo* is also translated to mean "true" kin, rather than ceremonial or classificatory ones. But this usage also denotes "true" as opposed to fanciful, or non-sensical, discourse (*kakaowi*). The implication is that siblings lashed together by food are more moral, their communication more real, reliable or objective than that which passes among noncommensal relations to whom an actor is less "bound" and who, as we shall see in Chapter 6, are masked and comical. Finally, I must note that the *nogo ya'ut*, the "creeper vine name," or "true name," is that name ego receives from his or her mother.

Like other Polynesian systems of sibling terminology (Firth 1970), the Murik *nog* is divided by terms distinguishing birth order seniority and sex (see Figure 5). "Elder siblings," both male and female, are referred to as

Figure 5: Sibling terms

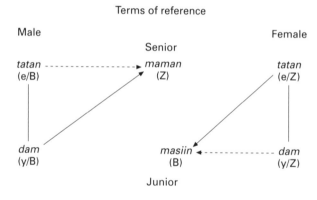

Terms of reference

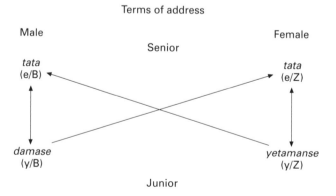

Terms of address

tatan and addressed as *tata*, the diminutive of which is *tatanse*. Younger brothers are referred to and addressed as *dam* or *damase*. There is, however, a separate term of reference and address used by the elder sister for the younger sister (*yetaman*). The birth order terminology (*tatan-dam*) is applied not only to commensal siblings but also collaterally. So long as the name of a common senior relative is known, siblings will inevitably call each other "elder brother" or "younger brother" according to the birth order of their common parent, or more remote ancestor. To know the relevant personal names and to be able to reveal them at an appropriate moment is decisive.

But genealogical knowledge is kept secret and not freely shared. Nor is it immune from deliberate elision. As such, sibling identity is fraught with ambiguity and optation, since birth order, especially as it branches out to collateral siblings, who can be identified through both mother's and father's siblings, is not clear-cut. As he introduced me to "our younger brothers" throughout Murik society in 1981–2, Murakau Wino, the man who adopted me as his "younger brother," would boast to me how my seniority in his sibling group was equivalent to his. This went on and on for many months until one of these "younger brothers" quietly suggested that I ask Murakau about his birth mother. When I did, Murakau admitted that he had himself been adopted into "our" sibling group and was "only a taken son" (*ma kuja sangait goan*) who remained eternally insecure about his senior status in it. From the point of view of discursive practice, that is to say, the use of sibling terms is not unmotivated. Nor is it unrelated to the cultural dialogue with and against the maternal body.

Nor is it entirely gender-free, but nearly so. Younger sister may address elder sister as "*tata*," just as younger brother may address elder brother (see Figure 5). In Polynesia, according to Firth (1970), there is nothing unusual about this. But what is "extremely rare," according to M. Marshall (1983: 11), is the use of an age discriminator irrespective of sex. If it is in their interest to do so, junior siblings, male or female, may and will call an elder sibling "*tata*," regardless of his or her sex. Younger brother may call elder sister "*tata*," or choose to use the cross-sex terminology (the "sister" term is *maman*; the "brother" term is *masiin*). In other words, no gender distinction encroaches in Murik age-terminology. Brother and sister may address *each other* by personal names, cross-sex terms, or appropriate elder/younger distinctions depending on the context. Both men and women, in short, can serve as "elder siblings." Today, it might be said that the main point of Firth's argument is that Polynesian sibling terms provide a gender-exclusive framework for moral action: the terms support a masculine structure of

Table 3. *Norms of Murik siblingship*

1. Elder brother is morally superior to, and has authority over, younger brother.
2. Elder brother should nurture younger brother, accede to his every request and avoid taking resources from him.
3. Younger brother should act with deference and restraint to elder brother: he should by loyal, willing and ready to work for him.
4. The ethos of siblingship should be modest: open reference to sexuality, violence and emotion by the younger brother is forbidden in the presence of elder brother.

politico-jural authority. In Murik, the terms and moral ideals of sibling-ship apply equally to both genders, both domestically and with regard to politico-jural matters. Nevertheless, since this ethnography largely focuses on Murik men, I shall phrase these ideals in fraternal terms.

The *tatan* is the firstborn of his sibling group and sometimes is called its "canoe-prow" (*gai'kev*) or "head" (*kombitok*). Elder brother is first in pres-tige, wealth and authority (cf. Bateson 1932: 287; Harrison 1982: 155). The corporate estate of the sibling group is vested in him. However, what is cul-turally stressed about the role is not his authority, wealth, or property rights but the ideals of indulgence, self-denial and quietism to which he should aspire. He ought to nurture and protect junior kin, his younger sib-lings and their children. Elder brother should withhold nothing from them (Table 3). Elder brother should always "give in" to younger brother or sister (Barlow 1990). Elder brother should never refuse their requests for tobacco, betel nuts, food, tools, or the use of lagoon canoes. As the cliché goes, he should allow younger brother to go through "his basket and take from it what he wants with his own hands." Younger brother should have complete and open access to all his resources. In Sahlins' old terms, reci-procity between elder and younger should be generalized. Exchange in siblingship is "a sustained one-way flow. Failure to reciprocate does not cause the giver of stuff to stop giving: the goods move one way, in favor of the have-not, for a very long period" (Sahlins 1972: 194). In Murik, this sustained one-way flow is one element in the schema of an abundantly nurturant mother "who" is an androgynous rejoinder to the eroticized, uterine mother.

An ideal elder brother should be proficient at advancing his interests through generosity and caretaking. His expectation is that, should he perform his duties well, his younger brothers will eventually reciprocate services and resources freely out of gratitude. In order to build a dwelling, male cult house, carve a canoe, or help mount a trip to an urban market or to the islands, elder brother may request that younger siblings work in the name of their sibling group. But in an effort to maintain his indulgent image, he should be too ashamed, even humiliated, to ask for anything from his junior kin directly and may go to great lengths to avoid making such a request. "The achievement of control and power through giving, feeding and manipulating others' dependence is the openly acknowledged prerogative of senior siblings" (Barlow 1990: 103). Voluntarily, the younger brother ought to defer to the wishes of his elder brother. He should "work" for his elder brother, carry his messages, do his bidding in the male cult house (see Chapter 8), fish and trade for him, go to market for him, and help him in ritual projects. Being subordinate to and dependent upon his *tatan*, he ought to seek him out and not require the elder brother to dishonor himself by having to ask for a favor (cf. Shook 1985: 6). Relations between younger and elder brother should be deferent and circumspect.

Although elder brother possesses resources and authority, he is nonetheless thought to be "enslaved" by his younger brother, who must, for his part, walk a "fine line [between] . . . exploitation and gratitude . . . More envied than admired," the role of the younger brother combines the freedom to make requests for goods and services with the obligation to support the interests of the sibling group as a whole and its leader in particular (Barlow 1990: 104). Younger brothers wish to appear both competent and grateful for their elder brother's efforts to help them. But a threat also sanctions their conduct. Elder brother may taboo his *dam* from using the sibling group's property, its outrigger canoe, coconut stands, or crab channels. "We are afraid," one man told me, "of disobeying the word of our *tatan*." But more commonly, elder siblings use a tactic meant to expose the failings of junior siblings called "setting an example" (*moan sikemo*, literally "to show a thing"). Should a younger brother neglect to fulfill an unspoken expectation of an elder brother, the latter may try deliberately to humiliate him by *performing the task himself in public*. Now to be chastised by an older sibling is one of the most horrid experiences imaginable. In a minor incident Barlow analyzed, a younger brother forgot to shove down a canoe into the water before the tide receded, preventing the family from going off to fish. Saying nothing, expressing no rancor, the elder brother simply *did the job himself*, which shamed the younger brother. The silent act of ridicule, which publicly mimicked what ought to have been done, demeaned him as unreliable and infantile.

Generally, however, it is expected that a younger brother will refuse to do what he is directed to do unless the elder brother giving directions is also willing to participate. Influence in Murik must therefore be ready to demonstrate a willingness to fulfill its own warrants. This is further reason why requests for cooperation between elder and younger siblings should only be made indirectly. Barlow has summarized the dynamic in their relationship eloquently.

> Parents actively teach children what their responsibilities and prerogatives are. The injunction to feed and give food is strenuously enforced with even very young children. Aggression . . . is not tolerated, but the young . . . are encouraged to stand up to the older ones. As a result, the older sibling learns to . . . obtain his/her cooperation in specific ways – through distraction, anticipation of needs and wants, exemplifying desired actions . . . to imitate, companionship rather than direction, play rather than assertion . . . Eventually children become proficient at caretaking and achieving cooperation through indirect means. The techniques become a repertoire for handling others which expresses the general attitude that *the most powerful position is that of giver.*   *(Barlow 1990: 113, italics mine)*

Elder brother should also resort to an indirect style of command because, in addition to the self-denial, the acts of nurture and generosity, for which his role is known, he must dissociate himself from violence, conflict and his own passions. Younger brother, meanwhile, is expected to behave as his moral contrary. While the character of the former is stereotypically poised, reserved and reticent about making requests, that of the latter is rash, demanding and ultimately violent. Still, it is elder brother who sets the ascetic, emotionally repressed ethos of their relationship.

"Elder brother," as the adage goes, "should not see the tears of his younger sister." Brother and sister may nonetheless have a more emotionally comfortable relationship than elder and younger brother. The crucial difference between cross-sex and same-sex siblingship is that, irrespective of age, a brother can always expect his sister's nurture (cf. Burridge 1959: 130–1). While elder brother should avoid asking favors of younger brother, a brother may always approach his sister for a meal. However, should the sister be firstborn, she may then expect his labor in return. Elder brother, who is otherwise "like a mother" to his younger siblings, remains forever "like a child" to his sister, whose nurture he jealously guards against the claims of suitors and whose sexual relations scandalize him. There is a story Mareta of Darapap once recited to Barlow about two sisters and their brother in which the tension between a brother's claim in his sister's nurture and her sexual autonomy is well told.

> The elder sister marries and bears a son. But the younger sister is prevented from marrying by their brother. He beats her brutally and incessantly for flirting on the

beach with her boyfriend. Finally, she goes to the beach and presents herself to the boy, who refuses her by saying "You already have a husband, your brother." The younger sister then decides to commit suicide. First, she makes new sago fringe skirts and several other things for her elder sister, in order to placate her for the trouble she feels herself to be causing. She hangs herself from a tree above a pool where her elder sister usually goes to wash. The next day, the sister discovers the dangling body and calls her husband to come cut it down and bring it back to the village. "Such are the consequences," everyone said. "It is the trouble with her brother, of course. He was always beating her." The brother flees in disgrace.

When does the brother begin to attack the incipient sexuality of his younger sister? Just as his elder sister gives birth, which is a moment at which a marriage usually begins to solidify and his claim in her is diminished. He tries to assert rights in his younger sister and control her desire. But as the sister's suicide suggests, the brother's claim is limited by her preferences. Implicit in the story is that when the sister begins to have sexual intercourse, marries and then becomes a mother, a brother's ongoing claims in her are irreparably damaged (Meeker, Barlow and Lipset 1986).

Siblingship and marriage

Marriage in Murik begins with the public recognition of sexual relations and the establishment of co-residence. Therefore it is of no small moment that there are two rival claims, not only about who has the right to transfer a woman's sexuality but also about postmarital residence. According to the first, the right to transfer the sexuality of a woman is vested in the self and marriage is entirely determined by personal preference. Marriage may begin in flirtation, seduction, or abduction, which is then legitimized when the couple moves in together, eventually to build a house and have children. Brideservice to the spouse's parents or elder brother is expected of the husband, but no brideprice is owed. Postmarital residence is bilocal, determined entirely by situational factors, such as the availability of land to build a house, childcare, and crab channels (Table 4). Now according to the second claim, marriage is defined as the elder brother's or father's right to transfer the sexuality of an unmarried woman, who ought to be eventually reciprocated by the transfer to a wife's brother of a woman related to the husband, i.e., sister exchange. Whether or not sisters are exchanged, brideservice is still expected of the groom by the wife's brother and parents. In this second system of claims, residence is virilocal and the husband is viewed as "a flying fox [a bat] who steals the sister and takes her to a foreign land where she bears her seed." At marriage, the brother ought to feel the loss of his sister acutely, which is a loss not only of her nurture

Table 4. *Postmarital residence in eastern Murik villages in 1982*

	Darapap	Karau	Mendam
Virilocal	45	12	13
Uxorilocal	14	4	12
Neolocal[1]	9	6	29
Bilocal	1	0	0
Other[2]	1	3	30
n	70	25	84

Notes:
[1] These couples were resident in towns and cities.
[2] I do not have data to classify these couples.

Table 5. *Eastern Murik village exogamy in 1982*

	Darapap	Karau	Mendam
Village (%)	28	27	15
n = 68	27	7	34

but of her reproductive capacity (she will bear her seed in a foreign land) (Table 5). The wife's brother remains the jural rival of her husband for claim in his sister's children (which still remains the case in the absence of sister exchange).

In the 1980s and early 1990s, the practice we saw mixed the two claims. Sister exchange was reported in Murik genealogies but only occurred in a minuscule number of living couples. In none of the marriages whose inception we observed did it occur, or was it claimed. Brothers were commonly called upon, or called themselves, to protect their sisters from abusive husbands. But what claim had they in their sisters' sexuality? Marriage was a matter of personal choice and, except in intervillage cases in which it was usually virilocal, residence was not a matter of prescriptive concern. In no case was bridewealth ever demanded, much less given.

Marriage is independent of ceremonial exchange: it goes virtually unmarked by ritual. No irrevocable rights arise from it. Within the village, marriage is not regulated; nor is it subject to prescriptive rules or norms. Polygyny is considered to be a hereditary right, permitted to the man whose father had more than one wife simultaneously; polyandry occurs but is a subject of dispute. The one incest proscription asserted by parents is that siblings whose grandparents had the same mother or father, or who grew up eating together in the same household, should not marry. Nevertheless, due to overlapping, bilateral kinship relations, sibling endogamy is frequent. Collateral consanguines commonly marry. Murik couples eventually value their marriages and develop emotionally close relationships; first marriages are brittle and typically end in "divorce," divorce simply being the return of one spouse to the house of his or her family of orientation. Several first marriages, which last only a few months or less, inevitably precede a more stable, long-term relationship. But the latter may continue to be disrupted by old, or the start of new, extramarital relations. A Murik husband reserves exclusive use-rights in the work, nurture and sexuality of his wife. And reciprocally, a wife reserves the same claims in her husband. But, for a number of reasons which I shall discuss in Chapters 7 and 8, neither spouse monopolizes their spouse's sexuality. Jealousy, suspicion and domestic violence are commonplace in married life, particularly among the young (see Counts 1990).

When a couple "marries," they begin to observe a hierarchical pattern of courtesy with their respective affines. The modesty which is called for takes its strictest form between younger brother's wife and husband's elder brother.[8] A senior man and his younger brother's wife substitute an avoidance term of respect (Murik: *wandiik*; Tokpisin: *tambu*) when communication between them becomes absolutely necessary, or else the junior spouse may be addressed as "wife of Kaparo" (*Kaparo'na neman*) or simply "Kaparo" (see Figure 6).[9] Because of the indirect, indulgent quality of his influence, elder brother is forbidden to make claims upon "the resources" of his younger brother. In everyday life, fish, sago products and rice are all cooked by women. Elder brother must be particularly careful to avoid food cooked by his younger brother's wife. Shortly after the younger brother marries, however, he may go ahead and "test" whether what she cooks will make him nauseous. If not, then he may elect to eat food prepared at her hearth. In any event, in his shame, he may not speak her name, may not talk to her, or even look her directly in the eye. Elder brother must avoid her body and person. More: he should avoid space identified with her. Therefore, he may not enter his younger brother's house.

Since elder brother is jurally superior to younger brother, the major responsibility for avoidance rests with the junior wife. She should keep a proper distance from him. Should she encounter the husband's elder brother on a footpath in the village, she should "break bush" by quickly jumping out of his way. During feasts, when younger brother's wife is expected to help elder brother prepare food and may have to enter his house, she should stay close to the door and should not leave her knees in his presence. And I have watched women "breaking their knees" as they move about the floor of the house of senior male affines. The culmination of these avoidances is a very strong emotional taboo between elder brother and younger brother's wife. They must behave with inhibited decorum in each other's presence. Should younger brother's wife sit on a bench made of old canoe sidings (*banumtom*) in view of her husband's elder brother, she should not dangle her legs, lest she appear coquettish. She should not carry firewood on her head in his view, lest she flaunt her breasts before him. Nor should she allow smoke from her canoe stove to rise within sight of the village, lest he understand that she is signaling a lover. Reciprocally, should elder brother become involved in an adulterous relationship and news of his dalliance reach the ears of his younger brother's wife, she ought to feel shame about it and keep the gossip from her husband, his younger brother.

Figure 6: Sibling and affines

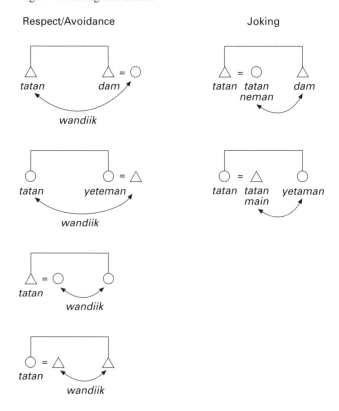

By contrast, few avoidances apply between a woman and her husband's younger brother. A husband may refer to his wife as his "mother" (*ngain*) as well as "wife" (*neman*). He may try to claim exclusive rights to enjoy his "mother's" sexual services. But he must share her nurture and reproductive capacity with his younger brother. For elder brother's wife (*tata neman*) is also termed the younger brother's "co-wife" (*neman gainaro*) and she is expected to act "like a mother" to him as well. These two have no avoidance relationship. Addressing each other by personal names, they may eat and speak together. Elder brother's wife should feed her husband's younger brother at his request: her hearth should always be open to him. He is a classificatory "father" of her children. Should an elder brother die, widowing the woman and leaving her with children, then the levirate may be claimed. Younger brother may marry his elder brother's widow and adopt the children. At the death of a younger brother, elder brother may not marry the widow but should merely "look after her" until she is ready to remarry and has found an appropriate man. His judgment or evaluation of the new husband is then said to be secondary to the preferences of younger brother's widow.

Classificatory younger brothers and elder brothers' wives who do not regularly eat together may enjoy a joking relationship if they like (see Figure 6).[10] They may tease one another about sexual immorality. An ethos of lascivious, mildly aggressive, humor may preoccupy their relations. When younger brother calls this woman his "co-wife" and reciprocally, when she calls him "co-husband," rights in sexuality are being satirized. The "joke" is that the younger brother's "unlimited claim" on elder brother's resources should include the "co-wife" but does not. Elder brother's wife thus becomes an "immoral woman" (*numaro mwaro*) whose vagina "is far too big for one man" but is not available to younger brother. The contradiction grasped and comprehended by their mockery is that, being a "co-wife," she should be sexually available to him, but being elder brother's wife, she is not (M. Douglas 1975: 98). Here we hear another rejoinder to the ascetic, maternal schema which elder brother should espouse. While elder brother is pure and chaste, his wife and younger brother understand sensuality and vice.

An alternative outcome to the joking relationship between elder brother's wife and younger brother takes place at the turning point in the most important legend in the culture. In the last chapter, I discussed the "Two Brothers" story as a representation of the Austronesian migration along the North Coast. Now I turn to the rendering of siblingship in the tale. The day after we arrived in Darapap in 1981, I sat in the male cult house and asked the men convened there to tell me a story. Without a moment's hesitation, Marabo Game immediately launched into it in Tokpisin.

The two brothers

We call the elder brother Andena and the younger brother Arena. Andena married first. His younger brother, Arena, had not yet married. Andena decided to initiate him and the rest of his age-grade into the male cult. Andena sailed to the islands to get the necessary goods and left Mwed, his wife, in the care of his younger brother.

The two set to work to prepare the food for the guests who would attend [the initiation rite]. Mwed paddled down a little channel off the river and dropped lines, Arena stayed at the mouth of the channel and pounded sago pith. After a while, he finished working.

To pass the time, he picked up a piece of white sago bark and began to engrave one of his designs onto it. Absentmindedly, Arena then dropped the piece of bark into the water, and it floated down the channel where it reached Mwed, his elder brother's wife. The woman noticed the design on the sago bark and greatly admired it. Pulling up her lines, she paddled back to Arena. "Is this yours?" she asked him, holding up the design. "It is very beautiful. Would you tattoo it upon me?" She placed the bark all over her body, on her breast, face, underarm, trying out possible places until she had wedged it squarely between her legs. "Would you put it here?" She opened her legs to Arena and he had sexual intercourse with the woman. Afterwards, while she slept, he tattooed the design across her vulva.

Mwed's pain was severe. When her husband, Andena, returned from the islands she did not come to meet his canoe to help him unload it. Andena suspected something was amiss. Being a spirit-man, he knew everything. On the way upriver, he had spotted a coconut husk which his younger brother had used to clot his wife's blood. So when Andena insisted that she help him pull up his canoe, he was not surprised to see her grimace as she entered the water. But he did not respond overtly: instead, he set to work building a cult house for his younger brother's initiation. Inviting men from nearby villages to carve designs onto the centerpost of the cult house, he planned to expose his younger brother by matching his carving style with the one on his wife's genitals.

After each man finished working, Andena fed and sent him home. Finally his younger brother carved a section of the post which matched the design on his wife. But again, Andena hid his anger and went on speaking and eating with his younger brother. When work on the frame of the building was completed, Andena had his younger brother dig a big, deep hole to sink the centerpost.

If Andena was powerful so was Arena. He anticipated what his elder brother was up to. Just before he told his elder brother that the hole was ready, he dug a sidetunnel in which to hide himself. Andena invited surrounding villages to help erect the centerpost. The men came and uplifted it.

"You climb down into the hole," Andena told his younger brother, "and guide the post." Arena had been chewing betel nuts and his mouth was full of red juice. When the men lowered the post into the hole, red juice came shooting up, which looked like blood. The men looked at each other in horror. Andena began to scream: "This man raped my wife! He raped my wife! He raped my wife! He is no good. Quick! Fill up the hole."

At dusk that afternoon, Arena was still hiding in the sidetunnel when his mother came to cry over her son's murder. He came out of the hole to comfort her. "Uhh!" she exclaimed. "Who are you?!" After assuring her that he was not a ghost, Arena requested that she get him something to eat and retrieve his stone axes because he wanted to go cut down a tree for a canoe. Next day, he set to work hollowing out the log.

Meanwhile, Andena, his elder brother, went to hunt for wild pigs with his dogs. Andena came upon his younger brother at work on the canoe hull. "You look like Arena, my younger brother," he said, as if not recognizing him.

Arena shook his head. "People look alike. What happened to him?"

Andena told his story. Admiring the beautiful work of this stranger [his younger brother], he asked him to make a second canoe. Arena agreed. He told his elder brother to beat plenty of sago [pith] so that both of them might go to the islands [to trade].

The elder brother [then] created the jungles of wild sago swamps which line the banks of the lower Sepik River. When the canoes were completed, the two spirit-men pulled them down into the river and loaded them up with sago. They paddled down to the mouth of the river. Arena lashed outrigger floats to the hulls. For his elder brother's canoe, he used weak beach vine (*wasami*) for the lashing, while for his own canoe he used a much stronger strong beach vine (*nog*).

At dawn the next day, the two vessels set sail. The elder brother's canoe went first. The lashing on his canoe tore apart. The outrigger split up. The canoe sank. In the water, Andena called for help.

"No! I will not help you. I am Arena, your younger brother, whom you tried to kill! Now you will die!" The younger brother sailed away in an easterly direction. But the elder brother did not drown. He drifted with the wreckage of his canoe to Kaup [the westernmost Murik village]. He married two sisters there and went on to have further adventures among the inland villages.

The younger brother sailed toward New Britain. Some men say he may have ended up in the Talasea area. Others believe that he found his way to America, where he married many women and had many intelligent white descendants. These "younger brothers" returned to colonize Papua New Guinea, and took whatever they wanted from natives, the descendants of Andena, the elder brother, who could not refuse them.

Although versions of the "Two Brothers" story have been collected all along this part of the North Coast (see, e.g., Hogbin 1935; Lawrence 1964; Pomponio 1992; Smith 1994), no analysis of which I am aware addresses the manifest content of the story, which in Murik seems to concern sibling authority. Given the norms of siblingship in this culture, it is easy to see that the two heroes of the tale are contrasted, just as siblings should be in everyday life. On the one hand, elder brother is associated with the creation of the regional landscape and moral-political institutions. His nurturing powers commence its action (his overseas voyage to provision younger brother's initiation, cult house construction and marriage). On the other, younger brother is a deceitful, disguised, trickster figure, a transgressor. The moral identity of one is firm. The other is unstable, requiring repeated assertion. Rather than the sponsorship of collective institutions, the magico-aesthetic design of the younger brother is phallic/erotic, arousing desire, conflict, and ultimately causing the dissolution of society. The one subverts the other. In particular, the younger brother's seductive power over women's emotions disrupts acts of nurture instigated by the elder brother.

Recall the burlesque of sexual flirtation which may go on in relations between younger brother and elder brother's wife. Arena and Mwed also exemplify this comedy. When Mwed, the woman, spreads her legs and beckons Arena "to put it here," she is referring to his design, which is and is not his penis. This kind of flirtatious joking demarks a moral boundary which distinguishes their grotesque marriage from conjugal marriage. The tale thus becomes all the more disquieting for not straying far from reality, but depicts the murderous potential of crossing beyond mock-flirtation into seduction. In keeping with his role, elder brother does not openly express his emotions. Instead, he sponsors a ritual event to use a moral pretext for revenge. In deference to his authority, the younger brother comes to work, to dig his own grave, as it were. During the feast, elder brother then tries to crush him into it with the centerpost of his new building. In response, the body and identity of the younger brother become ambiguous. What appears to be his blood is not his blood. His canoe lashing (*nog*), otherwise a quintessential symbol of commensality and association, becomes an instrument of vengeance and treachery (cf. Lessa 1966: 57).

As a representation of a "primal crime," this story recalls Freud's famous "Just So" story (1946), in which the murder of the father by his sons creates ineradicable guilt in them that gives rise to totemism and the incest taboo. Here, obviously, fratricide has replaced patricide. Here, the consequences are not clan exogamy, religion or the development of individual conscience, but sibling fission and the institution of a regional world (which is ultimately extended to include the World System).

Nonetheless, the disquieting moral is unmistakable: coeval men inevitably come to blows over disputed sexual rights in their "mother." In its realism, the story seems to suggest that society is eternally susceptible to this divisive force. After all, the two ancestor-spirits are "brothers." They do, yet do not, represent a single jural "body" *vis-à-vis* elder brother's wife. In spite of affinal courtesy, in other words, this intragenerational kind of Oedipal jealousy (see Chapter 7) creates an irresolvable dilemma. The relationship between sexuality and the reproduction of society must create ongoing rupture in cultural order. And this imagery exemplifies precisely what Collier and Rosaldo (1981) and later Collier alone (1988) argued about the unstable, highly charged, multivalent import of sexuality in brideservice-based societies.

What is Murik kinship all about?

On the one hand, the physiology of the maternal body is partially denied, dislocated and lodged from women. On the other, external, custodial practices of motherhood are honored and admired. Processes which go on inside the mother's body – conception, pregnancy and birthing – are stigmatized as unclean and must be quarantined from the community. The mother is not the "true mother," but only a "canoe-body," or an intermediary for the fetus: the real mother is the "bat mother." Admittedly, siblings do share maternal "blood" and paternal "semen" which have mixed during coitus. But they are not of "one womb" as the Tangu say of siblings (Burridge 1957a: 61 n. 25), since the womb is not part of the mother's body. Nor are they of "one breast" as the Dobu call their lineages (Fortune 1935), for the breast may have been "cut" by the payment of a childprice. Rather, siblings are "creeper vines" lashing a "canoe" together under the direction of their "prow" figure, the firstborn. It is true that the food they eat together – sago pudding – is the "bones of the ancestors." But eating these "bones" does not create affiliation through a symbolic incorporation of the substances of the ancestors contained in the sago puddings. Eating these "bones," acts of foodsharing under the auspices of a common mother, affiliates siblings to particular "creeper vines" (*nog*) and distinguishes them as sago-importing "Beach Murik" rather than sago-exporting "Bush Murik." This is a dialogue about the moral reproduction of society in which social aspects of motherhood contest interiorized, uterine and erotic, images of motherhood. The peace and satisfaction and dependency of the suckling babe, as well as the generosity and self-denial of a new mother, also appear in norms of the commensal, sibling group. The leader of this group, the firstborn, is no figure of force but a figure of indulgence. His influence is achieved through a generalized form of reciprocity of goods and services which he provides, as the people will sometimes say, "like a mother." Loyalty of junior kin is won by means of tactics which infantilize them and sustain their dependency. Younger brother is not merely subordinate but ethically deficient, "like a child."

While elder brother plays a virtually prudish, ascetic role, observing as he does careful avoidances with junior affines, younger brother appears to be a trickster, whose cosmic desire may disrupt the unity of the sibling group. As the representative of the jural estate and projects of his sibling group, the role of elder brother is morally incompatible with and opposed to passion and aggression, emotions located within the body – in the stomach, to be exact – which inevitably subvert order. At the same time, younger brother, who is dependent and shiftless, specializes in the very fleshly powers renounced by elder brother. Having no title to resources, he cultivates magical charms and lures that aim at seducing the sexual attentions of women. Despite an ascetic ethos, and despite the avoidance relations with junior affines, the interiors of women's bodies, in particular their sexuality, resist elder brother's attempts to defend and promote legitimate order, not only from others, but even from his own emotions. Elder brother is unable to remain a chaste mother. He is no less embodied than everyman. Sexually attached to his "wife/mother," he is subject to the same jealous rages. The embodiment of woman-the-lover dialogically challenges the ethical identity he personifies as woman-the-mother, a challenge which is not resolved but recreated in the organization of marriage and siblingship. The backdrop of moral relations and collective institutions in Murik is an endemic potential for violence motivated by Oedipal jealousies. While the hierarchical norms of siblingship stress nurture, respect and dependency, interior attributes of the body – such as sexuality and aggression – are no less prominent features of communal life. For the Murik belief is that everyone is highly sexual in motivation and action. Like elder brother's wife, women are understood to be sexually aggressive and no man is expected to refuse, or even to be capable of refusing, a woman's advances (see Chapter 7). Both men's relationships with other men and women's relationships with other women are constantly riven by suspicions and allegations of infidelity, and death by sorcery attack is perpetually traced back to Oedipal triangles. While the exteriorized, visible schema of motherhood constitutes the unity, peace and security of society, an internal, erotic image of capricious and flirtatious women and passive men provokes its collapse. The success of the institutions organized by elder brother, which seek to displace birthing, participate in nurture and inhibit passion, is only partial. A child's "true name" remains the name given him by his mother. And the sexuality of women (and men) inevitably inflames society, undermining the solidary values and symbols engendered by elder brothers, upholders of the quietist maternal schema (see Plate 1). But the relationship between these two images of the body is dialogical rather than simply hostile. They are contrary sides of a single argument. The next chapter, which begins with a brief comparative discussion of two other Sepik systems of male authority, continues to focus upon this schema in the context of Murik leadership.

4

The heraldic body

Among the Manambu people of the middle Sepik River whom Simon Harrison studied (1989; 1990a; 1993b), the polity is differentiated by totemic privileges and appurtenances. Each clan is headed by senior, male magicians who claim and assert rights over a finite set of powers which control the cosmos. Rights to these powers do not arise directly from the production of material wealth or demographic success but from a stock of personal names Manambu leaders give infant children. Babies given the same name cause disputes. In public tests, orators then contest genealogical knowledge (cf. Valeri 1985: 155ff). At the periphery of their debates – carried on in stage-whispers – defiant women dance angrily and fight among themselves. The loser must rename his grandchild. The point Harrison makes of this image is that it illustrates what he calls a realist but not a nominalist folk model of Melanesian inequality (1990a). According to the former, society is a totality of relationships that transcends the existence of its citizens. According to the latter, society is a contingent phenomenon that emerges out of political activity. In the western highlands, to take Harrison's example of the latter model, the clan is said to be composed of consubstantial entities, goods and persons, which men produce and deploy in order to make the claim that the polity "grows" on account of what they but not women do (see especially Meigs 1976; Biersack 1982; A. Strathern 1974). Along the middle Sepik, the reproduction of the body politic is imagined in less physical terms.[1] Among the Manambu, as well as among their Iatmul neighbors, a child's personal name is the decisive value. The body is here a site of a linguistically and totemically differentiated relationship between a status and its occupant. This more reified, hierarchical concept of society is a step removed from the hurly-burly world of material exchange. Why? Because Manambu politics, as Harrison concludes, are "less immediately engaged with relations of production" than in the western highlands (1989: 14).

Like its predecessor (Lawrence and Meggitt 1965), Harrison's dichotomy obscures the ambivalent relationship between Melanesian men and the maternal body which conditions the discursive form taken by their rivalries. The construction of inequality in this region takes multiple genres: realist values (e.g., personal names) are used to assert reproductive agency in society, and so are nominalist ones (e.g., pigs). But either way, sight should not be lost of the ways that the spectacles of inequality Melanesian men seek to direct constitute equivocal answers to troubling questions posed by the maternal body and "her" procreative force. Must not the Manambu debates, staged in hushed whispers between male elders debating the genealogies of infant namesakes, while surly women dance about, suggest such a masculine rejoinder? Iatmul *naven* rites, in which images of male motherhood are provoked by developmental displays of initiative and competence by children (Bateson 1936), also come to mind. In both Manambu and Iatmul cultures, symbolic acts of mothering are performed in which relatively masculinized, but decidedly androgynous, forms of collective expression displace, or at least respond to, the maternal body. In the 1930s, Wogeo men were both aware of and quite candid about the dialogicality of this relationship: the reason they played bamboo flutes, the men told Hogbin, was because women bore infants (1970: 101).

Just a few miles to the east of the mouth of the Sepik, another genre of masculine dialogue with the maternal body went on in Kaian village (Meiser 1955). The polity there was differentiated by material culture: each clan had its own distinctive clothing, color, house style, canoe, drums, dances and sacred spear. Most importantly, the adult men of each clan built their own wooden "platform" upon which they (but not their clans-women) might sit down and talk. Although clan built, membership in a platform group "apparently [had] . . . nothing to do with blood relation-ship" (Meiser 1955: 266). Ancestors and other spirits were said to dwell in them and were sometimes known to communicate as tutelaries to the men. The location, name and architecture of a platform demarked political boundaries within the village. Each platform, moreover, had "his" own moral reputation and code. The men of one were homicidal. Another was a den of thieves. Out of a third, indeed, came the hereditary leadership of the entire community. Although uterine motherhood seems to have been denied from the cultural "construction" of the platforms (the clan built them but their membership had "nothing to do" with consanguineal relations), the nurture of women was not. The women approached these structures to bring "food for the men" (Meiser 1955: 258).[2] Whether society is construed in realist or nominalist modes, and no doubt the Kaian case would have to be considered as another realist polity, my point remains that the symbols in terms of which Melanesian politics are conducted consist of deeply gendered, deeply ambivalent, dialogues about a single referential object, namely, a "mother" to whom men respond in a variety of instrumental genres. "This is a conversation, although only one is speaking, and it is a conversation of the most intense kind, for each present, uttered word responds and reacts with its every fibre to the invisible speaker" (Bakhtin 1965/1984a: 197).

This chapter will begin to investigate sacred emblems of hereditary authority in the Murik Lakes. In ritual arenas, moral order again recreates attributes of custodial motherhood which respond dialogically to "her" interior body. Here, quite literally, the reproduction of society is at stake. Ritual elders bestow insignia to children in order to recruit them as members of their descent groups, claims which are legitimized by competitive exchanges of food. Their gifts are not construed as consubstantial exchanges, but as acts of maternal nurture. This is not a public domain, however, in which men coopt women's labor or deny them citizenship. Both genders possess equal rights to hold the offices represented by, and perform the duties associated with, the hereditary regalia. Men and women both use the maternal imagery to negotiate their way through this competitive, androgynous field.

People of the ornament

Recall the dazzling appearance of Andena, the elder brother ancestor-spirit, when he came ashore in various parts of the North Coast in search of his lost brother.

> His skin was flaming red. A net bag with cassowary feathers hung on his chest. He wore a bark cummerbund about his waist, wicker bands on his wrists and ankles, and a beautiful loincloth. Boars' tusks and shell rings he tied into his armbands. He wore a dogs' teeth headband, *Nassa* shell bandoliers, a wicker hairpiece. The basket he carried bore his own name.

The polity in Murik is differentiated by ensembles of jewelry like that of the culture hero. Rights to transact and wear such outfits are held by named, ceremonial groups (*poang*), sections of which are domiciled in each village, but are residentially dispersed throughout the Lower Sepik. In the weave of a Murik basket, the *poang* are points where the threads meet and criss-cross to form the four corners at the base, each corner also being called a "breast" of the basket. As status groups, the *poang* are composed of overlapping, bilateral networks of men and women, the "creeper vine" groups (*nog*) discussed in the last chapter. *Poang* elders usually trace common bilateral ancestry back to an apical sibling pair who migrated into the Sepik estuary and founded their descent group in a Murik village.[3] Use- and transfer rights in the ceremonial property belonging to these groups are claimed by being able to call upon genealogical knowledge, which amounts to knowing several networks of personal names that denote, in the first place, common ancestors, trading partners as well as the sacred insignia (*sumon*) of the group. Although they are not armorial, I prefer to view these emblems as "heraldry," because they are systematized badges of collective identity which are associated with genealogical rank, the cultural attributes of which are held to afford protection to the groups for which they stand. Insofar as they are not armorial, the *sumon* are more similar to Japanese heraldry than to the bearings of European nobility and knights. I should admit, however, that I have never been able to elicit a satisfactory folk translation of the term. Matthew Tamoane, a native speaker with a BA in linguistics, once told me that he thought *sumon* meant "authority," a gloss of which I am skeptical.[4]

The heraldic basket (*sumon suun*) of the Mindamot *poang*, for instance, is called Wankau, which is another name of Andena, the elder brother ancestor-spirit. The basket of the Sait *poang* is called Yamdar (see Figure 7). The

Figure 7: The heraldic basket of the Sait (Yamdar *sumon suun*)

Table 6. *Insignia of the Kaun descent group*

(a) *mari marin sumon*: Braided with cords, with shells set in and four boars' tusks in the middle (two on top and two below), it is worn across the forehead and under the hairpiece.
(b) *mwara aran beron*: A tapa loincloth with red rattan woven into it and shells fastened to it in many places. Four boars' tusks and feathers are attached to it.
(c) *paripat*: A piece of possum fur with three boars' tusks on it is bound to the front of the hairpiece.
(d) *amwan urob*: A long necklace (shells strung in rows and boars' tusks bound in between).
(e) *murup*: A string bag decorated with knots, on each side are three rows of shells and twelve boars' tusks.
(f) large *gimiik*: A woven piece thickly set with dogs' teeth and four boars' tusks (two in the middle and a boar's tusk on either side). It is worn across the forehead and bound under the hairpiece.

Source: Schmidt 1933: 669.

identities of these groups, in other words, are discriminated aesthetically and linguistically by a semiotics of baskets and ornaments which solicit their assembly as a visible, ceremonial "body." Listing the *sumon* belonging to descent groups in only the two villages of Big Murik and Karau takes up many pages in Father Schmidt's pre-war ethnography. The emblems listed in Table 6 make up the insignia of a group which will play an important role in the extended case study I present in Chapter 8.

As in Kaian village, each descent group possesses its own outfit of these kinds of heirlooms. Leaders of the Kaun would bestow them upon first-born children during moments of high ceremony such as initiation. In minor rites, however, only two emblems are ordinarily given, the heraldic basket with its distinctive ornamentation and woven tartan, and a pendant of a circular pair of boars' tusks with some brightly colored bird-of-paradise plumage bound to it by a vine (*nog*). The *sumon*-holding groups are also called the "platforms" or "hearths" (*maig*) for their membership, as in Kaian, but culturally, the most prominent and potent images of their collectivity are the sacred regalia, or, more exactly, their prototypical images. The images, rather than the objects themselves, stand for the corporate estate, jural identity and moral identity of each group. Father Schmidt also recognized the distinction between emblem and image, while disparaging the "secularized" Murik of the 1920s and 1930s for having lost their sensitivity for the sacred value of it.

The *sumon* [spirits] dwell in several objects . . . The *sumon*, and the objects in which the *sumon* reside, or to which they are related, are various for each *poang*. Thus it can be shown that the path from superstition to economical thinking is but one short step – for today *sumon* signifies nothing so much as the productions which are the exclusive property of a particular . . . [descent group].   *(Schmidt 1933: 667)*

Notice how unstable the relationship of the *sumon* regalia to what is sacred about them appears in this passage. The *sumon* seem to be spirits "living in" or "occupying" a "house" or a "body" in Schmidt's initial phrasing. But then they become "related" in some more generic sense to ornaments. Neither metaphor specifies that the *sumon*, as Schmidt understood them, stood for embodied, totemic substances, however.[5] From my 1980s and early 1990s perspective, I would agree that these sacra are not signs either of cosmogonic power or consubstantial ties to an originary species. But certainly in Schmidt's day, as in mine, their use and abuse was believed to be closely monitored and sanctioned by both ancestor-spirits and rival ceremonial leaders. Their deployment and display were therefore embedded in minute protocol and carefully observed taboo.

The rituals (*gar*) during which *sumon* heraldry is displayed celebrate the recruitment of persons (or the consecration of a new house or an outrigger canoe) into particular descent groups. The insignia are not merely "exhibited" at such moments. They are "conferred" upon recipients. Such transactions require the verification of the *bona fides* of the donor, the local holder of his or her group's *sumon* insignia. In the case of important "property" not collectively owned, such as the financial remittances of children, or magic acquired individually during a temporary period of work outside the region, the dispersed *poang* members are not called upon to gather, and are irrelevant. But to bestow heraldry upon individuals in a high ritual setting, such as initiation or succession, a group of elders must be convened to share a pig feast and assess the sponsor's genealogical credentials to transact the *sumon* in question. The guests should bring jewels with them, boars' tusks and shell rings, literally to use in the assembly of the heraldic emblems which the recipient will wear during the climax of the ceremony, when he or she appears in public bearing the regalia of their group. In one instance we observed in 1981, a descent group leader sponsored a succession rite during which he received the right to hold and transact the *sumon* of his group.

A senior man named Kanari succeeded Pa'iin, his elder sister, as *sumon*-bearer of their sibling group. The day after the ceremony, insignia were disassembled and a group of about thirty senior members of the *poang*, both resident and visiting, were present to retrieve the boars' tusks and bird-of-paradise feathers which they had contributed to the outfit in which the retiring incumbent had appeared.

Men sat on one side of the house. Women sat opposite them (younger women stayed near the door). Kanari's wife and daughters had readied a meal of fish and pork for everyone. The two most senior dignitaries, guests visiting from the villages of Karau and Darapap, received the liver of the pig set atop a large plate of sago pudding. Kanari also dispensed Bacardi rum and some other sweet liqueur I did not identify.

Ginau, an elder visiting from Darapap, stood up and praised the celebration for having gone on without conflict and for having "raised the name" of the Sait *poang* to "cover up" all the rest. He then made mention of common ancestors, Nibu and Kerok, two brothers who had migrated to the Murik Lakes five or six generations ago: their sons' sons and daughters, whose names he also called, had since resettled in different Murik villages. Kanari was a descendant of Kamanda, their daughter, while Ginau himself was descended from Katem, Kamanda's brother. "The names are true," he concluded, speaking directly to Kanari, "Your claim to the *sumon* in Big Murik is just. Use them as you please." Ginau then sat down to his plate of liver.

While the *sumon* were dismantled, wrangling arose about who had brought which boars' tusks. Jaja, an elder *sumon*-bearer visiting from Karau, explained to me that the purpose of Ginau's speech was to give Kanari to understand that his rights over the *sumon* were now complete and he should feel free to decorate children, or display the ornaments whenever he wanted.

The idioms of association used in this meeting, the group of men and women assembled in a domestic space, the liquor and the meal, the naming of common ancestors of both sexes and the retelling of their migration history, the ritual accomplishment and, the basis of it all, their common entitlement to the heraldic appurtenances, made no reference to substances, fluids, or to any aspect of a common anatomy they might share by virtue of holding common *sumon* regalia. In 1981, neither the symbols nor the continuity of this group, nor the title to which the heir succeeded, were understood or expressed in terms of a uterine body, or any kind of totemic refraction.

Perhaps Father Schmidt was right. Perhaps in premodern understandings of *sumon*, some sort of spirit did "dwell in" or was "related to" the regalia, belief in which then atrophied under his ministrations and the other colonial forces. If so, we might well ask, what gender did such a spirit possess? Or, did it depend on the gender of the recipient of the ornaments? Or, was it always male or always female? Certainly, the *sumon* spirit could not have been a genderless spirit, given that all the rest of the spirits in Murik religion possess one sex or the other.[6] Except to speculate, I cannot say what may have happened to the meaning of these objects during the twentieth century. The contemporary deployment of these images certainly adheres to both male and female forms. In this episode, men and women gathered together in the dwelling house of their host rather than in his male cult house, the retiring incumbent had been an elder sister, and the toast praised the inclusiveness and success of the whole heterosexual group, whose ceremonial activity had "covered up" their rivals in the community. The claims made about this moral body asserted neither the dominance of male over female nor the hereditary authority of one group over the rest of the community. The spirits of the *sumon* have neither a fixed gender nor a single descent group to which they grant legitimacy.

The import and influence of any one descent group at a particular instant in a particular Murik village, or elsewhere in the Lower Sepik, depends entirely on the level of ceremonial activity and regional exchange relations maintained by its most senior sibling group in whom its heraldry is vested. No single *sumon*-holding group presides over others on the basis of genealogical credentials. No linguistic terms exist to discriminate and privilege an elite from commoners in Murik as they do among the Manam Islanders offshore, where a single hereditary lineage rules an entire village (Wedgwood 1934: 384; cf. Lutkehaus 1982). Each descent group is politically, economically and ritually equal. The Murik villages are acephalous. "There are . . . no chiefs," said Father Schmidt. "The people are organized in several *poang* [descent groups]. One *poang* does not stand above the others" (1926: 41). Or, as Murik elders like to put it by way of contrasting themselves to the Schouten Island leaders: "There, only one man holds the boars' tusks. Here, we all hold them" (see also Lutkehaus 1990).

Each Murik *poang* is internally stratified, however, by age. Its constituent sibling sets are ranked *vis-à-vis* each other according to the seniority of the parent through whom that group traces its hereditary claim to the group's *sumon* heraldry. The ceremonial leader of each sibling group, its firstborn, is called the *sumon goan* (literally, "heraldic son") or *sumon ngasen* (literally, "heraldic daughter"). The senior *sumon*-holder occupies the kind of office Sahlins once dubbed "non-chiefly rank" (1963: 294 n.). Now it is true that the ambitious *sumon*-holders do try to outgive their rivals in order to defeat them in the ceremonial arena. But I never once heard of any ambition to establish a hereditary dynasty out of their achieved position as preeminent feastgiver in their descent group.[7] Except in one case: in a mood of revitalization during the run-up to national independence, Michael Somare's autobiography *Sana* (1975: 17) made a strategic exaggeration when it translated *sana* as the name of a chiefly title. For "*Sana*" was nothing but the personal name of Somare's grandfather, who was a senior insignia-holder in his descent group in Karau village. Like all such leaders, his authority extended over the section of his group domiciled in his village but did not extend either over the rest of his village or over anyone else living in the Lakes, the estuary region or along the coast. If they are not Melanesian chiefs, the Murik standardbearers nonetheless remain "princely" titleholders whose status is not coterminous with their position in the kinship system.

Ceremonial reproduction

Being reckoned bilaterally and constantly subject to endogamous marriages, the genealogical boundaries of Murik descent groups are extremely ambiguous, to say the least. Extending the network of heraldic claims while depriving rivals of them is one explicit goal of ceremonial leadership. The preeminent entitlement of any insignia-holder, Murik men and women agree, is "to decorate children." The issue of decorations demarks a "child's" exit from the domestic group, with "her" explicitly matricentric representations and values, into membership in the androgynous, politico-jural community. *Sumon* ornaments may be conferred during life-cycle rites in early childhood, initiation, succession and death. Rarely, however, is any individual subjected to every rite in this sequence. Rites are rather mounted selectively to claim persons whose identity is particularly contested due to partial adoption, divorce, endogamy or descent group overlap (see Barlow 1985a). Typically, staging a single rite is a sufficient claim. As the principal idiom and goal of this system of status attribution is to differentiate the jural status of children through a principle of primogeniture, the heraldry would seem to be part of the dialogics of motherhood which I have been developing. Through an imagery of ceremonial reproduction, figured in terms of the adornment of the exterior body and competitive ceremonial exchange, heraldic elders take their turn at being aesthetic "mothers" of firstborn children.

In order to confer insignia upon such a child, a pool of elders – made up of the great-grandparents, grandparents, father, mother, mother's brother, or elder sibling – compete to obligate hereditary courtiers, feasting partners who come and render services during the rite in return for banquets.[8] When boys are decorated, these feasting partners come from the ranks of the child's classificatory mother's brothers, while those who serve girls are classificatory father's sisters. The matrikin personify male cult spirits (*brag*); the patrikin personify female cult spirits (*samban*). Claiming the heraldic body, in other words, is the result of a relationship between ascribed categories, birth order and gender, and reciprocal transactions that go on between the heraldic patron and the feasting partners on whom the honoree depends. What is interesting from my point of view is that the valuables exchanged do not evoke or confirm consubstantial ties of a phallic or uterine sort but are understood to square ceremonial debts.

The protocol for each successive rite demands a more elaborate feast, entailing a higher order of coordinated labor, travel and regional trade. The firstborn child is thereby associated not only with an emblem but also with an abundance of production, food and exchange. The ceremonial system challenges female procreative powers, with their stigmatized basis in processes going on within the maternal body, and turns them into morally unequivocal processes in which honor, aesthetics and the settlement of debt are played out upon a visible, androgynous body. The metamorphosis of dependency upon uterine motherhood into dependency upon heraldic, or jural, motherhood is symbolically unmistakable throughout the ritual cycle which, for the sake of felicity, I shall again describe for a firstborn son, although it can be and ordinarily is staged for firstborns of both genders (cf. Pomponio 1992: 78ff).

Three early childhood rites explicitly celebrate the emergence of a person oriented toward the polity and away from attachment to the maternal body. Barlow and I observed the first two of them in 1981 and then I saw the second one again in 1988. We did not see the third.

Until six months or so, the firstborn son is secluded with his mother inside her house until he begins to laugh and recognize people at a distance. The rite called *sumon arekomara* (literally, "insignia walk around") may then be staged. A child is carried through the community bearing *sumon*. The ornament in this instance usually consist of little more than a pendant of bird-of-paradise plumage tied to a pair of boars' tusks (see frontispiece). Next, when he begins to cut teeth, the *aragen* ritual may be mounted for him. On this occasion, the firstborn receives his first taste of the most preeminently social food in the culture, which is a thick, sweet-tasting porridge called *aragen,* made of fresh coconut milk, scraped coconut meats and sago breads. The same or another *sumon* is displayed at this time. Lastly, when a boy becomes mobile and begins to climb up and down the house ladder independently of his mother, the "ladder" ritual (*waik*) may be staged "to demonstrate" to him how to climb up and down it. A *sumon* emblem is given him by his heraldic sponsor.

While birthing is shunted to the periphery of communal space, marginalized and negatively valued, the heraldic "birth" of the firstborn may become a central event. As the child gains social skills – to see and respond to others, to eat cosmic food, and finally, to move beyond the space dominated by motherhood – his body becomes defined by and obligated to a prestigious matrix of ceremonial actors. The bestowal of heraldry honors and differentiates the firstborn child as the *sumon goan*, the "heraldic son," the leader of his sibling group as well as the heir of his patron and *pater*, the ritual leader of the particular descent group into which he has been decorated. Although other events, such as a first visit to a particular village or cutting down a first canoe log, sago palm, or planting a first coconut tree, occasion ceremonial exchange, insignia are not conferred again until initiation. A detailed examination of multiple stages

Plate 10: George Bai with plate-carrying women

of male initiation will be found in Chapters 6 and 7; here, my focus will remain only on the moment when they are deployed and displayed during this sequence.

At the end of a period of seclusion in the cult house, during which boys have sat passively incorporating sexual virility from wooden statues of their spirit guardians, classificatory mothers' brothers paint their bodies with red ochre. They have also readied new tapa loincloths for them, which is why this phase of initiation is called the "Loincloth Rite" (*nimbero gar*). One or more insignia-holders now enter the cult house and bestow *sumon*, the ornaments, baskets and other appurtenances, upon the initiates, who are then presented to the community.

Multiple cognatic claims are made to give heraldry to a youth. But to initiate a son, people tend to grant prior claim to the mother and her brother; reciprocally, the right to initiate a firstborn daughter they grant to her father and his sister. Others say that whichever parent, or whichever *sumon*-holder who represents the parent, is first to acquire the requisite food wins the right to serve as ritual sponsor of a youth. Alternatively, the preparation might be done cooperatively. A youth would then receive insignia from several *sumon*-bearers at once, each one representing their commitment to, and claim in, his jural identity. Whichever filiation is claimed, "the gift" which asserts the claim is always fixed at a rate of one pig per child. Decisive in this system of social reproduction is therefore not the identity of the mother or father, but the identity of an exogenous bene-factor: the hereditary trading partner who sells or donates the pig which is cooked and fed to the initiate's attendants.

Younger siblings may also be initiated at the same cost, but they do not receive an identical array of ornaments. Only the firstborn holds complete jural personhood in the descent group and only the firstborn receives the complete heraldic ensemble – beginning with the *sumon* basket – which adorns him when he walks through the community.

In 1982, the initiation of three brothers took place in Karau village. Two of the youths, who were employed elsewhere in PNG, were absent. Carved figurines (*kandimboang*) stood in for them. Only Bruno, the last-born son, was then resident in the village. After the classificatory mother's brothers painted his body red inside the men's cult house, three of Bruno's most senior mother's brothers appeared, representing different descent groups. Each man presented him with heraldry. "You receive only one boar's tusk" (*Kuja usiig ave*), one of them said to the youth with discernible scorn in his voice. "You may not carry the *sumon* basket. You are just a last-born son" (*Sumon suun ungoende. Mi goan kaikoro!*).

Bruno was then paraded through Karau (see Plate 11) and other Murik villages to "jingle his anklets" (*gigiinbo arekomara*, literally, "jingling walking around").

His *sumon* were repeatedly honored as he walked by both male and female kinsmen who broke open mature coconuts, swigged the juice and sprayed it at the young man and his heraldry.   *(see Lipset 1990)*

To expel the juice of a mature, "dry" coconut honors the *sumon* and anoints the recipient with attributes distinctive of a person. If the mature rather than the juvenile coconut evokes a withering maternal breast, then the violent act of splitting a coconut in half, drinking its fluid and forcefully expelling it is a public assertion of a jural relationship, via a nullification of suckling (which indeed is how Murik people literally drink coconut milk).[9] The heraldic displays thus overlay maternal dependency with ceremonial acts of reproduction. More: during ritual contexts, the resources of elders and the charms of youth are made public. As the former deploy their resources, they claim the power from rivals to reproduce society through the acts of nurture they sponsor. At the same time, through the

Plate 11: Bruno, a third-born son, was not permitted to carry the heraldic basket because of his junior status in his sibling group

display of their charms, youths make a seductive claim on those who possess the power to nurture (women). So the ceremonies which express the ideals and values of the community are at once a representation of both reproductive and seductive powers. These feasts are strongly marked by competition among elders, between elders and youths, as well as among youths themselves. These rivalries, in turn, disclose a system of values, the seductive powers of men, which oppose the moral order. The initiates' bodies are swathed not only in jural adornment but selfhood. The emergence of youth from the cult house bristling in *sumon* is not only a moment of high morality, it is also a sexually charged moment during which jealousies are aroused and the prospect of violence runs high. If the heraldic rites of childhood seek to claim the person as his social competences develop, they also claim him as his dependency upon his mother's body begins to diminish. Initiation, in turn, honors the bestowal of heraldry, but also masculine sexual identity, the magico-religious capacity which permits a man to replace "his" mother with a "wife/mother" of his own selection. The relationship between this ceremonial system and the maternal body is thus ambivalent rather than one of denial.

On the one hand, the *sumon* emblems are imagined to be independent of the maternal body. The ornaments thus lack gender. They lack totemic substance. They are not issued to honor the development of bodily functions. They are initially deployed to celebrate the emerging social capacities of the firstborn child: to honor his ability to laugh and recognize the other, to eat in and move about the community. The "Loincloth Rite" is no puberty rite but may be mounted irrespective of age. Later, the ornaments may be bestowed to legitimate jural authority within a descent group. *Sumon* are deployed to mark political succession. The career of an insignia-holder may begin and culminate during a pre-mortem tributary feast staged by an heir "to retire" an elder from incumbency of his or her title and office. An acquisitive, or transactional, and spatial metaphor, rather than a physical one, is used to speak of the heir's promotion: "He takes his elder brother's place" (*Tata kaban osanget*). Actually, a younger brother or sister, eldest child or eldest sister's child may succeed to a *sumon*ship. Nevertheless, the principle of succession remains that of primogeniture, primogeniture that is independent of gender or lineality. The appearance of *sumon* validates the transfer of the jural authority of a title-holder to his or her heir, and confirms that their kinship may be renegotiated by means of visible maternal actions and values. As part of this ambivalent dialogue with uterine motherhood, the *sumon* heraldry enable the detachability of body and person.

Thus, immediately following death, a new ghost should "carry" the spirit image of his or her *sumon* basket off to the numinous world. The ancestors there are then expected to recognize the basket and admit the newcomer to their community. Relevant elders will soon come to the house where the body is being mourned for and hang one or several heraldic baskets from rafters over the coffin (see Plate 12).[10] After the burial, nothing but a heraldic metonym for the person remains: his *sumon* basket. But corporeal death is of no little moment. The passing of a *sumon*-holder, a man or woman who has had a successful career as a feastmaker in the community, may transform the moral body into a contrary image of the

Plate 12: At the death of her grandson, Kanjo displayed heraldic insignia and discussed his descent so that his ghost might carry their spirit to show the ancestors in the after-life

maternal schema. The exteriorized, heraldic body, that is, fails to extinguish its defiled, interior counterpart. The relationship between the two is more complicated than mere repudiation. Shortly after the burial of an insignia-holder who has died of old age, a rite of reversal, called the "fight of the sibling groups" (*noganoga'sarii*), may break out in the three eastern Murik villages of Mendam, Karau and Darapap.[11] In this peculiar context, classificatory elder brothers' wives and husbands' younger brothers and elder sisters' husbands and wives' younger sisters (see Figure 6) may attack each other in certain ways. The village divides up by sex: individual women stalk and chase individual men, and vice versa (see Plate 13). Both male and female may carry coconut half-shells filled with a filthy concoction and wrestle each other down to the muddy ground trying to stuff a mixture of animal feces, mud and ashes into the mouths of their alters. The fighting goes on as long as it takes to involve everyone in the community in this grotesque nurture.

Noganoga'sarii turns the ethical system for which the deceased insignia-holder stood during his or her lifetime inside out. Instead of the purposeful and earnest project of decorating commensal kin, completing their identities by sponsoring tableaux of beautifully decorated bodies in the center of the community, classificatory kin takes center-stage and a contradictory kind of bodily liberation has its day. Instead of affirming gender-inclusiveness and equality, male and female divide up and fight each other. Instead of celebrating a child's upright mobility and his autonomous movement out of the domestic group and into the community, adults fight to push each other down "like children." Instead of "mothers" graciously feeding men and children pure foods, the sexes try to force-feed each other a filthy mixture of animal feces, mud and ashes which is unclean, decomposed and is associated with death.[12] The relationship of this rite to the bodily death of a *sumon goan* is not straightforward. True, it may break out whether or not the insignia-holder has been retired from his or her title. But protocol requires that it cannot take place unless a specific service has been performed by a classificatory sister's son (or brother's daughter) of the deceased during his or her final illness. Depending on the sex of the insignia-holder who lies dying, a canoe full of firewood or green coconuts must be brought to the house by one of these functionaries. Moreover, such gifts should only be brought to titleholders who have been active feastmakers. Nor is the fight random: specific affinal joking partners who otherwise trade mock abuses in daily life are pitted against each other. The threat *noganoga'sarii* poses to the moral order, in other words, is carefully crafted. It is a contrary version of the moral body that is neither wholly seditious, nor subversive. *Noganoga'sarii* does not break out every time a *sumon*-holder dies. Indeed, in recent memory *noganoga'sarii* has broken out only twice. Nonetheless, for many years afterward, women laugh and bemused men, in particular, shake their heads about specific indignities to which they submitted as a result of the gift of firewood or coconuts made to an elderly insignia-holder during his or her final illness.

In his book on Rabelais, Bakhtin sought to interpret the poetics of medieval folk humor and marketplace language within and against the dominant political contexts of their times. In ecclesiastical art, one body appeared oriented upward, in contrite supplication to the heavens. In carnival, by contrast, the orientation of the body was downward, toward its own "material, lower . . . stratum" (Bakhtin 1965/1984a: 151). In the medieval, ecclesiastical body, all aspects of "the unfinished world . . . as well as all signs of . . . inner life [were] carefully removed" (ibid.: 320). In carnival, "the life of the belly and the reproductive organs . . . relating to acts of defecation, copulation, conception, pregnancy and birth" came alive (ibid.: 21). This body was at once degraded but inexhaustibly generative. It was "not . . . closed . . . it outgrows itself, transgress[ing] its own limits" (ibid.: 24, 26–7). Its proportions were not refined but exaggerated.

The topography of this body was turned upside down and downside up. Buttocks and genitals became head and face and vice versa, head and face became the buttocks and genitals. Comic images of feces appeared in the billingsgate. As matter that was intermediate between earth and body, Bakhtin argued, feces were images of decomposition that also fertilized the earth. In this defiled image, both old and new were seen dying and pro-creating. Not purely filth, feces played a special role in overcoming fear of authority (Bakhtin 1965/1984a: 197). Throwing feces was an act in a mocking funeral of the legitimate order. Pervading all the trivia, abuses, scatology and farce that took place during carnival was a body whose logic Bakhtin called "grotesque" (ibid.: 62). The relationship between the grotesque and the moral body was not just one of reversal, however. It was one of ambivalence. Its distinctive mood was "contradictory and double-faced" (ibid.: 62), which is precisely the relationship that Bakhtin meant to foreground through his metaphor of dialogue.

By suggesting that an exegesis of popular culture in the European Middle Ages might shed light upon a Melanesian rite of reversal, I have two goals in mind. First, I want to advocate the methodological utility of Bakhtin's framework for the elucidation of this specific rite of reversal and for the analytical translation of this kind of phenomenon in general (see

Holy 1987: 1–21). But second, granting that this is not an exercise in controlled comparison whose purpose is to test a functional hypothesis cross-culturally (Radcliffe-Brown 1951; Eggan 1954; see also Evans-Pritchard 1963), I want to apply Bakhtin's notion of the grotesque with the discipline Weber demanded of ideal types: to distinguish "divergences or similarities" (1949: 43) in the Murik phenomenon to which I am comparing it. Obviously, a Melanesian grotesque must differ from one in medieval Europe. The moral figure it contests is domestic, androgynous and this-worldly rather than centralized, phallocratic and otherworldly. The metaphorical structure of the moral body against which it is struggling has given rise to a different topography. The crucial dimension of moral contrast in Europe was up–down. In Murik, it seems to be out–in.

While he was clearly aware of the procreative references of feces, presumably for reasons having to do with the repressive political conditions in which he lived, Bakhtin emphasized the license expressed by the grotesque body during carnival rather than gender. Dundes, a psychoanalytic anthropologist, made explicit the gender of the ritual use of feces (1976; cf. Epstein 1979). Following Bettelheim (1954), Dundes argued that when cultic initiators smear feces upon novices, they reenact birthing. In two cult initiations on the south coast of New Guinea he cites, a masculine, anal fertility thereby displaces the fertility of women. Novices emerge from a rotten world smeared with dung, mud and ashes, the remnants of their symbolic passage through the "posterior parts" of their male god (Van Baal 1966: 480; see also Williams 1936: 29). In another Aboriginal case, initiates are even made to drink urine and eat feces which represent "the milk" they would otherwise suckle from their mothers (Roheim 1942: 371). Why should men try to do this? They are jealous, says Dundes, of the uterine body and seek "to live without recourse to" it through ritual attempts to recreate a symbolic fertility in the masculine anus (1976: 234, 227).

The value of these two views of feces for developing an understanding of *noganoga'sarii* may appear unduly straightforward. For quite explicitly, this rite is a mocking funeral which overcomes loss. The startling image of women fighting men, force-feeding each other a fecal concoction is also, as Bakhtin would have it, an image of the ambivalent metamorphosis of the lower body that is both defiled yet generative (see Plate 13). Yet the meaning of feces in *noganoga'sarii* is unrelated to Dundes' startling image of infant-novices smeared with feces following their anal birth, suckling mud at the breast of an aggressive male mother. Watching and participating in it, I got no sense that the mixture of mud, ashes and feces with which the genders fought was an attempt, in either an explicit or an entailed way, to coopt the uterine procreativity of women by means of imagery of anal birthing. The reversals which ensued in the name of this rite rejoined the ethic of maternal nurture, dependency and solidarity that defines the ethos within domestic and ceremonial groups (see Plate 1), the very same ethic in which the deceased had specialized during her or his career as an insignia-holder. The object of the fighting in *noganoga'sarii* was to stuff filth into one's partner's mouth and to wrestle him or her down. For the Murik, *noganoga'sarii* is a highly entertaining form of slap-stick during which the moral ground upon which kinsmen of both sexes normally stand vanishes for a few days following the death of a ceremonial leader. It extends affinal joking relations which go on routinely. It does not parody uterine force. The imagery of feces in it is *not* a representation of womb envy. Rather, it inverts other aspects of mothering. Women and men fight to feed, and smear, the body with feces and mud during *noganoga'sarii*. The image to which they are responding is not that of a birth-mother, but a pre-Oedipal mother, the mother of infancy and early childhood; the beautiful woman who loved, nurtured, protected and kept them scrupulously clean. It is "she," in "her" ceremonial guise, who has

Plate 13: After a senior insignia-holder died, a ritual fight (*noganoga'sarii*) broke out. Men and women tried to force-feed each other a mixture of mud, animal feces and ashes and/or wrestle each other to the ground

been responsible for the reproduction of society through the staging of feasts and the bestowal of "her" heraldry: "she" has now died. It is from attachment to "her" mothering, rather than to her birthing or uterine capacities, that society must now separate itself. Upon the death of an insignia-holder, a prominent "mother," a prominent figure of dependency, has been lost: the ethics and values of moral life give way to archaic impulses. Relationships become infantilized, thrown back to an infrasocial time when the boundaries between inner and outer body remained incomplete, when orifices were uncontrolled, open and not closed, before resuming a "body" in which "mothers" feed children without passing on, or suffering, defilement, while insisting that they "stand up" and remain scrupulously clean.

How did *noganoga'sarii* conclude in 1982? Combatants exchanged dishes of sago pudding garnished with seafood. Rather than a masculine attempt to birth, the symbolism of fighting with feces, ashes and mud in *noganoga'sarii* represents a grotesque "weaning" played out in a mode of aggressive slapstick by heterosexual society as a whole. Mead would seem to have lent a certain amount of credence to this interpretation when she noted how an Arapesh mother used to wean the child she was nursing when she became pregnant again. "This is done by smearing the nipples with mud, which the child is told, with every strongly pantomimed expression of disgust, is faeces" (1935: 38). Like the *sumon* emblems whose holders define jural identity but possess no gender or biogenetic substance, the license permitted during *noganoga'sarii* recreated no uterine presence. It contradicted oral dimensions of dependency upon the maternal body through a defiled, aggressive representation of foodgiving.

The role of a *sumon goan*

The metaphoric opposite of a *sumon*-holder is not said to be a commoner, but the son of a "fruit bat" (*naboag goan*), an illegitimate, bastard child, who, lacking parents, is likened to a thief who comes in the night to steal from other people's gardens. Eating other people's food, he is chronically indebted, and being indebted he becomes inferior, if not completely subordinate. By contrast, a *sumon goan* both settles debts and creates indebtedness by offering food to others. He is a man who "has plates" for guests. Like an elder brother, the insignia-holder is more of an indulgent presence than a forceful one in society, or rather, the force of his presence is his nurture, which gains compliance through the strategy of manipulating the dependency and sentiments of junior kin. The role of the *sumon goan* is taken up with ceremonial and daily foodgiving. Pig mandibles dangle from his porch beams, the proud record of the many feasts he has sponsored. He should be a man "of tobacco, betel nuts, coconuts and *aragen* porridge." His hospitality attracts a steady stream of guests who come to his house for a meal and become the focus of his fastidious attention. Only chewing betel nuts while entertaining, a *sumon*-holder should never "eat in public" but should only eat privately, either before his guests arrive or after they leave. Never straying far from women who personify the schema of values he also seeks to espouse, he only eats food prepared by the women, his wives and sisters, whom he trusts to protect his health from contact with menstrual fluids, or sorcery theft. Together with daily and ritual nurture of children, junior kin and trading partners, finally, a *sumon*-holder should also excel in minor exchanges with feasting partners to be discussed in Chapter 6. When one of them, for example, who has been visiting kin is ready to return home, he should present him with some small departure gift (cf. Malinowski 1935: 345 n.; and Powell 1960: 126). This courtesy sustains the feasting partner's indebtedness and obligation to perform ritual services when called upon.

A successful *sumon*-bearer and his womenfolk should remain hard at it, fishing, manufacturing trade goods, trading, cooking, feeding and planning feasts. Their lives turn upon the demands of hospitality. Relations within this group, particularly between the *sumon*-holder and the many industrious women with whom he lives, may therefore be complex, at times strident, but generally held in high regard. For different reasons, Collier and Rosaldo included such a positive valuation of marriage as a feature of brideservice-type systems in which husbands do not depend upon surplus wealth produced exclusively by women to win prestige and do not need to define their intentions and identities as completely opposed to those of their wives. In Murik, precisely because husbands depend upon their wives' productivity and wives depend upon husbands' productivity, they are not rivals. "The husband is a canoe hull," as the Murik aphorism expressing such interdependency goes, "his wife is the outrigger float." A senior man once praised the basket-weaving skills of his first wife to me, a woman who was quite a shrew, saying that his trading partners could "come all at once and try to attack me [by calling in debts] . . . But I always win. I give them everything. No debt can stay with me. I always pay [them back]. This is the way of my first wife . . . which is why I keep her." It follows that lacking the labor of his wife or co-wives, the career and ambitions of a widower *sumon goan* will be curtailed, if not completely obstructed (see Chapters 5 and 8; see also Plate 10).

Larger feasts, whether to celebrate an initiation, succession, or the end of mourning, usually involve a dance which begins at dusk and lasts until dawn. The beginning of the performance is formal, performed by a troupe of dancers who have rehearsed a set piece for the occasion under the tutelage of a *sumon goan*. However, in order to keep up the singing and dancing all night, others may join in to perform favored pieces that everyone knows. Mothers, or anyone who admires a dancer, may honor him by sprinkling white lime or talcum power on his back, forehead or chest (see Meeker, Barlow and Lipset 1986: 23). As the evening lengthens, and everyone begins to show signs of fatigue, the *sumon*-holders start to assert themselves, leading verses of songs of their own composition and dancing flamboyantly. Early in the morning, well before dawn, the host insignia-holder should serve an elaborate pig feast to all the dancers and singers, to compensate them for their performance. Then, near dawn, the original troupe concludes the celebration with a final performance, and another meal is laid out for all those who have danced until sunrise. If the dancers are still inspired, and food remains, they may go home to sleep and rest until late afternoon when they will begin dancing again. The celebration should continue until everyone has had all they want of it. For several days afterward, many people can be seen hobbling stiff-legged around the village.

If he is known for acts of abundant nurture and flamboyant entertainments, the *sumon goan* is also a visible image of sacred purity, quietism and self-control. A fight may be stopped through the intervention of a *sumon*-bearer, or, it is important to add, an initiated woman. Like a *sumon*-bearer, such a woman may stride into the middle of a fracas; when the antagonists see her holding up her hand, they must desist from their violence (cf. Valeri 1985: 149). During the great feasts, the *sumon* which signify the jural identity of the host group also taboo conflict. A number of *sumon* are displayed during these festivals. The baskets, armbands, headbands and shell vests worn by the dancers, the slit-drum around which they perform may all be *sumon*. The scene is rife with these symbols of moral order. For good reason: the dancers have been secretly doused with love magic. Feelings of sexual jealousy and rage are thus not far to seek. The potential for conflict is high. Before the dancing begins, a host will usually make an announcement encouraging everyone to enjoy and celebrate, but entreating them to avoid fighting. High ritual is rent by rumors of sexual intrigue and potential conflict. Elders may intervene beforehand with words of caution, threats of punishment, or by forbidding the attendance of certain members of their family. Should conflict break out and a woman or a *sumon goan* is forced to intervene, the festivities must be immediately terminated. The *sumon* themselves are then said to be "broken" (*sumon onagaga'iro*). Slit-drums are "turned over" so their "mouths" face down to the ground. The participants must remove their decorations and immediately depart for home. Having been violated by personal injury, a loss of blood, or even the very observation of violence, the *sumon* become taboo. They, and the body of their holder, become defiled. The title of the *sumon goan* is suspended; his status is temporarily nullified (see Pomponio 1992: 92). Children may not be decorated, mortuary rites may not be performed, new outrigger canoes may not be consecrated: rival insignia-holders, however, can go on making their overlapping, rival claims to these persons and vessels. Following an injury, or any sort of direct contact with communal violence, a descent group's leader should stop all ceremonial activity until his body and insignia are ritually "stood up" (*sumon odekara*) from having "fallen" in battle. An act of ceremonial compensation (a pig feast) must be sponsored by the culprit. When the indemnity feast is served to him, the holder's insignia are said to be "put back together" (*sumon kotoboare*).

Paying this penalty, for violating what is today called the "law of the *sumon*," is considered expensive. Titleholder and heraldry are protected from personal injury not only by this great debt, but also by a mystical penalty associated with neglecting to repay it. The compensation, and related reconciliation feast, stop fellow titleholders within the victim's descent group from conspiring to ensorcell the assailant (see Chapters 5 and 8). As his ambition is to keep his insignia in circulation and to keep himself as well as his group in the visible center of society, the *sumon goan* will also try to avoid compromising the purity of his sacred paraphernalia and body. To get their fragile insignia out of danger, *sumon*-bearers will hasten to disassemble and put them back in the rafters of their houses as quickly as possible after the ritual in which they have been deployed has been completed. A fistfight which takes place in a dwelling house, or inside a male cult house belonging to an insignia-holder, will taboo the *sumon*. But whether the *sumon* of a man not wearing the regalia become "broken" should their holder suffer deliberate injury outside of his house was a matter of dispute in 1982 (see Chapter 8). In either case, my point is that the Murik heraldry are considered to be mystically vulnerable to conflict and physical violence. Harrison (1985) has described an important string bag Manambu men receive when they enter the third grade of their male cult. This bag is also only carried on ritual occasions and otherwise kept hidden in the house of its owner: to carry it around in everyday life would constitute a public menace. The Manambu initiatory string bag is feared to contain a dangerous, mystical agency which threatens moral order. By contrast, the Murik heraldic basket, like the maternal schema of which it is an evocation, is a delicate image of interdependency that is threatened by conflict.

In formal meetings, the voice of a *sumon goan* ought to be "strong." He should speak often and, having an acute sense for a sequence of events, with clarity. He should "crush the bark floor" now and again, stomping his foot to dramatize his rhetoric. But otherwise, the *sumon*-holder ought to keep near his house. A poised and discreet man, a *poap nor* (literally, "a man of the locality"), he should be loath to take sides in the ubiquitous sexual intrigues which are so close to the Murik heart. It is said that when carrying the *sumon* basket, he ought not even spread gossip about adultery. Even if he hears that his own wife, or the wife of one of his younger brothers, has taken a lover, he should simply "put" the gossip in his basket, and "leave it there." Michael Somare has written that

when people come to fight us, we call them to eat first. We sit down together. We talk. We eat. Then we say to them: "All right, if you want to fight, take your spears and stand over there. We also will take our weapons, and will stand on this side." But we believe that after eating, their minds will be changed. They will not want to fight us any more.   *(1975: 23)*

The tactics of nurture and generosity, that is to say, should be used as part of an overall strategy to defuse conflict and maintain the unity of his insignia. Heraldic elders thus personify and embody a nurturant order.

But for all the self-restraint and solidary values associated with them, their *sumon* remain daunting emblems. Giving rise as they do to jealousy and resentment, the mystical temperature of the *sumon* insignia is said to be "hot." Not only does their display intensify preexisting Oedipal rivalries, they are also held to elicit sorcery from the wider descent group. Elders resident elsewhere in the estuary region are invited to attend major heraldic transactions in order to certify the genealogical credentials of the feastmaker. The threat of sorcery (*timiit*) is their sanction. A man, for example, who is not an eldest sibling, or who is a senior firstborn but has not mounted the requisite succession exchanges, or whose adoption has weakened his claim to seniority, is seen to be risking his life, should he pursue illegitimate *sumon* transactions. Such pretenders will eventually die, it is believed, from mystical causes: a rival *sumon goan* within his own descent group will ensorcell him by stealing his foodscraps during the illegitimate feast (see Chapter 5). Boars' tusks and dogs' teeth ornaments are emblems of title in the Schouten Islands as in the Murik Lakes. An expedient, competitive, political life also surrounds the island chiefs, but commoners hold them above it, because "to be without a leader would be worse than having to submit [even] to one who [is] angry" (Hogbin 1978: 143; cf. Lutkehaus 1982). Murik society, in short, is less centralized. The monopoly of force is more diffused. In Murik, the bodily presence, rather than the absence, of a splendidly arrayed titleholder creates a potentially dangerous situation. Earlier in this chapter, I described assurances offered a newly installed *sumon goan* which confirmed his ritual autonomy. Implicit in these assurances was a veiled reference to the dangers of being an active *sumon goan*, for whom the fulfillment of duties must inevitably jeopardize the career, not to mention the life, of the incumbent.

Bigman, elder sibling, chief, or great man?

Functional analyses of inequality in Melanesian societies pose several questions about the status and leadership role I have been examining. To what extent is becoming a *sumon*-holder a personal achievement? To what extent is heraldic authority vested in a hereditary status? And to what extent is it caused by the production of wealth rather than the control of ritual status and knowledge? What, in short, are the sources of political inequality in this culture? Or, to submit these questions to more conventional categories: is the Murik *sumon goan* a bigman, elder sibling, chief, or great man?

It is self-evident that the *sumon goan* is no "bigman," if holders of that status wield provisional power and prestige that are won and lost through action, exchanges of wealth, the influence of their oratory, etc., and gain nothing at all from membership in ascribed kinship or descent categories.[13] The Murik political system is clearly not a meritocracy but a gerontocracy. The *sumon* are preferentially held by the eldest sibling in the senior generation of each descent group. Nevertheless, Murik primogeniture is negotiated through competitive ceremonial exchanges. The sacred insignia are transferable. Should the firstborn fail to perform the requisite ritual work, the *sumon*ship which ought to be his by birthright may be assumed by a younger sibling, if the junior man is able to mount a succession feast for the elder sibling as well as any intervening siblings who have prior claim to the status (see Lipset 1990). To maintain a heraldic title, a man must then remain continuously active in the prestige economy. Rights to, and the quality of, his ascribed title are contingent upon ritual agency, i.e., upon the level of feastmaking he can reach and sustain. Although the *sumon goan* is not a bigman, his ambition – pig exchange – is the same as that sort of leadership.

Is the *sumon goan* merely an elder brother, or is he a Melanesian chief?[14] His rights and duties and those of the domestic "big brother" status are virtually indistinguishable. Their role is formulated in the same imagery. Both are expected to provide indulgent nurture, protection and instruction. Both stand for modesty, self-control and order. Fortes once argued (1962) that two criteria can be used to discriminate hereditary offices from kinship statuses. The first is that an office may not go unoccupied because, should it do so, the very existence of socioeconomic life would be threatened. The second is that an office must be perpetuated, not merely by inheritance, but by the staging of mandatory succession or installation rites. Property or status may be claimed through inheritance. But assuming an office requires a more formal ceremony in which symbols of the perpetuity and rights associated with it are conferred upon the new holder. By these criteria, the Murik *sumon*ship is an office, albeit a weak or minor one. When unoccupied, the maintenance of jural order in, and the social reproduction of, the descent group are obstructed. Moreover, an outbreak of *noganoga'sarii* following the death of an officeholder may disrupt maternal values. But the point is that becoming a titleholder requires sponsoring a succession feast, while becoming a firstborn sibling only requires filiation and actions commensurate with this status. The title held by the senior *sumon goan* is thus to a nonchiefly office; the ceremonial group it represents is the political equal of its rivals. The relative statuses of the *sumon goans* who hold insignia are all the same. The prestige of their insignia is differentiated only by individual success in ceremonial exchange. Moreover, to return to the Fortesian criteria for a moment, neither domestic reproduction nor production are contingent upon their perpetuity. Bodily injury mystically taboos the insignia and their holder from competing in the prestige economy until a compensation feast has been staged. In this interval, the holder does not lose his title, so the office does not go unheld, but, at the same time, he may not fulfill his ceremonial duties. He may not invest firstborn children, novices, his constructions, or the dead, with his insignia. He may not remarry. But while jural reproduction is tabooed, other jural relationships go on unaffected; other *sumon*-holders decorate children, build canoes and sponsor mortuary rites. It is only the victim's heraldry which is "broken." Fortes was thinking of a more highly centralized political system when he argued that jural authority constitutes the domestic order to the extent that an office lacking a tenant will cause society to collapse.[15] In Murik, the domestic order is far less contingent upon the *sumon*-holders. *Noganoga'sarii*, the rite of reversal which metamorphoses maternal nurture and authority, is circumscribed: it only begins if a particular ceremonial service is performed during the titleholder's terminal illness and only lasts until everyone in the community has tasted the filth. What is more, it may only break out following the death of an insignia-holder who has died of old age. Such a *sumon*-holder would have already been retired from his or her jural position. The burlesque of the maternal schema it permits, while being related to the loss of an insignia-holder, is therefore not directly caused by an empty office.

If I may be permitted one admittedly farflung contrast, perhaps a bit more light may be cast on how relatively tenuous is the link between domestic and politico-jural orders in Murik. The culminating act in Ashanti installation ritual is the moment when the chief-elect sits by himself on the "supreme ancestral stool" in front of his advisers and is enjoined to observe taboos and have a moral career (Fortes 1962: 59). Among the Murik, the heir is not raised before society; rather the retiring incumbent is so honored. When the Ashanti chief sits before society for the first time, one man sits alone. When a Murik younger brother succeeds his elder brother (or sister), not he, but his wife, sits down on a platform next to the retiring incumbent. Surrounded by collateral kinsmen and kinswomen, the two of them chew betel nuts together. The night before she has made love to him: their ceremonial act of commensality the next day signifies this intimacy. With the elevation of her husband to the genealogical status of the incumbent, affinal avoidances have been lifted. Having been displaced from his politico-jural status, the husband's elder brother is no longer the wife's senior affine from whom she had hitherto to maintain avoidance relations. The shift, or transfer, of the title is represented in terms of its impact upon affinal inequality, which is to say it is represented as effecting status differentials within a particular sibling and insignia-holding group. It does not create a unique figure of authority in and for a whole community.[16] Murik heraldry does not represent royal blood, of which there is none in the culture. It rather makes up an egalitarian system of titles which requires continual affirmation through the ceremonial redistribution of food. Now despite the literature on Melanesian political leadership, which has led many to assume that cultures in this region insist upon an "essential identity of men and of groups" (Forge 1972: 539) and therefore lack symbols of legitimate authority, it seems to me that another way of conceptualizing the Murik polity and system of leadership could be to conclude that, since every group holds its own distinctive heraldry, and therefore possesses its own ritual-political autonomy, there are no "commoners" in the society. Murik is an egalitarian political system, that is to say, composed exclusively of elites.

Using a distinction that recalls the brideservice–bridewealth dichotomy of Collier and Rosaldo, Godelier (1986; see also Godelier and M. Strathern 1991) has rethought the causes of political inequality in Melanesian societies in terms of a contrast between status differences caused by wealth and those caused by cosmology. Bigmen, he has argued, gain power from the production and exchange of nonequivalent, material goods. They transact objects "to buy" women whom they or others marry, to make compensation payments or to offer sacrifices to their deities. By contrast, "great men" make equivalent exchanges. Marriage, for them, requires some form of woman-for-woman exchange and the primary response to injury or homicide is similarly eye-for-eye. Control of material resources does not give rise to political hierarchy. Instead of wealth, great men master ritual by which they gain magico-religious control over the continuity of society. Obviously, Murik ceremonial leaders are not great men. Although, ethnohistorically, sister-exchange was claimed to legitimize Murik marriage and contemporary elders continue to compete to control a ritual hierarchy, the whole ceremonial process of exchange is contingent upon a combination of an exchange of *both* equivalent services and nonequivalent things. The right to confer his emblem, which stands for jural identity, birth order, etc., upon individuals during life-cycle rites, depends upon a nonequivalent exchange: one pig = *sumon*. All the heraldic rites depend upon reciprocal exchanges of services provided by representatives of the male and female cults. In the past, these transactions were supplemented by a one-time transfer of sexual access in the heir's wife during succession rites, which was viewed as an equivalent exchange in return for the same service that the incumbent's wife had once rendered the previous titleholder. The *sumon goan* are deeply enmeshed, not only in material production and asymmetrical exchange of the kind Godelier would associate with bigmen polities, but they are also involved in the kind of symmetrical exchange of actions he associates with great man polities.

If the *sumon goan* is not a bigman, big brother, chief, or great man, or rather, if he is partly a bigman, big brother, chief, and great man, then what are the causes of inequality in this system and what kind of a leader is he? And what might that type be called? Although the Murik system of primogeniture vests authority in firstborns in every sibling set, I hesitate to call him "big brother," or even "heraldic big brother," since the title is not gender-specific. I am also loath to call him a "heraldic mother." This label at least makes proper reference to the larger schema upon which his role is modeled, but obscures the fact that his title is largely managed by men. Perhaps "head person" might do; this phrase at least suggests that the attributes, qualities, entitlements, etc., of his office are hierarchically distributed throughout society in a gender-neutral way. But it conveys nothing of the dialogics in which the position is cast, so I find it wanting as well.

Dialogics of inequality

In order to understand and appreciate the dialogics of inequality, I have been arguing that we must look beyond "prestige goods [whether they take] . . . material or intangible forms" (Harrison 1993b: 156). We must look to a culturally particular maternal schema from which they take their cogency and motivational force. Certainly, the meaning of the principal modes and tactics of Murik leadership – generosity, quietism, instruction and so forth – would seem to confirm my argument. In Murik, as elsewhere in Melanesia, dominance nurtures while subordinance consumes. Hierarchy is constituted by exchanges of food.[17] In Murik, a donation of wealth, which takes the form of a feast, is validated by the transaction of an emblem, the *sumon*. There is a shift from the feast to a sign for the feast, and the latter realm is that of bodily purity, honor and prestige, which is associated with elders of both genders. In Murik, what we learn is that foodgiving, a mode of female/maternal agency, is reflected in a sign, an image of honor. But the sign stubbornly refuses to be single-sexed. Men and women share rights to wear, hold and confer *sumon*. What we learn is that, although the moral order epitomized by *sumon* may sometimes be managed by men, these emblems do not evoke any exclusively male body. The values and conduct that an insignia-bearer should personify – the heraldry he competes to bestow upon children, the grand feasts he seeks to mount, the peace he should protect and promote, his sexual modesty and emotional self-control, the instruction he is ready to offer – are virtues abstracted from an image of a mother of infant children. The Murik more or less openly acknowledge the salience of this prototype when they say that the elder brother and the *sumon goan* ought to act "like a mother" (*sumon goan ngain enambo*). This maternal analogy is again evident in the metaphoric expression for the guidance an elder may offer a younger man who is organizing his first feast. Such advice is said "to give" the junior man "strength" to stage-manage the delicate event, but idiomatically the elder is said to be giving him "to suckle from his breast" (*ningiinominga*), suckling him with his counsel rather than fluids.

Since I have been arguing that the dialogics of inequality rejoin an interiorized, uterine image of the maternal body, perhaps this suckling image is more contradictory than vindicating. If we return to the Manambu and Kaian concepts of society I introduced at the beginning of this chapter, perhaps this contradiction can be resolved. Although personal names and platforms also appear in Murik culture, obviously, the polity is not principally represented by them but by ensembles of heraldic ornaments. In Murik, jural order is aesthetically and linguistically differentiated by bodily decoration. This mode of representation makes a different sort of claim about agency in society than do the images of names in Manambu and platforms in Kaian, where nature and society are both constituted by cosmic powers vested exclusively in men. In Murik, among the cluster of rights which adhere to the display of *sumon*, a primary one is the social reproduction of children made through ceremonial exchange. When children are decorated by heraldry, they receive identity and entitlements in particular jural networks. Lacking grandiose reference, the Murik emblems signify "nothing more" than this. The agency of the heraldic masters and mistresses of society is thus far less magisterial than that of Kaian or Manambu men.[18] The power represented by their *sumon* is less totalistic. The display of *sumon* does not mystically recreate the cosmos or cybernetically staunch its inevitable decay (cf. Errington and Gewertz 1987; see also Penn and Lipset 1991; and Harrison 1993b). Their display makes no chiefly claims. The display of *sumon* is rather a heterosexual and androgynous response to a feminine center. Through ritual action, the *sumon*-holder seeks to create a dependent other – a vulnerable child – in need of the nurture and protection afforded by his emblems. He seeks, in short, not to coopt, but to participate in, the image of woman-the-mother. The image of the *sumon goan* "nursing" a young feastmaker with guidance rather than breastmilk suggests this quite clearly. But while men may be responsible for the management of their heraldry, as I have said, they do not seek to remove these symbols from women's hands. Rights to display *sumon* must be validated by an exchange of food, an act to which both sexes must contribute work and for which both may take and receive jural credit. As senior firstborn sisters, women may formally participate in the system *on a par with their brothers*. They may and do instigate ceremonial exchange and receive benefit from the labor of husbands, brothers and sons in these projects in the same way the rites men sponsor benefit from the labor of women (see Chapter 5 and 8). The dialogics of this system of inequality are clearly audible: for women, the people say, may hold *sumon* "just like a man." In the next chapter, I make use of an extended case study of heraldic succession to investigate the pressures which have been impinging upon it, and the changes that they have brought about during the 1980s and early 1990s.

5

Who succeeded Ginau?

Having newly arrived in Wewak, the capital of the East Sepik Province in February 1981, Barlow and I went looking to meet urban Murik. Among the first people we encountered was Ginau, a lanky, rather elderly man, with angular features and penetrating eyes. "I am the government of Darapap," Ginau announced to us in a raspy, somewhat thin voice that seemed worn out. For many years, he told us, he had been the local government councilor in Darapap village as well as one of its leading *sumon goans*. At that time, however, what we saw was an unkempt man with thick, bootblacked hair, straggling beard, who was dressed in a torn singlet and a threadbare black waistcloth. Black bandanas were neatly wrapped around his ankles and wrists (see Plate 14). He was in mourning, we later found out, for his wife and youngest child, both of whom had died some years earlier; so both the man and his insignia were living in defilement. Ginau was neither grooming himself nor purchasing new clothes; he was forgoing the display of his heraldry. For the past several months, he had been staying with a sister's daughter at Kreer camp, a ramshackle, beachfront community made up of urban Murik and other regional groups, while waiting to go back to the village to celebrate his "Washing Feast" (*Arabopera Gar*) that would end his mourning taboos. "As soon as I go ashore," he kept insisting to us, "my party will begin." Much time came and went, however, before Ginau finally returned to Darapap, where no great feast greeted him. The events, local tensions and foreign pressures which I shall discuss in this chapter confounded his expectations. Finally, two years later, the feast was successfully staged. To be sure, exposition of this case will serve to index or gauge the autonomy of local authority during this time, but will also reveal an utterly crucial idiom in the dialogue between the uterine body and the maternal schema through which

Plate 14: Ginau of Darapap

senior men and women reproduce moral order in Murik. The somatic prototype for the latter is a hollow vehicle, a "canoe-body" that is endowed with both male and female agency.

An heir chooses his father (cf. Hogbin 1940)

The sponsorship and organization of the Washing Feast were the responsibilities of Ginau's junior kin living in two Murik villages. In Darapap, a man named Bate imported a large quantity of sago from his trading partner in Mabuk, an important "trade mother" village on the Sepik River. Women cooked sago breads to serve during the feast and to be sent out as invitations to it. Still remaining to import were plaited canisters of *canarium* almonds from the Schouten Islands, garden produce and, of course, the pigs from a number of regional locales, as well as sugar, rice, tins of meat and cartons of beer from supermarkets in Wewak.

Why did Bate acquire the sago? Because he considered himself to be Ginau's heir. One of Bate's younger brothers explained their relationship to me this way: "Ginau has the *sumon* Bate wants. There is a debt yet to be repaid. Ginau 'gave paint' to Kapram. Now Bate is trying to pay this debt." Although Kapram, his father, had been one of Ginau's elder brothers, which would, according to kinship terminology, make Ginau one of Bate's "fathers," and although Ginau succeeded Bate's father as *sumon goan* of their sibling group (see Figure 8), when I asked Bate why he was planning to honor Ginau, Bate denied that this was his goal at all and abruptly broke off our conversation. I quickly returned to my original informant, his younger brother, to ask about Bate's peculiar response. After his wife died, I was told, Ginau had renewed a longstanding love affair with Bate's wife, about which Bate remained unspeakably bitter.

Bate was not preparing for Ginau's feast in Darapap alone. Ginau had not only succeeded Kapram, his father, but he had also succeeded Kerok and Mwangar, the father's younger siblings (see Figure 8). In order "to repay" Ginau, offspring of the latter two were also helping Bate provision the feast. Sauke, the firstborn daughter of Kerok, had decided to give a pig to Ginau "in the name of [Bate] her elder brother." And Kanjo, Bate's wife, had also vowed to give a pig to Ginau in her husband's name, because Kerok had been her "mother's brother" (see Figure 8) and had sponsored her initiation. Bate, his wife and "younger sister," had each purchased one pig from We and Koil Island trading partners.

In addition to these three, two women from neighboring Karau village were planning to honor Ginau. Sarakena and Tangiin were daughters of Mwangar, the elder sister whom Ginau had succeeded (see Figure 9).

Figure 8: Sait *poang* in Darapap

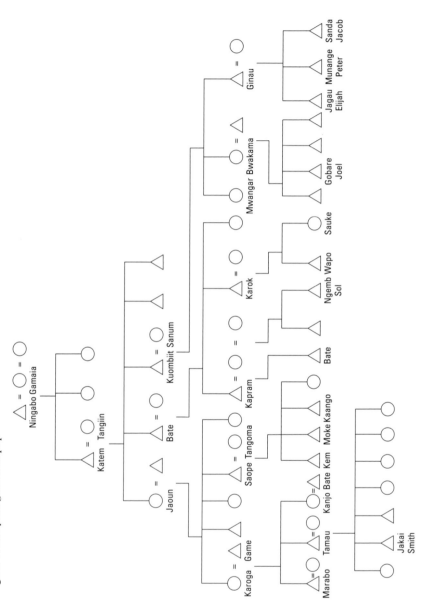

Their husbands had already taken them to purchase pigs from trading partners among the Bush Murik of the Marienberg Hills. The husbands imported *canarium* almonds from the Schouten Islands and had distributed several buckets of nuts, together with a gift of sago and bananas, to Ginau's feasting partners to obligate them to officiate during his rite.

The total number of pigs – five – was equal to the number of pigs Ginau had given the heirs' parents to make his succession claims. When these pigs were reciprocated, the ceremonial and fosterage debts of this rising generation would be squared. In the village of Darapap, Bate would receive Ginau's title as senior titleholder in the Sait *poang*, while in Karau, Sarakena, his most senior firstborn daughter living there, would be installed in the *sumon*ship.

Heraldic title is won at the fixed rate of one pig per heir, irrespective of the gender of the donor. Sarakena, the woman in Karau, was seeking to become a *sumon*-holder. Her husband, brothers and sons were working for her by traveling throughout the region to trade to secure the requisite foodstuffs. While the privilege of either gender to gain a title in this system

Figure 9: Sait *poang* in Karau village

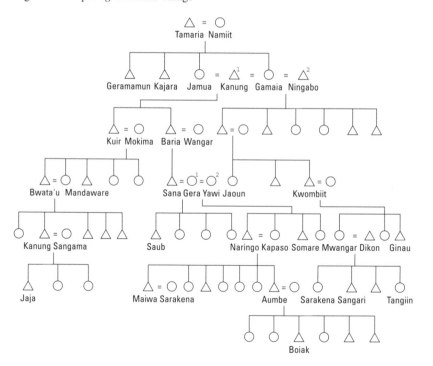

is unambiguous, claiming the rights and fulfilling the duties associated with the title are sometimes phrased (by both sexes) as a masculine capacity. Both men and women agree that the woman who becomes a *sumon ngasen* can trade, instigate and attend ceremonial exchanges after having succeeded her elder sibling, father, mother or mother's brother. Bate used to complain to me that since work for one's *sumon* should only make use of an outrigger canoe which belonged to the descent group of one's ornaments, his preparations were blocked. He owned no vessel. The politico-jural domain, from his point of view, depended upon a mastery of extraterritorial vehicles and spaces that only men construct and manage. This side of heraldic reproduction requires a masculine and not a feminine activity. But the safe passage of any canoe on the open seas ultimately relied, at least in the days of sailing outriggers, not only upon male skill, but also upon the magico-religious effects a woman's actions might have on a voyage. What, then, is the gender of this process and title? Is it male, or female, or both? A woman may hold a title "just like a man," it is true, but a man may hold a *sumon*ship and must adhere to its maternal ethics.

Notably, illicit desire divided incumbent and heir. An interesting parallel obtained between the love affair and the politico-jural stakes between the two men. With respect to both affection and heraldry, an object of desire was sought from which a third party felt dispossessed. The "son" was going ahead with his preparations to succeed to the title which he felt Ginau had "taken" from his father, but denied doing so, apparently because Ginau, his heraldic "father," had also "taken" his wife's affections. The difference, as Simmel put it (1955), between jealousy and envy is that the former turns around the possessor, while the latter turns around the possession. In this instance, Bate was both envious of the possession and jealous of its possessor. There would seem to be a homology between the *sumon* ornaments and the contested woman as objects of desire. In any case, desire, an emotion located within the body, ruptured their relationship, while the exterior, heraldic body united it. Refusing to acknowledge the possessor, Bate nonetheless admitted that the point of the feast was to succeed to Ginau's possession, the Sait *sumon*ship. I shall defer discussing the obvious Oedipal dimensions of this kind of sexual rivalry until Chapter 7. Here, I need merely observe that the attitudes, beliefs and practices pertaining to *sumon* insignia are, in Freudian terms, highly cathected, being permeated by and arousing powerful jealousies and rages that evoke a feminine presence (see also Chapters 8 and 9).

Three months following the arrival of the first shipment of sago, Bate was continuing his preparations for the feast. A group of twenty men and women crossed the lakes and paddled up a channel to a sago stand owned by their "trade mothers" to process more starch. They had been encamped for several weeks when a healthy young woman in her thirties suddenly died. The work party immediately returned to Darapap with the corpse. Kin of the deceased woman poured into the village from throughout the Murik Lakes to mourn (and consumed most of the sago which had been brought back).

As people keened over the corpse, a sympathetic gathering attended by senior men from Darapap and Karau convened in a nearby unfinished house. Bate was present and the discussion quickly turned to his feast. One Karau elder spoke of his concern that the woman's death had been mystically caused by ancestors or sorcerers critical of the preparations for this feast, since she had been working on it. All the delays, he said, were dangerous. "You people of Karau," Bate quickly countered, "own a [copra] boat. Why don't you think about helping me? My sago is now ready. All that remains to do is to go to the Islands and get the pigs and the garden food." One of Bate's affines shook his head and apologized. In the past, he said, his copra boat could have made the trip but it had long since broken down, so "now we don't have a 'path'." One of Ginau's collateral brothers lamented more general problems. Profits from fishing business now went to pay for education fees and village council taxes rather than to provision feasts.

Before, the ancestors used to have time to mount very big feasts. In those days, they thought about feasts and nothing else . . . Things are different now. Now it is better to complete a smaller feast and return to thinking about business. The idea behind Ginau's feast is to make one in the old style, to bring in a lot of people in a big way. This is a big job, requiring a lot of work and a long time to prepare. Look what has happened: a woman has died.

A few days later, Bate paddled to Karau to borrow gas from his sister's husband living there to go to the Islands on behalf of the mourners. In the male cult house, he announced that since the dead woman was only a second wife, the ritual ablutions were going to take place immediately without any lengthy interval of mourning. Elders rehearsed the details of the pending Washing Feast and then set a schedule for the upcoming period during which final preparations would also be concluded for Ginau's feast. The Karau copra boat would first bring women to market in town so they could sell fish and buy the necessary rice and liquor. The boat would then go to the islands, pick up Bate's pigs and produce, return to town to get Ginau and bring him back to the village for the festivities. The meeting went on in a desultory fashion after this timetable had been set out, but I, at least, noted that no one had discussed it with the boat captain, who had been last seen taking a siesta in his house.

A short time afterwards, Bate left for We Island on a loaned Darapap outrigger powered by a borrowed outboard motor to import food for the end-of-mourning feast. He took lower Sepik sago and Murik seafood to trade for vegetable produce. He returned in the middle of the night three days later, having run out of fuel near Big Murik, where he had to borrow some more gas to get back to Darapap, several miles to the west. Next day, the Washing Feast was held for the widower and the rest of the mourners. While he did succeed in staging this ceremony, Bate's plan to return to Koil Island to pick up his pigs for Ginau's feast never materialized. Engine trouble held up the Karau boat in town for months. Meanwhile, the sago breads rotted. Upon hearing a rumor that two of Ginau's employed sons had remitted their father about US$500 with which he had purchased tradestore goods, Bate left for town determined to persuade one of his employed sons to buy him a new outboard motor.

Ginau then decided to sponsor the Washing Feast himself. Word reached Darapap that he was getting ready to return in order to stage two preliminary exchanges in preparation for it. One was meant to oblige his classificatory sisters' sons to perform a folk opera. The other was to oblige a fellow *sumon goan*, the head of the Mindamot descent group, to allow him to carry the latter man's heraldic basket when Ginau paraded through the village at the conclusion of his Washing Feast. (He had received this basket during his initiation as part of his subsidiary membership in the Mindamot descent group for which it stood.) But this *sumon* had been "broken" during a brawl (see Chapter 8). Mwaima, the *sumon goan*, had been injured and his insignia remained taboo. Titleholders in the latter man's descent group had yet to receive a taste of pork, provisioned either by his assailant or by a fellow *sumon goan* in his generation. By serving a small pig feast, Ginau meant to "to repair the broken" heraldry and oblig-ate Mwaima to bestow the basket to him at the appropriate moment during his upcoming celebration.

No little annoyance greeted Ginau's plans in Darapap. The village was struggling to stave off the collapse of its fishing cooperative (see Lipset n.d.a). Many elders were irritated that Ginau was burdening them with ritual duties when they had business to look after.

But by far the most serious reservations belonged to Mwaima's heirs, Minjamok, his collateral firstborn daughter (FZDD), and James Kaparo, his firstborn son (see Figure 10). Both daughter and son had been thinking about staging a succession rite for their father and were worried that

Ginau would claim their title by repairing their father's *sumon*. All three denied that their heraldry had been broken. When the injury occurred, they argued, the old man had been wearing no ornaments and had not been carrying his heraldic basket. Mwaima told his family that he would flee to town should Ginau "truss a pig for him" (and indeed he did leave for hospital treatment soon afterwards).

Ginau countered that the *sumon* and its holder's body were not detachable. "The *sumon* are on the skin of the *sumon goan*," he maintained to me, "whether or not they are worn, because in the past, the *sumon*-holders always wore plaited wrist, arm and ankle bands." In early August 1981, seven months after we first met him proclaiming his imminent return to the village, Ginau finally left town on an outrigger loaded down with cartons of beer, tinned fish and bags of rice. On the way to Darapap, his canoe stopped at Big Murik to drop off a gift of rice and beer to Ginau's feasting partners there in order to oblige them to perform a dance during his feast.

In the several days following his return to the village, Ginau stayed near his house and did not enter the halls of the male cult. After an extended absence from his community, he told me, a *sumon goan* must not return to public life brashly, but should mark the transition by sponsoring a small feast. Ginau felt shame, he added, about the offense that had been taken at his proposal to repair the broken *sumon*. He was no less troubled about his loss of influence in the village. In the past, he said, leaders made feasts. "Now a man must make money." One of his rivals, the leader of another *sumon*-holding group, who had been a village official during the colonial period, had never made any feasts. But the name of his son was now on

Figure 10: Genealogy: Mwaima–Minjamok–James Kaparo

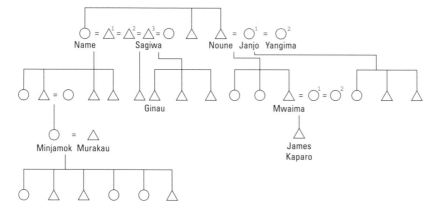

everyone's lips because he had founded the village's fishing cooperative. "All of my children have quit the village. There is no one left to replace me. What has happened is that the son of this lesser man is rising up, using money" rather than feasts. Ginau was referring to Matthew Tamoane, a linguistics graduate of the University of Papua New Guinea.

With the help of a widowed sister and the wives of several younger brothers, Ginau laid out a small feast in the male cult house. When the meal was over, nothing was said to him, no speeches were made, although such a repast is supposed to occasion a summary of events which have transpired during the absence of the returning *sumon goan*. The men of the village were exhausted from fishing business which had kept them up throughout the night. Eventually the youths slipped off to play soccer on the beach. Next day, another meal was staged in the cult house to settle an unrelated dispute connected to the fishing business. After this feast, one of Ginau's collateral firstborn sons spoke up about the lack of discussion the day before.

Food has meaning! Yet no one said anything. People just ate and left in a hurry. Ginau wanted to hear about what had gone on in the village during the many months he was in town and he had something to tell us as well. But everyone flew away . . . Shame should have been aired yesterday. Ginau waited and waited and waited. He was the very last man to leave the cult house . . . Now there are new ways. In the past, each [descent group] had a *sumon goan*, and when one of them laid out a meal of sago pudding and people ate it, they had better rinse their mouths with saltwater afterwards to clear their heads of the power with which the food had been bespelled. This is not a lie! It is true!

Despite the plea and veiled threat in the speech, village attention continued to ignore Ginau. Middle-aged and senior men began to discuss a dispute between a sago-supplying village and another Murik community (see Lipset 1985: 75–9). Youths again drifted away, leaving the elders in the cult house, sitting idly beneath their spirits stored in the dark rafters above them, listening to a lengthy story. Many men smoked; others fell asleep in the dimming, overcast twilight.

Ginau mounted no other feast in 1981. As time passed in 1982, other rites of lesser import, but more urgency, were staged in the two villages of Darapap and Karau, which further deflected attention from him. Meanwhile, the money Ginau's sons remitted him was spent on subsistence. Neither did Bate succeed in putting together the goods. As he told me over and over, he owned no canoe. "The men will talk if I use someone else's canoe to go to the islands . . . So first I must buy my own motor." Moreover, the house his younger brother had vowed to build for Bate to live in after he assumed the title, had not been completed. The house was to be a sacred, named dwelling (*pot iran*). And Bate felt that moving into it was an indispensable part of succeeding Ginau. "I can't simply live in any house," he told me, "after I become *sumon goan*. It must have a name." But such a house also requires a consecration rite before it can be occupied, which rite is also provisioned by overseas trade partners. Early in 1982, Bate wondered out loud whether one of his collateral brothers who was eligible to succeed Ginau might not beat him to the feast. "I just don't have a path! (*yakabor ungwe!*) I don't have a canoe. My brother, Kem (see Figure 8), already has 'given paint' to his father. Who [of us] will win?"

Ginau's Washing Feast
We returned to Darapap in 1986, and found Ginau changed. Clean-shaven, and sporting a new green Hawaiian shirt, he wore nothing around his wrists and ankles. His end-of-mourning feast had taken place the year before in Mendam, the easternmost Murik village, rather than in Darapap, as Bate and the others had planned four years earlier. His sponsor there had been a collateral firstborn son named Dana (see Figure 11). No pigs had been given by any heir in Karau or Darapap villages. Only Peter, Ginau's second-born son, had remitted money to pay for food and liquor. Ginau recalled the feast to me with great satisfaction.

Men from three Murik villages spent the night in the Mendam cult house singing the sacred spells of the war spirits [of our descent group]. At dawn, the performance ended, and my sisters' sons washed . . . and shaved me and cut my beard. They decorated me with paint and gave me the *sumon* baskets, Yamdar of Sait and Wankau of Mindamot, and the two Sait walking sticks that are studded with boars' tusks. They took me down [from the cult house] and walked me to the outrigger canoe. Dry coconuts were broken in my path. My son Peter took photographs.

Before, we used to send women to the *sumon*-holder or to his [classificatory] mother's brother [to accommodate him sexually]. When [either] man took the woman, he [the titleholder] gave everything to the young man. Now [we pay] money: today money has replaced "skirts" [e.g., the sexual services of women]. Dana gave me K40 [= US$45] . . . He took my place. I am finished now: he has kicked me out.

I went and sat down on the platform of the canoe with the *sumon* baskets to eat betel nuts with Dana's wife . . . Gera. Dana is my [firstborn] son and wanted to take my place so he sent his wife and all the *sumon* went to his skin. Gera climbed up onto the platform and sat a little below me. We ate betel nuts together . . . She had become a big woman . . . I told her: "You get up and take the [*sumon*] basket and carry it on your head. You can't look from side to side. You look straight ahead and go to your house" . . . She left and then I returned to the cult house where my [classificatory] sisters' sons looked after me. A big feast was served.

Next day, [my son] Peter took me back to Darapap. Then we went to Big Murik with the *sumon* [insignia] and baskets of food to give to my [classificatory] sisters' sons there.

At the same time as Dana succeeded Ginau as *sumon goan*, Ginau cleansed himself of death pollution. Sitting atop the platform of an outrigger belonging to his descent group, he displayed *sumon* baskets and other paraphernalia belonging to the two descent groups, the Sait and Mindamot, into which he had been initiated as a youth. Resplendent in this heraldry, he toured other Murik villages and received gifts from dispersed kin in recognition of both of these transitions. To register the transfer of his title to Dana, his "son" in Mendam, affinal avoidances between himself and the "son's wife" were lifted. Dana "sent" the woman to give money to him, as a substitute for "offering" him sexual access to her. Ginau accepted the money and the two appeared in public together eating

Figure 11: Genealogy: Dana of Mendam village

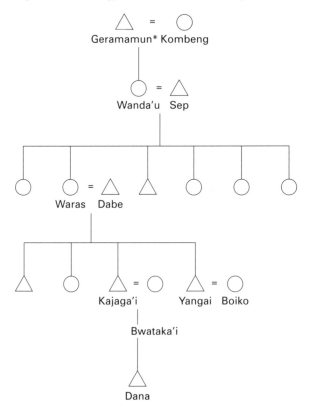

* See Jaja in Figure 9.

betel nuts. This act of affinal commensality stood for the husband's status shift and succession to the *sumon*ship. His promotion meant she was a son's wife no longer. She herself had become a "big woman" (*numaro apo*). As the new *sumon goan* in his descent group resident in Mendam, the heir simply inserted a boar's tusk into his armbands. Ginau's retirement there had no effect upon his incumbency in the villages of Karau and Darapap because these latter heirs had not reciprocated the pigs he had previously given their parents when he succeeded them. Ginau went on to discuss his expectations for them.

Now I am waiting for Sarakena [his eldest daughter in Karau village] to get pigs [see Figure 9]. Peter [his second-born son] has already bought pigs at Mangan village [in the Marienberg Hills]. Kanjo and Tangiin also have pigs [see Figure 9]. They are waiting for Sarakena. When she gets two pigs, my sisters' sons from Big Murik will come and decorate me again. I already sent a pig to them which they have eaten. So, if Sarakena and her younger brother get the pigs, they will come, and she will kick me out and I will no longer be the government. Then, she will be able "to work" with the *sumon* as she wishes.

Heraldic title is vested in the most senior sibling group resident in a particular village. The extent to which the authority of a particular titleholder radiates outside of a single community depends on the migration and genealogical histories of the insignia-holding group and on the titleholder's success in ceremonial exchange. Ginau's position in the Sait descent group extended to several Murik villages because these villages had split, as had the descent group. It was also due to his own relentless competitiveness. "In terms of pigs," he used to boast to me, "I always won." Ginau had been raised as a last-born son by a Sait insignia-holder. In order to legitimize what he viewed as an attenuated status, which he felt was doubtful on account of his adoption, Ginau succeeded each of his elder siblings and had received jural rights in their progeny (see Figure 8). The project undertaken by Ginau's "heraldic children" – to retire him from his title as their "heraldic father" – was therefore a complicated multivillage process, the most significant exchange in which would take place in Darapap, Ginau's village of residence, because yielding his title there would most curtail his authority and prestige. But while openly referring to the Karau women who were preparing a feast for him, Ginau refused to acknowledge Bate, or, for that matter, any other heir in Darapap. Ginau's second-born son, Peter, once told me that his father wanted to keep the *sumon* "in the family." Here again, the dialogue between moral order and the interior body is audible: ceremonial exchange answers feminine sexuality and procreative force through an idiom of generational succession.

When visiting the village on home leave in 1983, Peter saw Ginau in his mourning rags, unshaven, slovenly, all the while meticulously covering up his wrist- and legbands. Feeling embarrassment about his father's predicament, the son vowed to stage the requisite feast for him. But Ginau objected that a second-born had no right to do so prior to senior kin. Should none come forward, then Jagau, Ginau's firstborn son, would have the opportunity, and only failing him might Peter take the initiative. Angry at what he considered to be his father's rigidity, Peter sulked as he left the village to go back to work in the Gulf Province. By the following year, Ginau had abandoned his objections. The senior heirs had died in the interim and his firstborn son had shown no interest. Peter purchased five pigs (one for each of his siblings) and spent about another US$3,200 on food and liquor. In response, Ginau began to talk about initiating him so as to establish his heraldic *bona fides*. But this plan so upset Jagau, the elder brother, that Peter declined on the grounds that he was living too far away from the village to manage the insignia and fulfill duties associated with them. In 1986, we found Ginau, clean-shaven as he was (see Plate 15), trying to persuade Peter to stand for MP in the national elections so that he could return to live closer to home and receive the title in which he still was hoping to invest him.[1] The heraldic system, as this case clearly

Plate 15: Ginau seated next to the Sait heraldry

demonstrates, does not coopt the maternal body but revises and aug-
ments "her" through its ornamental paraphernalia.

The dwelling, which Bate's younger brother had begun to build in 1982,
was nearly completed by 1986, and Bate told us of plans to use it as a female
cult house for a large women's initiation rite he was preparing to sponsor.
The same younger brother had also refurbished an old outrigger for him,
and his children had purchased two outboard motors. Bate was organizing
minor mortuary exchanges and the construction of several new outrigger
canoes when he died of tuberculosis in early 1987. By the time Barlow and I
returned to the Murik Lakes in 1988, Ginau had also died of kidney failure
and prostate blockage, as his daughter-in-law, a nurse, told me.

Neither Bate nor his younger sister in Karau ever did mount a succession feast for Ginau. So an obvious question to ask is who in fact did
succeed to Ginau's title in these two villages? And the answer is that, in a
sense, no one did. Being last-born, Ginau had no younger sibling, and
Bate, his elder sibling's firstborn son, died before mounting the requisite
tributary rite. No one immediately stepped forward to take the title.
Ginau's three sons were living elsewhere in the country: two worked for the
state, the third was employed in the private sector. All three had met with
varying degrees of success in school and were villagers no longer. Elders
complained that the cash economy cut into the monopoly ceremonial
exchange used to hold on to the investment of the surplus resources of the
society. Perhaps the more significant challenge to the reproduction of
heraldic order is the drain of talented and genealogically positioned youth
(see also Gewertz and Errington 1992). No heir trailed in the wake of
Ginau's highly competitive *sumon*ship, only a batch of unsettled ritual
debts. For safekeeping, Marabo, his eldest sister's firstborn son, took
charge of his heraldry, but did so only provisionally and illegitimately,
since he had not validated his right to the title by sponsoring the appropriate prestations. In Karau, Songwari, the younger brother of Sarakena,
Ginau's firstborn daughter, donated two pigs which Ginau's feasting partners ate during his funerary rites. The one belonged to his elder sister; the
other pig, as he told me, was his own. But when I asked him who was
looking after the insignia, "True," he answered quickly with a broad grin,
"I covered up Sarakena [by giving that pig]. But I told her, 'Never mind!
You hold the *sumon*. You are the firstborn.'"

Succession and the hollow body

This case study depicts not only some of the obstacles impeding the
autonomous reproduction of the heraldic body but also symbolic changes

which this body is undergoing. In the premodern era, succession was separate from mortuary ritual. By the 1980s, installation rites had been merged with end-of-mourning ablutions. Alternatively, succession might follow an incumbent's death, when an heir sponsored the funerary exchanges and validated rights to the title of the deceased in so doing. The pre-mortem prestations had divorced body and person, leaving the retired incumbent jurally dead, while remaining physically and morally alive. The folding of succession into mortuary exchange should not obscure the distinctive relationship of political status to the body in Murik culture (see also Smith 1994: 40). At death, heraldry is brought to the corpse; it is neither immanent in the corpse, nor does it represent the body, like a consubstantial relic. Ginau argued that the heraldry of his fellow *sumon goan* was mystically damaged by violence because premodern *sumon goan* always *wore* their insignia, not because their bodies were spiritually possessed by heraldry. Similarly, it is only when an individual undergoes Van Gennepian transitions of state and status that ornaments are conferred. But these rites honor the emergence of social competencies rather than physiological changes within the body. Ensembles of Murik heraldry condense particular migration and genealogical histories. They validate successful ceremonial exchanges. They stand for the property of a descent group, the ceremonial entitlements of its leadership, its ethical system, and, most importantly, they discriminate individuals as jural persons. But while the *sumon* emblems evoke the body, being worn on its exterior, and are even said by the Murik completely to transform a person's appearance, these valuables are not in and of themselves consubstantial signs of the body. They are not, for example, associated with semen, blood or bones, but rather with exchanges that square social debts, the ceremonial services provided by a senior generation.

This transactional inflection is also apparent in two other prominent images that denote installation rites, namely, paint and sexuality. Among the most common Murik metonyms for the succession feast is "to give paint" (*waikur mariabo*). This idiom refers to the cosmetic red ochre with which the incumbent's skin is bedaubed as he (or she) is made up for his (or her) last formal appearance in the community. Among the Abelam, Forge (1962) reported a different construction of ceremonial paint, regardless of hue: it is a "hot," magically powerful substance, considered to be a source of several kinds of male potency. As an imported substance, the "source of . . . [a big man's] paint . . . [is kept as] *a jealously guarded secret*" (Forge 1962: 10, italics mine). Now in Murik, ceremonial paint is valued because it is *given* to a titleholder by an heir, and it is the right to make this exchange which is jealously guarded. Paint possesses neither mystical temperature nor potency. That is to say, like heraldry, the significance of paint lies not in its intrinsic attributes, or its origin, but in the fact of its exchange. Decorating the body of the incumbent titleholder, paint is understood to be a cosmetic given in conjunction with the transferral of rank rather than as a "substance full of power" (Forge 1962: 12). Similarly, the sexual access to an heir's wife that used "to be given" to his incumbent "on a one-time basis" was not viewed as a physiological act of procreation or bodily augmentation. Nor was it viewed as an expression of emotion. Sexuality in this context was a quantified value which was exchanged to settle a debt. Intercourse was a reciprocal service, a husband's act of giving. Being a "token" of exchange, it bears the mark of a currency. The Murik themselves seem to be saying as much, having substituted money for sexuality (see also Chapter 7). Nevertheless, the adhesion of aesthetics and exchange to sexuality must remind us that the *sumon* possess profoundly emotional charges and impulses. Heraldry must be bestowed, rather than withheld, to win prestige for its holder and his group. The *sumon* are valued, not for their durability, but for their detachability and exchangeability (see M. Strathern 1981; 1988a). The *sumon*, it is true, are said to be "hot." As nodes of conflict, they are brittle, eminently breakable objects, whose display solicits the gathering of a similarly fissile group of dispersed kin. Their "heat" is immoral; it comes from the mystical rivalry they provoke. As moral emblems of collective order, in short, the *sumon* are perceptible objects afforded surface rather than interior qualities. They possess no fluids extracted from within the body but are simply worn upon it.

A hollow template has structured the meanings of paint, sexuality and heraldry. Ginau's argument that insignia were always subject to mystical defilement whenever their titleholders suffered physical injury confirms that both should avoid contact with interior qualities of the body – menstrual fluids, violence and passion – in order to maintain their integrity (see also Valeri 1985: 148). Physical injury might entitle the aggrieved to compensation, but such an exchange is not construed to be "blood money." Bodily injury is of secondary consequence. The indemnity repairs the *sumon*, cleanses it and allows heraldry and holder to return to action in the ceremonial arena. The relationship of heraldry to the body became clear in 1988 when, during the Washing Feast for Kanjo, Bate's widow, two brothers got into a fight while Kanjo was sitting on an overturned canoe decorated in *sumon*.

In a drunken state, Marabo, the elder brother, fell into a rage because his younger brother and not he had assembled the walking stick studded with three pairs of boar's tusks (*kajimon sumon*) for his younger sister. Grabbing the stick from his sister, Marabo swung it about in the air, screaming that his entitlement as firstborn had been defied. One of the boars' tusks fell to the ground. The heraldry literally had been "broken." A feasting partner recovered the walking stick from him and took it home for safekeeping. The Washing Feast was not completed. Kanjo had yet to tour her own village or the other Murik villages in regalia. The rite ended abruptly and the guests, many of whom had come from other Murik villages as well as elsewhere in the lower Sepik, "left hungry" (cf. Spradley 1969).

Several of Kanjo's collateral "brothers" immediately began to compete to stage the compensation exchanges to restore the unity of the tabooed heraldry. Among them, of course, was Marabo, the culprit who had broken the *sumon*. He told me about his plans:

I will go down to Blupblup Island and get a pig there. Then I will go back to Darapap and call for my feasting partners to bring back the *sumon* walking stick. We will fill up a basket with food, and set the porkbelly on top and give it to them. They will put the stick in the *sumon* basket and we will take it back. The pork will be distributed to all the Murik villages, to all the people of this Yamdar ornament. If I do not do this, I must die because the men of this *sumon* will be angry and make a secret pact to kill me. Now it is less certain. But in the past, a person in my situation did not have a chance. If you made a mistake with the *sumon*, you would die immediately [of sorcery].

In this episode, when the *sumon* emblem was "broken," no one had been physically injured. Marabo expressed no residual shame about the fistfight he had gotten into with his younger brother. He was only concerned to repair the heraldry by mounting the appropriate exchanges in order to forestall the threat of sorcery, a threat, I should add, which he did not view as emanating from the mystical potency of the ornaments themselves but from elsewhere in his descent group.[2] Here again, we see the relationship between powerful emotions and secret powers which emanate from within the body and a moral body which takes a perceptible, exteriorized form.

Among the Mountain Arapesh, a father might phrase his claim to authority over a son by reminding him of the "many pigs [the father] . . . fattened from which [the son] took [his] growth" (Mead 1935: 77). In Murik, authority over junior kin is also expressed in terms of pigs and indebtedess. But a Murik father, or heraldic father, would never think to claim that his work or food made his son grow. Murik pigs do not augment bodily growth. Ginau had gained rights to the "children" who considered themselves to be his heirs as part of the entitlement he received from their parents, his elder siblings, in return for tributary feasts he mounted to honor them. Thinking of the pigs Ginau had given their fathers and mothers, his heirs wanted to repay this ritual "work," and retire him as their heraldic *pater*. Except for Bate, the sons and daughters in the case openly acknowleged a desire to honor their "father" in this way. Each person sought to settle their ceremonial debt through the sponsorship of a conjoint pig feast during which Ginau would appear for a last time decorated in the regalia of their descent group. The pigs were meant to square a visible, aesthetic debt rather than one that affected the interior body.[3]

What reaction did the remittance of Ginau's second-born son provoke in the father? Instead of being automatically ascribed by a uterine event or filial relationship, primogeniture has been cast as a wealth-object that may be exchanged. The father interpreted the son's remittance as a claim upon his title. Primogeniture is politicized: possessing a hereditary title is subject to the capabilities and foibles of actors. Defiled and "broken" by violence, repaired through ceremonial exchange, ultimately yielded to the first person from a cognatically designated pool of junior kin to sponsor a tributary feast, the relationship of the *sumon* to its holder is eminently negotiable. What motivates an incumbent to pass his (or her) title to the "skin" of an heir is the ethical system of accommodating the requests of inferior, junior kin. But what prompts this sacrifice in a proximate sense is a feast, the prestation of foods that are largely imported, rather than locally produced foods, which are endowed, in other words, with visible, not interior, meanings. Succession requires two types of ceremonial exchange. The heir gives foodstuffs to the incumbent titleholder who uses it to fete his feasting partners and compensate them for their ritual services. This pair of local exchanges, in turn, has been provisioned by a great deal of regional trade.

The crucial relationship of regional trade to the reproduction of the heraldic body deserves comment. Preparations for *sumon* exchange plunge potential heirs into the canoe-dependent domains of overseas trade, petty capitalism and the remittance economy. In this case, foods were imported from the Schouten Islands and the Marienberg Hills. Sago came from a lower Sepik River village. The alcohol and rice were purchased in Wewak, the provincial capital. While business, education and loss of children to urban jobs clash with and divert attention from the reproduction of the heraldic body, remittances from employed kin, either directly in cash, or in the form of outboard motors, also facilitate it. But the remittance of Ginau's son, which financed his feast, reveals a very important point about the means and relations of production in this culture: Ginau had neither a

Plate 16: Sakara, a retiring *sumon*-holder, is seated next to his younger brother's wife on a platform. Note the paddles included in the tableau, suggesting that the platform is itself a canoe

canoe of his own nor a wife to weave baskets and attend visiting trading partners. While the ceremonial career of a *sumon goan* who does not own an outrigger may bog down, that of a *sumon goan* without a wife is virtually thwarted. Indeed, the extent to which a Murik title is culturally constructed in terms of canoes and wives is well represented in the *tableau vivant* which is staged to culminate the *sumon* transfer. The heir's wife, who has given sexual access to herself to the retiring incumbent in order to "make her husband a *sumon goan*," sits next to him and the couple chew betel nuts together. This reversal of affinal form takes place on the platform of the heir's outrigger canoe (or a platform built specifically for the tableau). When the heir retires the titleholder, the heir's wife becomes a visible sign of his promotion. The ex-titleholder and the heir's wife appear together – at nearly the same level in space – on the platform of the very vehicle that allowed the scene to occur (see Plate 16). The wife herself then becomes a vehicle for the signs of her husband's new title, literally carrying the *sumon* basket to him.

Rather than foregrounding the promotion of the heir, the exemplary representation – of a heterosexual, commensal relationship set upon an outrigger canoe – depicts the transfer of title as a lifting of affinal inequality and avoidances between the incumbent and his heir's wife. Offstage in his house, the new *sumon*-holder merely inserts a circular boar's tusk into each of his wicker armbands. This modest act of attaching the heraldry to his body signifies that ceremonial credentials have been validated by a series of exchanges provisioned by overseas trading partners accessed by outrigger canoe. It also signifies genealogical credentials, it is true, but not royal blood. Over and over, Bate complained to me that he "lacked a path." The path was not lined by his birth nor by the pigs he imported, or, as it turned out, he had failed to import. Bate meant that he owned no canoe. Indeed, I have heard it said that titleholders are themselves precisely this: "canoes" for their heraldry. The maternal body, the body of procreation, birth, blood, eros, rage, injury and death, provides few metaphors for moral personhood or political association in this culture. It has come under the design of an image of a hollow vehicle-body moving through foreign, watery spaces. I have made a material argument that the nonagrarian, entrepot adaptation has shaped the poetics of the moral body in this culture in custodial rather than physiological imagery. The idioms of ornamentation, generosity and commensality are not related to bodily growth, consubstantiality or emotion. The prototypical figure of the maternal body, in all of "her" emptiness, now comes into full view. "Her" body is a hollow, deeply androgynous, vehicle. The maternal schema is set in a "canoe-body" (*gai'iin*) endowed with female and male, rather than only female, agency.

Many gender transformations are associated with canoes. In inland relations, "trade mothers," who are accessed by canoes, give sago, the "bones of the ancestors," to their landless "children." In the overseas sector, hosts infantilize guests with abundant hospitality. In outrigger canoe construction, the hull undergoes a gender change in a masculine canoe-shed that is symbolically equivalent to a birth house, or womb. The hull emerges from this building amid images of parturition. While a canoe is at sea, like a vulnerable pregnant woman, the steersman's wife should remain chaste, like a husband. Most tellingly, of course, are pregnancy beliefs in which shapes of, rather than the substances inside, the foods a woman eats affect not the growth, but the bodily integrity, of her fetus. She herself both is and is not the "real mother" of the fetus, only its "canoe." One might speculate that this is why a knot is tied in a creeper vine to secure the soul to the "canoe-body" of a newborn, as an outrigger is lashed to its hull. No conjecture is necessary to understand the imagery by which siblings are known – as "creeper vines" or "canoes" – and to their moral center, the indulgent figure of the firstborn, who is their "canoe-prow." A man's wife, whom he calls his "mother," is idiomatically his "outrigger float." And, esoterically, a wife's body is also said to be her "husband's canoe" who has given herself to his ceremonial rivals. Finally, the insignia-holder, himself "a canoe" for his heraldry, ought to act "like a mother," through feastmaking and the bestowal of *sumon* upon children.

This chapter has focused on Murik notions of ceremonial succession and, willy-nilly, personhood. Its main point has not been to show how the attribution of maximal personhood is limited to a privileged elite, but to show that the idea of title is neither gender-exclusive nor rigidly ascribed by a principle of primogeniture. Title is a negotiated and transacted category: receiving it is a creative act which demands initiative rather than passivity, anonymity or death (cf. Fortes 1973; Geertz 1973). Succession activates a momentary, if quite delicate, network of trade, domestic and ceremonial relations. Without the ceremonial reproduction of the heraldic body, this wider reticulation need not occur. Concepts of politico-jural authority, no doubt, provide a measure of and for personhood (La Fontaine 1985). In Murik, each descent group possesses its own heraldic accessories and its own system of ceremonial leadership. The title and office of *sumon goan* is an accomplishment which constitutes the very field of its authority. The titleholder who remains inactive in ceremonial exchange, who "merely eats in his house," as the Murik say, not only reduces his own agency and the reputation of his heraldry, but reduces the influence of his sibling and descent groups within his village as well as the influence and reputation of his village in other parts of the Lower Sepik and North Coast region. As such, there is a strong element of contingency built into this prestige structure. If personhood is not unilaterally conferred upon actors by society, or achieved in feats of "strength," then what is the cultural relationship of person and society? Geertz once praised the notion of personhood as "an excellent vehicle by means of which to examine th[e] whole question of how to go about poking into another people's turn of mind" (1983: 59). In Murik, the notion of "vehicle" turns out to be an excellent way of poking into the question of what personhood means in terms of its relationship to the collective life.

Of course, however powerful a metaphor it may be, the notion of a canoe-body remains a hollow image which must fail to exhaust the whole range of human experience. The elements of interior life it helps to smother, namely, sexual desire, uterine fertility and aggression, nevertheless find relatively untamed cultural voices. The maternal canoe-body schema remains an ambivalent rejoinder to a body more carnal, with which it is locked in an eternal dialogue. In Parts II and III of this book, I shall go on to examine how male cosmology and conflict disarrange its quietism and ethics, only to cause them to be rehearsed once again in ritual performances.

PART II

DIALOGICS OF THE MATERNAL SCHEMA AND THE COSMIC BODY OF MAN

Plate 17: Pimi of Darapap with the mask of Emang, a spirit-man

6

A body more carnal

Cleansed, as they ought to be, of defilement and passion, the men and women who lead the polity seek to emulate sacrificial, quietist mothers raising dependent children. While the heraldic bodies they seek to reproduce are not single-sexed, there are two other sex-exclusive bodies in the society with which they are in dialogue. They are male, war spirits (*brag*) and female, sexual spirits (*samban merogo*; see Barlow 1992; 1995). This chapter examines the extraordinary discourse with which masculine spirits alternately satirize the moral order in everyday life and lend it a virtually unconditional support during ritual contexts. I first provide exegesis of their absurdist poetics as a caricature of the stoic, sacrificial body espoused by those who would uphold the maternal schema. I then discuss the quite moral role these same spirits are given to play in the course of the ritual cycle when the values they would otherwise ridicule they carefully affirm. The distinctive relationship between their mocking and moral poses, in other words, is neither mimetic nor functional but contrapuntal, doublevoiced and unfinalized.

Among the genres through which such a spirit voices objections to the maternal schema, perhaps the most basic one is expressed by his unambiguous gender. His body is not inhibited and androgynous: masculine passion and fluids overflow him. He is not just a spirit but a spirit-man (*brag*; or *pot nor*). Far from repressed, the glamour of his face (*brag sebug*), depicted in wooden masks (see Figure 12), is animated by blatant images of genital desire and aggression. His visage is dominated by a great phallic, beak nose. In wooden figurines (*kandimboang*), the top of the spirit-man's head often appears to turn into the face and head of a serpent (*wakun kombitok*) whose nose reaches down to his genitals. He can be no less beautifully ornamented than the heraldic body. But his decorations are intended

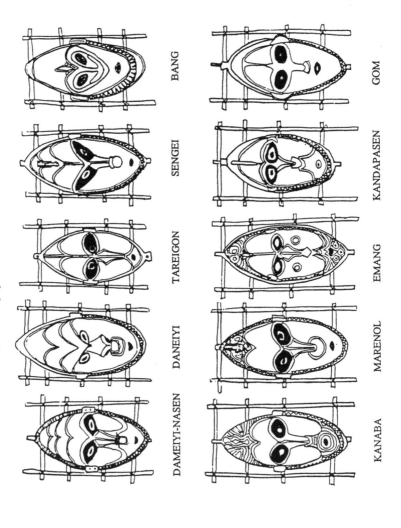

Figure 12: Spirit-men (*brag*) in Darapap

to seduce women rather than identify collective affiliation. The hollow wicker frameworks into which men harness themselves when they impersonate spirit-men are also called "canoe-bodies." Covered with a thick, leafy bunting, they are not empty but contain men whose canoe-bodies, in turn, carry the spirit which possesses them (see Chapter 7). His body is one of protrusions and orifices: it is an open, not closed, body whose boundaries are intricately connected to other forms of life. Zoomorphic motifs on the foreheads of the spirit-men are the particular "canoe-bodies" in which they appear to travel about in nature. Complex, Escher-like, figure–ground reversals abound in their iconography (see Figure 13). On the prows of lagoon canoes, lizard and bird motifs become noses on humanoid faces. Mouths gape open, as if to consume the watery space through which they travel (see Lipset n.d.c.). Their ubiquitous buffoonery, which transgresses the decorum of the sibling group, also erases limits between man and society and man and nature, to man's advantage. A tone of triumph characterizes their comedy, of the body overstepping itself, celebrating a victory over material deficits and asceticism as it does. Among these men, body and world are interwoven rather than distinct, merging rather than differentiated. Unlike those of the heraldic body, their bodies are anything but empty. Above all is their limitless passion, a point made clear in the story of the twelve "Fog Men" (*Wau Nor*), whose twelve wives were seduced by Sapendo, a spirit-man who lived under water.

The Fog Men[1]

All the wives of the Fog Men secretly met Sapendo, when they went to fish [by the river]. [One of the women named] Pandima [would] beat a slit-drum and Sapendo would come to the surface beautifully decorated in his ornaments. The women undressed and lay down on their backs. Prepubescent ones lay in a row; adult women lay in another. One by one, Sapendo had intercourse with them. Afterwards, each woman presented him with a dish of sago pudding she herself had cooked (see Bateson 1936: 192). They offered him only their very biggest fish to garnish the puddings. He ate and went back into the water. The women returned to work.

Because their wives only left very small fish for them to eat, the Fog Men began to become suspicious. Wok, their leader, told his son, who was also named Wok, to cry [deliberately] next morning, when his mother left to go fishing, so she would bring him with her.

[Next day, he did and was brought along.] The women got to the river [and went to work]; Wok's mother immediately caught a huge fish. "It's for papa!" the boy yelled, holding the fish. Wok played for a while in the canoe, until he fell asleep. The women cleaned all the fish they caught, including the big one Wok had picked out for his father.

Figure 13: Carved post in Sendam male cult house, Darapap

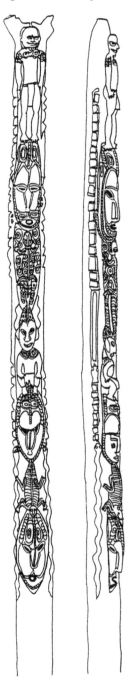

They cooked sago pudding. Then Pandima beat the slit-drum to call Sapendo, who was such an enormous man that his decorations drifted ashore long before he himself reached the surface.

Wok was still asleep. The women took off their skirts in anticipation of the arrival of the spirit-man. Their pile of skirts covered up Wok. The sun made him hot. He woke to see the twelve women lying on their backs. Sapendo had intercourse with his mother first. "My mother is being raped!" he burst out (*Mana ngain tokesonara!*). Sapendo fled back into the water. The women threw the plates of sago pudding after him. But he did not return. In their anger, they sewed Wok's mouth, eyes and ears shut with little fish bones. He lost consciousness.

The Oedipal troubles caused by the insatiable virility of the *brag* spirits are a major theme throughout the entire corpus of their tales.[2] The bodies of these masculine spirits are hardly celibate or virtuous. Instead of a quietist maternal body, they are uninhibited and lewd. We shall see how they disrupt the prudish decorum of the sibling and ceremonial groups by substituting a mood of indecency, mayhem, but, above all else, desire. However, this Melanesian liberation, a carnival in the midst of everyday Murik life, creates no "freedom" in any sort of voluntaristic or revolutionary sense. It succeeds in exposing the incompleteness of the moral body by countering its totalizing claims. The modes of passion these spirit-men celebrate are perhaps less explosive than those permitted during the Balinese cockfight (Geertz 1973; cf. Rosebery 1982; see also Dundes 1994). Their ubiquitous presence does not set them apart from the moral body as the cockfight is set apart from hierarchical Bali. Nor does the body to which they give license threaten to topple the sociopolitical structure. Just the same, it would be wrong to conclude that "nothing happens" (Geertz 1973: 443–4) because of what people do in its degraded guise. The *brag* spirit-men and the female *samban* spirits voice crucial, and yet nonetheless quite equivocal, aspects of experience. They are not simply an indirect instrument of social control (cf. Eco 1984) with which to "renew . . . the system" (Gluckman 1963: 112). They puncture the moral body in ways that rather "embrace both poles of becoming in their contradiction and unity" (Bakhtin 1965/1984b: 203). In this cosmos, motherhood is neither idealized nor vilified. Instead, "she" becomes the subject of one consistently contrary system of discourse.

The Spirits of Wealth

Commensal sibling groups give *mwaran*, literally "valuable goods," or ceremonial "wealth,"[3] to the men or women who personify these spirits in ritual contexts. The category of these relationships is therefore called the "valuables" or "wealth path" (*mwara yakabor*). In number, any individual actor, male or female, may have twenty or more *mwara* partners in his or her "wealth path." They may live in ego's own and in other Murik villages; or they may live elsewhere in the Lower Sepik. But the extent to which any individual *mwara* relationship is pursued depends upon propinquity, the ongoing level of ceremonial exchange maintained by the actors, as well as the personalities involved.

The principal *mwara* statuses are the classificatory mother's brother, the father's sister and a heraldic double. Here, I only intend to analyze the classificatory mother's brother and the heraldic double; that is, I limit my focus to the men's relationships in which I myself participated.[4] Since the classificatory mother's brother appears chronically in daily life, while his heraldic colleague becomes active only during rarefied moments of high ritual, I shall first turn to the matrilateral relationship (cf. Bateson 1936; see also G. Lewis 1980). A man, his daughters and his daughters' sons inherit *mwara* kin from a woman, her sons and her sons' daughters. The man and the woman are classificatory "siblings." The "brother" is therefore the "mother's brother" of the woman's sons, while they, in turn, become "brothers" to the man's daughters (see Figure 14). These relationships recall the alternation of the sexes in succeeding generations that the Mundugumor likened to "a two-ply cord" (Fortune quoted by McDowell 1991: 245; see also Mead 1935: 176). There are three important differences between the two systems, however, and one similarity. The first difference is obvious: the Murik metaphor for the set of relations is not a cord or a "rope" but the exchange of "wealth."[5] The second is that the alternation of the sexes in Murik does not register male–female rivalry and conjugal suspicion as Mead maintains it to have done among the Mundugumor. Thirdly, the Murik pattern has no positive relationship to marriage whatsoever (see McDowell 1991: 266–86). Common to both is that rights to cosmic powers, but not land, are at stake. In Mundugumor, actual paraphernalia imbued with cosmic power were passed "down the ropes," while in Murik the stakes are less literal.

The values and behavior associated with the *mwara* partnerships at once debase and humiliate the inhibited, maternal canoe-body with a forgiving kind of humor, while also restoring "her" generativity, endowing "her," to put it bluntly, with cosmic id. At stake is not the completion of a finite, heraldic body. At stake is the metamorphosis of that body, its endowment with gender. This metamorphosis is accomplished through an elaborate exchange of goods and services between the Spirits of Wealth and their rivals, the commensal "creeper vines" (*nog*). Part of it is accomplished through a restless satire, a dialogized opening up of the maternal schema whose prototype is not itself "reproduced, it is merely implied. But the entire structure of speech would be completely different if there were not this reaction to ['her'] . . . implied words" (Bakhtin 1965/1984a: 195).

The Murik carefully distinguish between the *mwara* mother's brother and his commensal (*nogo*) counterpart. The differences they emphasize reveal the former as occupying a nether position that is neither kin nor yet stranger/enemy. "If the mother's brother works, the sister's son eats," goes a Murik saying about the commensal version of this relationship. And the two do eat together on a regular basis. The *mwara* mother's brother, by contrast, does not routinely share food with his sister's son. Their world is one of banquets and drunkenness. Commensal kin know their common ancestors; *mwara* kin cannot specifically trace all the relationships in their common ancestry. Younger commensal kin know they ought to defer to their elders hierarchically. They should speak "honestly" to each other, primly avoiding reference to the interior body, to its passions and processes. They speak *nogo'iin*, which literally means "creeper vine talk," but

Figure 14: The Spirits of Wealth: classificatory *mwara* kin relationships of women and men

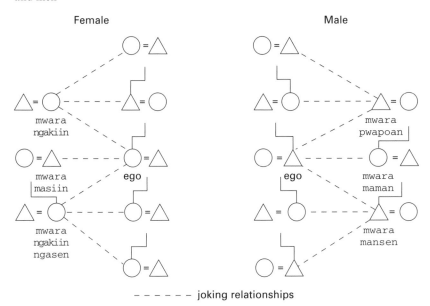

- - - - - joking relationships

ngakiin: **father's sister**   pwapoan: **mother's brother**
masiin: **brother**   maman: **sister**
ngakiin ngasen: **father's sister's daughter**   mansen: **sister's son**

is translated as "truth" in Tokpisin. *Mwara* kin, by contrast, joke with one another more or less as equals. They speak *gwaga'iin*, which generally means "play talk" but also connotes sexual ribaldry. Or, they talk "nonsense" (*kakaowi*) to each other. In their shameless discourse, the interior body is far from taboo. Instead, the explicit subject of their mockery and laughter is genital desire. Domestic and ceremonial exchange of goods among commensal kin is organized by genealogical age, birth order and the values of the maternal schema. Exchange between *mwara* kin is largely limited to ceremonial contexts during which it is competitive. The symmetrical term of address (*jakum* in Murik; and *katres* in Tokpisin, from "cutlass" or "cartridge") the *mwara* kin use to each other, regardless of age and sex, further distinguishes the rivalrous and aggressive qualities of their relationship from the hierarchical, modest and courteous decorum of the *nogo* relationships. Finally, unlike collateral members of the *nog* who may marry each other and still carry on "pre-affinal" relationships should they choose to do so, *mwara* kin should never marry. If a *mwara* brother and sister do marry, they should begin to feel too much "shame" to continue their joking relationship. Therefore, the rights and privileges they share do not mediate affinal tensions in a direct way as Radcliffe-Brown used to argue about Dogon joking relations (1965b and c; cf. Sharman 1969; see also Hammond 1964). Instead, the *mwara* relationship both constitutes yet violates a discursive boundary between the moral body and the world of nonkin immediately outside it. They surround domestic and jural kin with mockery, ridicule and laughter. But they also play a crucial role as ritual courtiers, or functionaries. When individuals exit the domestic group to become ritually differentiated as jural persons, the *mwara* kin look after and instruct them, not as rivals from without but, again, as "mothers" from within.[6] During ritual, they shift in and out of their satirical voice and in and out of a mode of unconditional support.

A classificatory mother's brother, for example, will transport his nephew across, and transform him within, the cosmic boundaries dividing jural groups in a community. When a retiring *sumon goan* leaves the men's cult house to parade through a village, adorned for the last time in heraldry, the first steps he takes should cross over the bodies of his mother's brothers and sisters' sons who prostrate themselves, like babies in a state of impotent helplessness, or like sacrificial mothers, in his path. Their duties recreate precisely this image: they do not obstruct the path to a new status so much as they constitute it. What is the "value" of such unconditional support? It is measured in *mwaran*. The sibling group and its heraldic extensions reimburse the *mwara* kin with food or other goods. Each time a *mwara* mother's brother performs a ritual function for his nephew, he also presents *mwaran* – e.g., a dog's teeth necklace, a boar's tusk, or a pittance of cash – to his protégé. While the services the mother's brother provides are reciprocated by food, both raw and cooked, his durable gift is not. This asymmetry serves as the rationale for the continuity of their relationship across generations. The *mwara* sister's son is left indebted by these durable gifts: he is therefore obligated to organize funerary services when his mother's brother dies and begin then to reciprocate both the life-long ritual role and gifts to the daughter's sons of the deceased, which work to indebt the latter youth. And so on from generation to generation. Their relationship is defined in terms of alternating exchanges of services and alienable goods. The *mwara* sister's son ends up owing neither his wife, nor his body, nor any of its fluids, nor the origin of his being, to his *mwara* mother's brother. He should reciprocate to the latter man's daughter's son exactly what he has received: the ritual services and the valuables which helped him assume the cosmic attributes and capacities distinctive of a man in this culture.[7]

A grotesque symposium

What is funny, when something is funny, and with whom it is funny are learned through the *mwara* brother–sister relationship. While her son is just a toddler, a mother should introduce him to her "wealth brother" (*mwara masiin*), the boy's "wealth uncle" (*mwara pwapwan*). No later than when he begins to talk, or even before, she should start to whisper some appropriate profanities for her son to yell back to him, when he walks up to them sitting on the porch of her house, and pulls at the boy's penis "to lengthen it," or when he charges his nephew with having committed youthful infidelities by mentioning the name of some purported girlfriend (usually an old woman) whom he has "seduced," on account of having inherited the "big balls" (*pa'iin apo*) of his mother's father. I once saw a young mother hold up her surely preverbal toddler to her face and sweetly coo to the little boy, "You fuck! You fuck!," in Tokpisin, just as her *mwara* brother walked by on his way to see off a canoe. A boy will also learn how to joke by overhearing abusive banter going on between his mother and her *mwara* brother during chance encounters. He will watch as this man walks by screaming about his brazen mother who has the nerve to go by herself to the bush "to get water" (water being a euphemism for semen), whose back, as he might say, is all too familiar with the beds of needles in the casuarina forests, and whose "clam" was broken there long ago, no doubt, by several men whom she, being the inexhaustible, wild woman that she is, must have seduced (see Tiffany and Adams 1985). Through her own and her son's voices, a woman simultaneously retaliates against her "brother" and teaches her boy to distinguish between the moral discourse expected of him within the commensal group and the absurdist parody demanded of him by the Spirits of Wealth. From observing and parroting his mother, perhaps by the age of three, a little boy should begin to tease his *mwara* mother's brother, groping him back (or, at least, knowing to raise his middle finger back to him as he passes by the house) and should begin to learn to hurl at him some simple abuse such as "Big prick!" (*Edug apo!*). In a gradual way, the *mwara* sister's son assimilates the bawdy genre: the repertoire of impropriety, profanity and aspersion, as well as the debased gestures which enact a body armed with fantastic, phallic protrusions, inexhaustible libido and gaping maw.

Few hard and fast rules debar *mwara* comedy from any particular setting in the culture. Joking in cross-sex groups should avoid explicit mention of male genitalia and should remain phrased more delicately through the use of periphrasis or euphemism.[8] The most prominent taboo against joking, of course, exists within the commensal groups which gather inside dwellings. But should a sister's son go to the house of his *mwara* mother's brother, the two may exchange mild abuse. The most unrestrained form of *mwara* joking takes place during same-sex assemblies in the halls of the male cult, especially as they begin, lull or end (see Chapters 1 and 8). However, joking is hardly limited to such venues, and will break out whenever *mwara* kin congregrate, to see off or greet an outrigger canoe, or when, as in Darapap, they used to wait for the regular Monday morning local government councilor's meeting to begin, or really, any time at all. We were walking past the house of his *mwara* mother's brother on our way to meet an outrigger canoe. Ginau called out the man's name: "Pimi-ay!" On hearing from his wife that he was asleep, Ginau turned upon the woman. "Are you holding his cock? Have you pulled back the foreskin yet?"

No little of the daily texture of public life is taken up by this sort of obscene, irreverent and somewhat strident travesty of the moral body. Like his phallus, the trail of lovers seduced by the *mwara* mother's brother throughout the region is lengthy. Being a *nor mwago* (literally a "bad man") of immense sexual appetites, he is sometimes likened to Sapendo, the mythic hero who made love to a dozen women, and then ate the plate of sago pudding each had prepared for him as a token of their postcoital gratitude. Following the return of his sister's son from a trading expedition, the *mwara* mother's brother will question him mercilessly about "how many plates of sago pudding he got to eat" while away.

Sitting together in the male cult house, two brothers were teasing a short, bony, rather unpleasant looking old man, who lacked an eye. "Last night," one told the other, "I dreamed that our mother's brother was in Boem Island fucking a woman from behind, standing up like a dog." In spite of his repeated denials, in the minds of his abusers, the elderly man was soon journeying to each of the Schouten Islands, between Manam and Kairiru, ending up in Wewak town, "eating plates of sago pudding" in each place before returning home by boat to start the circuit over.

Mwara kin adore such ridicule which, they say, "sweetens" life. "Oh! Affine! My apologies!" goes a standard greeting. The implication is that the one has just seduced the other's sister. Now the two men ought to give up their play, the play from which they take so much pleasure, and begin to observe the staid conventions required by affinity. But of course the greeting is ironic because it distinguishes their relationship as precisely "not affinal." The one in fact has not come from lovemaking with the other man's sister and so their joking relationship remains intact.

At the same time as they mock sexual fidelity and affinal decorum, the *mwara* mother's brother may act as an immoral go-between to facilitate the ongoing philandering of his sister's son (cf. Fortune 1939). Throughout the lifetime of their relationship, the *mwara* mother's brother should secretly arrange trysts with women (single or married) to whom his nephew is sexually attracted, and should protect the sister's son's privacy, while he meets his lover (*mari neman*) in the gardens, or in a canoe in the lagoons.

The example men give to illustrate this duty usually begins with the sister's son sitting together with his *mwara* uncle in an unwalled, gazebo-like structure (*banumtom*) when suddenly the young man gets an urge to visit one of his girlfriends. Casually, he shoves his basket over to his mother's brother, explains that he is off "to urinate in the bush" (urine and semen being synonymous) and asks him to look after it while he is gone (*Ma minjiro menowa. Suun dewan!*; literally, "I am going to urinate. There is my basket!"). The *mwara* mother's brother is thereby meant to know to collude with his sister's son when the latter's wife approaches jealously asking after the whereabouts of her husband. The uncle should deceive her with something like the following disclaimer: "No, no. He has not gone far. He was just sitting here with me. Look! There is his basket!"

The bawdy comedy and duties of the *mwara* kin suspend the discourse of conjugal morality, affinal modesty and sibling hierarchy which confine and enclose the body, ultimately enveloping it in an impenetrable heraldic facade (see Chapters 3–5). Their laughter and abuse is filled with a pronounced hyperbolism of bodily images. Equipped with a huge, insatiable phallus, a lascivious leer upon his face, and a predilection for ridicule and the ridiculous, the image of the *mwara* mother's brother is exactly not that of a mature, modest, indulgent elder brother carefully avoiding the younger brother's wife. No longer is his body a beautiful but empty canoe, moving carefully through space. No longer must he talk honestly. No longer must his passions be repressed. The chaste, androgynous body of a custodial mother to which so many emblems of interdependency cling – plates of food, baskets, heraldic regalia and hungry children – is here answered by a grotesque body, an erotic, aggressive body of a warrior. Although the latter upholds no sacrificial, maternal ethic, "his" is an equivocal body, a body which is no longer moral, but remains deeply positive in character.

Occasionally, a *mwara* mother's brother may come and tug at his infant nephew's penis when he encounters the child sitting with his grandfather in a public venue, while his mother and her sisters have gone to fish in the lagoons for the day. I first understood this gesture to be an attempt "to lengthen it." But it is also the beginning of socialization for horseplay, a great deal of which goes on during the many public meetings and events when joking partners gather. In the male cult house or in village avenues, *mwara* kin engage each other in obscene, slapstick comedy, the repertoire of which further liberates the moral body from the deference and avoidances that confine its domestic/jural form. *Mwara* mothers' brothers and sisters' sons will grope each other in the crotch and buttocks (see Plates 18–21). Walking by a sister's son seated on the floor of a cult hall, a mother's brother may casually lift his leg over his nephew's head, partially undoing his waistcloth as he does in order to flutter it over the younger man, so as to pollute him with the residual sexual fluids from the woman he has just seduced. He might even do his worst, and jump, crotch first, onto his nephew's face. In pantomine, he then forces these unclean substances into his partner's mouth. (Such an image appears at the beginning of Chapter 1.) For example: about twenty men convened to discuss the illnesses of two elders. The brothers and sons of the two men served a meal. Afterwards, Kaibong goosed his *mwara* mother's brother as he was about to stand up. Mariim, meantime, was squatting over a plate, washing it. His *mwara* sister's son reached over and quickly poked his buttocks causing Mariim to leap into the air from the squat. Landing on his feet, Mariim resumed washing the plate.

Discussion started (and then concluded) about the two men. Mariim was now lying flat on his back; his knees were propped up so that his buttocks were exposed. Sauma threw a little wooden skewer into his crotch and Mariim kicked his legs up into the air overdramatically. Sivik walked by and gently stuck his foot into Mariim's crotch. Sauma gingerly picked up the skewer and poked Mariim again. He jumped up wildly to the great amusement of the house. "You hole! You!" Jakai Smith yelled at Sauma, who was his *mwara* mother's brother, in defense of Mariim.

The *mwara* kin give license to morally inadmissible behavior. Their kinesic invasions of the genital and anal zones show little concern for the preservation of social boundaries (M. Douglas 1970: 98). But two carefully observed rules limit this debased drama. The first is that the slapstick between *mwara* brother and sister should be less aggressive than that permitted between the mother's brother and the sister's son. And the second is that a nephew should *never* lift his leg over the head of his uncle until he has given the senior man a "skirt" (*dag*). That is, the *mwara* sister's son must, on a one-time basis, yield rights in the sexual services of his wife to his *mwara* mother's brother. In return, he may then lift his leg over the elder man's head. An expensive penalty, enforced by the sorcery of the spirits (*brag timiit*), sanctions violations of this rule. If, in a moment of exuberance, a sister's son who has not done so, forgets himself, and lifts his leg over his *mwara* mother's brother, he must quickly compensate the latter man with a pig feast. Otherwise, the *brag* spirits will kill him. Their slapstick is thus an unruly, yet rule-bound, parody of the moral body.

Nevertheless, a dilemma is created and exposed in the ethical system of the sibling and descent groups, a dilemma that is posed by its celibate and modest virtues. For men, the process of marriage begins with an act of sexual intercourse legitimated by the taking up of coresidence, the initiation of affinal avoidances, and the fulfillment of brideservice duties, but perhaps most importantly, by reproductive success, ongoing childcare and commensality. Sexual intercourse, in a way, gradually turns a woman into a celibate mother, a mother of both her husband and children, who must feed them food free of sexual impurities. The conjugal body becomes moderated by obligation and inhibited by the maternal schema. For the Spirits of Wealth, the body remains sensual and irrepressible. Their sexuality leads to no co-residence, decorum, brideservice or reproduction, but results in the mockery of commensality ("How many plates of sago pudding did you eat?"). Siblings should avoid talk of the body, displays of emotion or sexuality by junior kin and affines, much less feign to grope their genitals. The *mwara* mother's brother relishes his immoral discourse. *Son nor!*

Plate 18: At work, cutting grass for Gobare, the village councilor (seen gesturing); men take a break

Plate 19: Gobare's *mwara* sister's son furtively pokes him

Plate 20: Another one of his *mwara* sister's sons turns on him as well. Gobare makes a mock attempt to flee

Plate 21: Gobare mock-curses his pseudo-antagonists

(Lecher!; literally, "Intercourse man!") is another one of the standard greetings which a *mwara* sister's son may use to salute his mother's brother. More than likely, his mother's brother will answer by calling the nephew a "dickhead" (*edu'poap!* literally, "glans penis"). Younger brothers learn to observe proprieties toward elder brothers. Affines defer to and avoid each other in space. The *mwara* mother's brother pokes the crotch of his sister's son. The spectacle of his slapstick and humor continuously violates the boundaries of personhood, substituting a riotous sensuality for the ascetic maternal schema. Remaining within bounds of kinship, and being itself informed by a normative idiom, this comedy is not unbridled. The discursive challenges posed by the *mwara* mother's brother are more like satire and burlesque than resistance or treason. Their comedy is no betrayal or denial of commensal morality, but a gendered rejoinder to its flawed, prototypical image. A momentary revolution of the senses, and an exhilarated feeling of pleasure and agency, sustain their ambivalent dialogue with the chaste, hierarchical voices who personify the maternal canoe-body.

Mimicry, hyperbole and social control
Young people are expected to call upon elders when they are in need of social or material support. Their dependency is encouraged by such kin, who seek to fulfill the maternal value of indulgent generosity. In addition to their grotesque travesty of the moral body, the Spirits of Wealth use hyperbole to pressure individuals *not* to show this dependent face outside the commensal, sibling and descent, groups. But can every moment of need or vulnerability be nurtured by the moral body? No. People are imperfect. By overstating and exaggerating accidents and miscues, the *mwara* kin merrily reprove those who commit offenses of social ineptitude. Their actions decree that no one should present his or her domestic face – the face that can admit of helplessness, indigence and inadequacy to senior "mothers" – in public. The Spirits employ two ironic devices, which I shall call "aggressive generosity" and "'sympathetic' mimicry" to sanction such inappropriate displays of dependency.

Aggressive generosity
Following his return from overseas trading, a *mwara* sister's son may want to hurl "spears" (*ningeg*) – minor gifts of food – at his classificatory mother's brother. This is one of many moments in their relationship when the wealth they exchange becomes coded as metaphoric "weapons" (see Codere 1951; Young 1971). Another is occasioned by the public display of indigence. "At all times, we must watch our *mwara* mothers' brothers and

sisters' sons very carefully," my adoptive elder brother used to advise me early on during our initial period of fieldwork. "When one of them asks one of us [in the sibling group] for something, then that person ought to go quietly to all our brothers, pool the thing that has been requested, and really *give* it him!" I had no doubt that the crescendo of stridency and boldness to which his voice rose when he explained this to me meant that making such a gift was akin to smacking the recipient with one's fist. Indeed, this sort of aggressive and humiliating generosity is explicitly glossed as a kind of metaphorical "murder." The same man used to boast to me that long ago he had "killed all our" *mwara* mothers' brothers living in Kopar village at the mouth of the Sepik, "killed them," in other words presented them with gifts they had failed to reciprocate. Whenever *mwara* kin meet – which is all the time – they continuously size each other up in a nonchalant way, hoping to seize upon an opportunity when their partner comes up wanting some ordinary thing a person should normally possess: a utensil, tool, betel nuts, tobacco, a cigarette lighter, or even firewood. When the sister's son appears to have forgotten his spoon during a feast and he is spotted applying to his elder brother for one, the *mwara* mother's brother will alert his siblings and assemble six or eight spoons. To the immense delight of the gathering, but the great humiliation of the forget-ful "mark," the donor will then make a grand gesture of presenting the gift to the sister's son, at the last moment encircling his head, clockwise, with the bundle of spoons. Although being caught short of practically anything will prompt the aggressive generosity of the *mwara* kin, the example offered to illustrate this sort of exchange is always "spoons." One need not stray too far afield exegetically to conclude that multiplying spoons trans-forms a utensil which would otherwise refer to the element of generalized exchange (Sahlins 1972) in the moral schema of mothering[9] into meta-phorical ammunition, a sign of combat, and a quantified good, a sign of regional exchange. The maternal schema is thereby moved into a different metaphorical environment, one in which it is immersed in masculine values and tactics, abstracted from the semantic fields of warfare and trade, and of course, it is moved into a field which is genealogically less intimate. The use of hyperbole in *mwara* comedy thus demands a bounded and more strategic presentation of self. It does not cultivate the dependent self, the sustenance of which is so basic to domestic and jural authority. But neither does it become disengaged from the object of its satire.

Two tactics follow from the premise that displays of dependency should be kept hidden from the *mwara* mother's brother. The first is that no peti-tions for goods or favors should be made directly to him and the second is that no requests should be made in his presence. A sister's son who does either obviously leaves himself open to displays of aggressive generosity. For example, a meeting was held in a male cult house to resolve a dispute within the Darapap fishing company. A meal was served.

Just as everyone was almost done eating, Gidion arrived carrying two plates of sago pudding and half a coconut, the meat of which he was nibbling. Kaibong, his *mwara* mother's brother, looked up from his plate and impulsively threw up his hand toward Gidion, as if to ask for the coconut. A moment passed: the two men looked at each other. Kaibong began to protest, vigorously shaking his head. With a mischievous smile, Gidion answered, "So you want to eat coconuts do you?!!" As Kaibong tried to deny the request, Gidion rushed out of the cult house, somewhat later to return with a dozen coconuts roped together, which he dropped down by the lap of his mother's brother, who slumped over. The latter man then divided the gift among his *nogo* brothers, fathers and uncles who were present.

The redistribution of a "spear-gift" (*orub ningeg*) integrates the commensal group because it obligates the recipients to help repay the debt when the opportunity "to retaliate" inevitably presents itself. A "spear-gift" of equal or, better still, of greater size will then be assembled and "shot back" at the indigent wretch, or at one of his kin, who has made a careless display of neediness. The pervasive threat of aggressive generosity heightens tension and boundaries between these groups, distinguishing them as adversaries, and differentiating the commensal contexts and their embodiments of indulgent motherhood, from cosmic contexts and their embodiments of grotesque motherhood. This minor rivalry does not specifically concern indigency as much as a more general contrast – a dialogue – between a dependent self which may openly express need for the other, and an autonomous and assertive self capable of "standing up" on his own. Misdirected requests are therefore not the only kinds of dependency which arouse the hyperbole of the Spirits of Wealth.

"Sympathetic" mimicry

During feasts they attend together, or anytime they happen upon one another, *mwara* kin lie in wait, looking for the opportunity "to ambush" minor injuries, accidents, or other misdeeds which may befall their joking partners. Falling down, cutting a finger, bumping a head, or just carelessly dropping something, will bring forth a mocking mimicry from available *mwara* kin, of whom an individual may have ten or more living in any particular village. From all corners, they will converge upon their "victim," to help "share his shame" (*yanga nabiin koteka*, literally, "everyone's eye take away"). One by one, they imitate his accident as precisely as possible. "Laughter," it is said about such incidents, "is heard in the houses."

A teenage boy wearing thongs was carrying a gunny sack full of sago starch that he had slung over his shoulder when he slipped and fell on a slick tide flat. Six or seven of his *mwara* mother's brothers and sister's sons quickly appeared to form a queue. With a single pair of thongs and a sack stuffed full of clothes, each man took a turn falling at the same spot where the youth lost his footing, passing on the thongs and the sack to the next in line to sling over his shoulders as he did.

The oaf should feel a double shame: shame for the embarrassment he has caused his siblings by being so careless and shame for the shame which his *mwara* kin have brought upon themselves through their sarcastically "sympathetic" mimicry. To compensate the latter, commensal elders should stage a minor feast (not requiring a pig), a meal which is said "to stand them up." The feast, in other words, does not restore relations of equality between *mwara* mothers' brothers and sisters' sons. Nor does it "buy their blood"; it rather "buys their shame."

A sprightly senior woman named Konombe was mounting a sharp incline at a bank of the Sepik River in Angoram, the district headquarters, when she slipped on a slick patch of mud and fell flat on her buttocks. Other people in her party, standing some yards away, saw the accident. A woman named Musiit, who was Konombe's classificatory daughter, called out in a teasing way that she would tell the woman's *mwara* nieces what had happened after they returned to the village. Konombe picked herself up and screamed back to Kaibong and me, who were standing next to Musiit, and had joined in her threats to spread the news. "No! You are my children! I bore all of you!"

Sympathetic mimicry is routinely offered for "falling down," which becomes a more general message about a state of being, that of incompetence and vulnerability, the display of which persons should try to contain within the commensal group which is normatively prepared to care for it. But like the performance of aggressive generosity, the sympathetic mimicry of the Spirits of Wealth has sociological effects upon commensal kin: it not only discriminates their contrasting moral ethos but circumscribes their membership. They must not betray each other's misdeeds or accidents to their *mwara* rivals. Moreover, they should collude to help each another during assemblies in horseplay and acts of aggressive generosity.

Barlow observed (1990) another object of their mockery which should not pass unnoticed here: the device of mimicry itself. The mimicry of the *mwara* kin burlesques yet another element in the maternal schema. Their hyperbole not only caricatures the display of helplessness in public contexts, it caricatures the domestic tactic of exemplifying desired actions (see Chapter 3). Wanting to appear patient and indulgent toward junior kin, and wanting to avoid the dishonor of making direct requests of them, elder siblings are taught deliberately how to perform the action they want junior kin to imitate. The failure of a younger brother to fulfill a duty for an elder brother may be reproached, not directly by anger or resentment, but by *showing by example* (Murik: *mwan sikemo*; Tokpisin: *soim pasin*). This "technique of demonstrating what should have been done is learned within the sibling relationship but finds powerful expression on other occasions as well" (Barlow 1990: 105).

Within the sibling group, displays of indigence, incompetence and helplessness provide opportunities for elder brothers and sisters to express compassion for, instruct, nurture and generally help or protect their junior kin. Seeking to fulfill the normative schema of mothering, the elder sibling will try to act like a custodian and teacher. The *mwara* mother's brother turns these same states into opportunities to indebt and defeat their rivals. Sympathetic mimicry and aggressive generosity pressure the individual to perform in a competent manner, to be in possession of appropriate utensils, and, at least, to confine requests for resources within the sibling group. The individual, in short, should appear as an autonomous, sociologically bounded, person who knows where and when to show dependency and vulnerability, helplessness and incompetence. When he fails, carelessly or accidentally, he can expect that the *mwara* kin will humiliate him by multiplying his failure many times over. The message of their comic hyperbole would seem to be: "Stand up like a man! But if you must act like a dependent mother's child, restrict your neediness within your sibling group where it belongs. We will take advantage of you should you not!"

In its comic moods and devices, the grotesque symposium of the Spirits of Wealth is part of a paradoxical dialogue with the commensal, domestic and jural groups. Their generosity and mimicry sanction public lapses of autonomy. When and where does their verbal abuse and slapstick break out? In gaps between contexts, during transitions, when interactions begin, lull, or conclude. Upon entering a meeting hall, or after the departure of an outrigger canoe, a person should immediately distinguish commensal from *mwara* kin.

Therefore in contexts outside of the dwelling house, the joking of the *mwara* kin both orients the self within, yet also disaggregates the self from, its most reliable sociomoral field (cf. Hallowell 1955). The *mwara* mother's brother simultaneously demands that the sister's son behave as a competent member of the groups which define his personhood in society and that he misbehave as trickster-adversary. His comedy is no utopian liberation from moral norms and inhibitions. It does not simply promote the possibility of a complete exit from the moral body and its maternal schema. The laughter of the *mwara* mother's brother does not just involve the sister's son in a subversive concept of the body. The deeply generative side of his position is made clear in the explicitly maternal role he takes during rites of passage when the maternal schema is in flux. I shall now examine this quite positive, altogether supportive, side of his ambivalence in the course of the life-cycle.

Childhood

The *mwara* mother's brother is not merely the first exemplar of community presented to a boy. In firstborn rites, he is the very means by which a child enters the community. From whom must the firstborn be hidden until he is able to identify others? With whom must he be "strong enough" to interact? With whom must he be able to laugh before he can be introduced to the community? The *mwara* mother's brother.

After a firstborn child has become able "to see" him at a distance, and after the necessary goods have been imported, the first duties of the *mwara* mother's brother are to donate an ornament (or some money) to his sister's firstborn son, remove him from his mother's dwelling and carry him through the village bearing heraldry (see frontispiece). In return, he receives several baskets full of imported delicacies: tobacco, sago breads, betel nuts, Pacific almonds and fruit. In a second rite, the *mwara* mother's brother literally feeds the child a first taste of *aragen*, the ceremonial porridge. Unlike everyday food cooked by women, this porridge is prepared by the male cult. Sweet and white, some people liken its taste to breast milk. But it does not come from mothers but from the spoons of the *mwara* kin, who come dancing up to the dwelling house of the firstborn and, in a stylized, rather aggressive way, line up "to teach" him to eat it. In "the ladder" rite (*waik*), the house ladder is decorated with imported foods, and the mother's brother "instructs" the sister's son how to climb up and down it.

In each of the three episodes, the part played by the mother's brother is purely formal. A boy will more than likely have begun to eat "social" foods, fish and sago pudding, by the time he is "taught" to eat *aragen*, the ceremonial porridge. And he may have already begun to roam about the village with his siblings by the time he is "taught" how to climb up and down the house ladder. The sponsorship of these ceremonies is often a major claim in an ongoing custody dispute between divorced parents over a firstborn child. The provision of food to the mother's brother then becomes a jural claim made by mother's or father's kin, a claim which is legitimized by whichever family's *sumon* is displayed. But whether taken developmentally or jurally, my point remains the same. As in Iatmul *naven* rites, the mother's brother is afforded the duty of honoring the child as he develops skills which eventually lead to his separation from the domestic group and its maternal values. The three attributes and capacities – politico-jural status, the ability to eat social food and the mastery of proximity – celebrate the progressive differentiation of the self away from the mother and toward becoming an autonomous member of the community.

As children grow up, many other moments of separation and jural incorporation are marked by the ritual guardianship of the *mwara* mother's brother. The first time, for example, a sister's son is permitted to go overseas with his father or mother's brother to trade, a *mwara* uncle is entitled to rush up to the outrigger and anoint the departing nephew with his spittle, by spraying saltwater into his face. This is meant both as a magical gesture of protection and as an act which places the youth in his debt so that, upon his return, he will be obligated to give his mother's brother something from the bounty of trade goods he has received. This may happen every time the youth goes to a new port for the first time. Fathers will sometimes try to hide sons in the depths of an outrigger as they leave the community so the *mwara* mother's brother will not notice the departure of his nephew. But having been blessed, the countergift is not simply an act of compensation; it is again coded as "homicide." Upon return, the sister's son should "shoot" and "kill" his mother's brother with a "spear-gift," an act of reciprocity which is part of their adversarial exchange I have called "aggressive generosity." From early on, the *mwara* mother's brother celebrates and abets the autonomy of the sister's son. Rather than accepting him as a helpless child, hopelessly attached to the indulgent nurture, instruction and protection of his many "mothers," he furnishes the child with a contrary set of capacities and resources which allow him to become independent of "her" infantilizing ministrations.

Male initiation

In the premodern era, male initiation used to consist of four phases which were either performed separately or were interconnected. The *mwara* mother's brother continued to attend to his sister's son like a courtier during each of them. His duty in the Loincloth Initiation (*nimbero gar*) was to dress the novice in a new loincloth, ornaments and *sumon* heraldry, preparing him for his first appearance in the community as an adult member of a particular descent group. In a second phase, when all adolescent youth collectively entered the cult of the war spirit-men (*brag gar*), he would begin to teach him how to play the bamboo flutes, the voices of the *brag*. When firstborn sisters' sons were initiated into the cult of the spear ancestor spirit-men (*Kakar gar*), he was responsible for feeding his charge. In the last phase, called the "Tree Sap Initiation" (*yarar arum gar*), during which all youths received a great dose of virility magic, he would smear the love potion upon the sister's son. I shall discuss his role during only the first two rites in this section, reserving the third for the chapter to follow. Since the Tree Sap Initiation had become defunct in the 1980s I will omit it entirely.

The Loincloth Initiation

As elsewhere in the insular Pacific (see Goldman 1970: 531), the ritual transition to adulthood in Murik is signified by new clothes. It used to be said that a boy walked naked through the community until puberty, exposing his "unripe" or "raw" penis. When a firstborn son reached puberty, or, if he had learned to shoot fish, carve and use a bow, his mother might then conclude that he had acquired the skills of manhood and might think about importing a pig and the other foodstuffs necessary to initiate him and the rest of his younger brothers into her descent group. While these preparations were under way, the boy's mother would ask one of her *mwara* brothers to make her son a new bark loincloth. On its pubic apron, he might paint the face of their *brag* spirit-man outlined in *Nassa* shells, positioned just so its long nose would dangle in front of the youth's genitals, like an enormous phallus.

In an initiation I observed, the young man's mother wept quietly as a group of her *mwara* brothers lifted Bruno up and bodily carried him down the house ladder and off to the male cult house. There, they "taught" him how to smoke and chew betel nuts, encircling his head with the cigarettes and nuts as they presented them to him. They gave him his "first meal" of sago pudding, crab and fish. He tasted a mouthful of each before dispatching the plates to his mother's kin. The mother's brothers had also prepared a red ochre (*waikor*) which they smeared all over Bruno's body. Singing love spells (*simog*), a few of them held the decorated loincloth they had made for Bruno above their heads, and danced gaily with it, before encircling his head with the garment and slipping it onto him. Each man readied *mwaran*, wicker armbands, dogs' teeth necklaces, leglets, boars' tusks, or some cash to give Bruno. As the several mother's brothers presented a gift or attached it to his body, a slit-drum beat was sounded (*mwara debun*) to signal throughout the eastern lakes that valuables had been exchanged.

A meal of nuts, breads, fruit, tobacco, sago pudding and crab was served by Bruno's family. The mother's brothers tasted it and redistributed the food throughout the village. As they ate, each man stood up with a single feather in his hand, red, white or black, and stuck it in Bruno's hair (see Plate 11). Naming a woman as he did so, the mother's brother explained to the youth that he had been in a particular village where the woman whose name he had just revealed had seduced him. In return for making love to her, she had given him the feather he was now giving Bruno. One of the uncles, who was his paramount mother's brother (*mwara pwapwan da'uraro*, literally, the "nose mother's brother") presented Bruno with a lime gourd (*a'ir*), which he also named after a woman who had seduced him.

After being presented with heraldry, the Spirits of Wealth led him down from the cult house into the village, where people honored the insignia by breaking mature coconuts in his path, and spraying the juice at him. Following the procession, they returned Bruno to his mother.

Before the beginning of an all-night dance performance later that afternoon, the mother's brothers were rewarded with a varying number of baskets of imported nuts, fruits, sago breads and uncooked strips of pork according to the size and quality of the gift each man had given the initiate earlier in the day. Bruno's paramount *mwara* mother's brother, who had been responsible for making his new loincloth and had given him the lime gourd, received the porkbelly and liver, the most prestigious pieces of meat.

In this instance, the *mwara* mother's brothers transported, instructed, donated "money" to and dressed their charge. After physically removing him from the feminized, domestic space and depositing him into the masculine, cosmic one, they "taught" him to smoke, chew betel nuts and eat in the new setting. By giving him alienable wealth, they presented him with the wherewithal to become an independent, regional actor. In so doing, they also revealed the secret, erotic rewards of overseas travel, for the feathers they stuck in his hair color-coded a geography of their lovers. Black feathers stood for Sepik River women, red feathers for inland women, white feathers for coastal or island women (cf. Harrison 1990a: 104). Painting his skin red, adorning him with "clothes" and jewels charged with images of cosmic virility and beauty, they enlarged the social significance of the youth's body, making him the brilliant subject of collective attention. "The radiations of adornment," Simmel noted, provoke "sensuous attention . . . [and] supply the personality with . . . an enlargement or intensification of its sphere: the personality . . . is more when it is adorned" (1964: 340). Escorting him through the community for the first time differentiated as a "heraldic son" (*sumon goan*), the youth was subjected to honorific gestures (see Plate 11). Then, in their role as cosmic nexus, they returned the initiate to his mother. In response, reciprocal acts of exchange were offered them: the sister's son donated gifts of food which distinguished two ranks of *mwara* mother's brother. The men who had metamorphosed his body and made it so much more social also became the object of social differentiation. The pageantry of the scene was punctuated by banquet imagery: carefully enumerated plates of food circulated within the cult house as well as throughout the community. And the *mwara* kin protected the unfinished bodies of their protégés as they transformed them. The values of the maternal schema were not far to seek.

Brag initiation

Boys, the saying goes, grow up "hanging from the skirts of their mothers," which dependency depletes their masculinity. In a second phase of initiation,[10] one of the duties of the *mwara* mother's brother is to move the youths out of their mothers' "stomach/womb" (*sar*) and into the devouring "stomach/womb" of the *brag* spirit-men. Now his charge is to cure enfeebled, feminized boys and turn them into virile men who are aggressive, invulnerable warriors. This process turns upon the voices of the *brag*, the paired bamboo flute spirits which are considered to be mystically toxic to women's health. The novices are brought to the male cult house where they are said to be "swallowed" up, digested and gestated in his "stomach/womb."

The boys entered the cult house, Father Schmidt observed, to find "spears . . . arranged diagonally across each other; also the men stand in line with their legs spread apart and the novices must crawl under the spears and under the men's legs" (1933: 57). One by one, the novices were laid face down over big slit-drums and beaten by their fathers with the butt of a big palm leaf or whipped with the tail of a big eel. Accusing the youth of having grown up indulged and lazy, they would sing:

> When you live under your mother's leg
> you do what you like,
> and you have what you want.
> You ask for food,
> you receive it.

You ask for water,
you receive it.
You come and cry,
you want to eat a big fish,
your mother cooks it for you.
But now,
now, you come under your father's leg.
Now you pay
for being treated softly at home.
A new time begins for you
when you must know
that you will be a man.   *(Somare 1975: 25–6)*

Youths were expected to take this hazing stoically. The commensal, rather than the *mwara*, mother's brother could protect his sister's son from the spanking by draping his body over the back of the youth. The duty of the *mwara* mother's brother was rather "to show" his nephew the flutes, which were performed by torchlight, and to begin to teach him how to play them. This lengthy and difficult process was started in a playful, teasing way. The nervous youth should try to control his anxiety by averting his eyes from his clowning *mwara* mother's brother, either shutting them completely, or fixing his gaze upon his feet. Until this point, the youth had been allowed to sit quietly in the presence of the *brag* spirit, head down, permitted only "to steal" glimpses of the flutes, should he have received permission to do so from his commensal mother's brother. At the end of the performance, the fathers threaten to ensorcell their sons if they reveal anything of the secrets they have seen to their mothers. The youths spend the whole night in the cult house learning to sing the spells of the war spirits (see Chapter 8). At dawn the next morning, the *mwara* mothers' brothers take the novices to the beach to teach them that they must "wash" in the sea following bodily contact with spirits. Each *mwara* mother's brother would also begin to teach his sister's son a technique for expelling the "black blood" which collects at the base of a man's spine as a result of contact with feminine sexual fluids. The accumulation of black blood, the youth would be warned, will inevitably doom a man's abilities to succeed in projects outside the community, by making him torpid, slow-witted and weak. The blood should therefore be discharged before any enterprise, such as going to war, traveling overseas, playing in a soccer match against another village, participating in a string-band competition, when hosting a visiting trading partner, and most crucially when carving a new outrigger canoe. The first time the *mwara* mother's brother tutored his sister's son in the bleeding procedure, he was said to be "ripening" or "cooking" the youth's penis, the penis of an uninitiated youth being called an "unripe" or "raw banana."

The uncle takes his nephew to a secret spot near the beach where he first shows him how to select only the freshest stems of a barbed grass (*gaisi-ikup; Coix lacrima*) because rotten ones are thought to cause deadly infections.[11] The mother's brother would then rub the penis of his sister's son until it became erect. Inserting the barbed stem into the urethra, he would quickly yank it out so that the barbs caught its walls, causing blood to be expelled from the penis.[12] Each mother's brother might do this several times "for" his nephew, during which process, I was told, weaker youths would succumb and faint in pain. But those who could withstand the operation ten times or so did great honor to their fathers, who, upon hearing of their sons' stoic performance, might happily exclaim something like, "Oh! My canoe has come ashore!"

Once this has been done, the sister's son would wash in the sea and then return to the cult house, where his group would reassemble to recover together for a few days. The youths were "fed" by their *mwara* mothers' brothers during this period of seclusion. After their wounds healed, they received a shell ring or a dogs' teeth necklace from them as well as new clothes. The *brag* spirit was then said "to spit" the youths back to their mothers.

Through an oxymoronic pedagogy, which was at once festive yet coercive, the secret illusions of the cosmic body were revealed. The *mwara* mother's brother clowned as he introduced the novices to the bamboo flute spirits, "his" voice. The youths were also ambivalently thrashed – the fathers sing as they administer it – and were made to suffer other bodily abuses before they were taught the hygienic techniques which protect their cosmic bodies from feminine sexual impurities and protect women from their cosmic bodies. Abuse became praise and social metamorphosis. The youths received a change of costume and new privileges. They became "entitled to marry" (Beier and Somare 1973: 8), although, since the *brag* cult initiation is not a puberty rite, many of them were either much too young to do so or else had already been married for years. This spirit-man had a devouring body. As a destructive and cannibal warrior, "he" swallowed and digested the sisters' sons. But "he" also had a generative body "who spat" them back to their mothers metamorphosed into adult men. The male cult house within which this process took place was the symbolic "womb" of a pregnant woman as well as a birth house. When the youths entered and left the building, these two contradictory bodies were combined into a single grotesque image of birth and warfare: the youths crawled through a tunnel of legs and spears made by a row of cult elders who held the weapons diagonally against their legs. The *brag* spirit-men "ate" the sisters' sons in order to render them maternal services – instruction, nurture, hygiene and grooming – before regurgitating them with the attributes and capacities distinctive of adult men. The *mwara* kin who personified these ambivalent spirits become cannibal-mothers, both devouring yet pregnant, punitive yet custodial. The hidden dialogicality of their relationship to the uterine body could hardly be rendered any more apparent.

Death

As part of this dialogue, the commensal group takes as its model or prototype a hollow canoe lashed together by "creeper vines" that is decorated by heraldry. The men and women who seek to become a moral "canoe-body," mother, but do not give birth. How does such a body die? What could a decomposing corpse possibly mean in terms of such a vehicular prototype? In what sense might Hertz's admonition "to follow the corpse" when analyzing funerary rites apply here? The import of these questions is not just for my interpretation of the dialogics of Murik culture, but for an empirical reason as well. Death remained by far and away the most common context in which *mwara* kin remained ritually active in the 1980s and 1990s. At death and during mourning, the *mwara* sister's son recreates an elaborate image of grotesque gestation and birth (see Barlow 1992: 74–9).

The *mwara* mother's brother has many duties during consecration ceremonies for the first house built by his sister's son, for a new outrigger canoe, as well as in cult house construction and conflict settlement (see Chapter 8). In each context, he donates valuables to the sister's son, indebting him. When the senior man dies, the sister's son begins to resolve and reverse his indebtedness by managing the movement of the man's soul to the ghostly community (*pot kaban*). As the *mwara* mother's brother oversaw jural transitions during his nephew's life, in other words, the sister's son must now take charge of this final status-shift for his uncle.

On the day of his death, a group of perhaps one dozen sisters' sons will come to wash and dress his corpse in new clothes which they donate. Each of them should also place one boar's tusk, or any other currency, upon his chest, a gift which obligates his kin to sponsor a festival of ablution called a "Washing Feast" (*Arabopera Gar*) to recompense them and celebrate the end of mourning of his wife, siblings, children and age cohort.

The following day, but sooner if the corpse begins to putrify, the *mwara* sisters' sons return to the mourning scene. Those kin who are still weeping over the body retire to the periphery of the house and turn away or gaze impassively at them as, now armed with hatchets and dressed in ornamental regalia, they burst into the middle of the room, not to nurture the mourners or attend the corpse, but to "play" (*gwaga*) with it. They shout the names of the *brag* spirits they personify and then they begin to call for the deceased to get up and join them. They taunt and insult him, demanding that his ghost return to its canoe-body so the deceased may celebrate too. "*O'dekara!*" (Stand up!), they yell, as they try to revive their joking partner from his passive, asocial state. Kicking and knocking the bier with their hatchets, shaking the body, prancing about, they threaten to lift up a floorboard and dump it into the mud below so dogs can drag it away.[13] Onlookers laugh and whoop. Their mock antagonism soon gives way to a more formal and deferent performance. Should the deceased have owned title to a folk opera, the *mwara* kin may begin to dance it outside of his house for all to see.

The mourners who have come into contact with the corpse become taboo and must submit to an equivocal caretaking, which is at once maternal yet abusive. *Mwara* sisters immediately tie creeper vines (*nog*) about the foreheads[14] of their brothers[15] and feed them food they may even premasticate. Suddenly, however, the maternal schema is turned topsy-turvy. The ghost now becomes "an initiate" who is subjected to hazing. Except in the case of the deaths of the very elderly, ghosts are assumed to be angry at having been suddenly cut off from attachments for which they continue to have feeling. Hovering about the company of the bereaved is a jealous ghost "who" seeks to charm them from their bodies. The *mwara* kin must therefore insist that the ghost go to the ancestral community (or go inside the cone of the volcano at Manam Island) and not interfere with the living. No one checks vital signs. Instead, the *mwara* sisters' sons place the ghost in their debt by giving it wealth and taunt it, happily demonstrating that it has made an irreversible departure from its canoe-body. While the person is subject to many kinds of asocial states which resemble death, no sentient Murik would ever fail to take up the mocking gauntlet thrown down by his *mwara* kin or reject an opportunity to dance with them. The first exemplars of the community now become its final signposts. How do the *mwara* kin certify death? By exaggerating the failure of the mother's brother "to stand up" to their travesty. Comically humiliated and defeated, the soul must come to accept that it is no longer "a passenger" of its canoe-body, no longer their rival. The *mwara* sister's sons put valuables on the corpse and go dance. Their succession as the new senior generation of mothers' brothers to the daughters' sons of the deceased is a spectacle tinged with mockery, laughter and an exchange of wealth. The death of the grotesque body does not provoke grief among them but a debased victory over a longtime adversary.

The Washing Feast

The mourners then withdraw into seclusion, which is ended, and a place for the ghost among the dead is secured, by the Washing Feast, which is a kind of secondary burial festival. In my terms, the Washing Feast is nothing but another form of aggressive generosity, a triumphant banquet image during which the cosmic body is rid of impurity and is returned to its pure, moral form.

The festivities begin at sunset, the time of day when shadows lengthen – the ghost is a shadow – and suddenly appear to detach themselves from people. The community summons the ghost from the outskirts of the village, where it is said to have been lurking during the mourning period, to join in the celebration. Small Washing Feasts then divide into two groups: men repair to the male cult house while the women adjourn to a large dwelling which serves for the evening as the female cult house. Larger ones take place outside in a central square where both sexes assemble around the heraldic basket of the deceased, which hangs from a post erected in the middle of the gathering. There, they dance and sing all night. An ethos of victory fills the air. The sibling group of the mourners try to overwhelm the *mwara* kin with their prestation. Twice during the evening, great feasts will be served. Throughout the night, individual sisters' sons will take their uncles by the hand to offer them the opportunity to drink an entire case of beer by themselves: again "to kill" them with the gift. We once saw an elderly man appear at intervals, literally to dance circles around his *mwara* kin while bidding them: "Have your fill, my sisters' sons! Have your fill!" In same-sex, indoor gatherings, lewd antics inevitably break out among *mwara* kin.

It was well past midnight in the male cult house in Mendam. The men had been singing the spells of one of their war spirits for several hours, accompanied by the heavy percussion of several slit-drums which were continuously beaten. Even so, one senior man had not been deterred from falling asleep. His *mwara* sister's son, one of the drummers, noticed him and, without missing a beat, pranced across the hall toward the man, who was curled up on his side. As he danced, he gestured with the lengthy drum-beater as if he was about to poke the man's buttocks. He toyed with his audience for several minutes, rhythmically approaching then retreating, until his old mother's brother finally woke up. A man then leaned over to me to say: "Never let your *mwara* mother's brother get away with falling asleep."

At sunrise, the exhausted community reassembles. Beaten across their backs with branches of juvenile mangrove trees by their cross-sex *mwara* siblings, the mourners are driven through the village to the beach (e.g., a *mwara* sister will beat her brother and vice versa). Both the mourners and their *mwara* kin squeeze inside a large conical monster (*gaingiin*) made of casuarina branches, and run into the sea (see Plate 22). Throwing the monster from their shoulders, they bathe briefly and cleanse themselves of the death pollution. After a few minutes, they run from the water and jump over a small bonfire that has been built on the foreshore. They then make their way back to the village. A few *mwara* kin of the deceased remain behind and fashion a model outrigger canoe (*bun gai'iin*), origami-style, out of coconut leaves. The ghost is meant to use the little vessel to voyage to its ancestral community. Having provisioned it with food, they cast the canoe into the sea. Should it return to shore, the ghost is understood to want to remain among the living for what are presumed to be its own malevolent purposes.

Plate 22: A Washing Feast concluded at dawn. Mourners were beaten and pushed toward the lakefront. They squeezed into a monster (*gaingiin*) which they threw off in the water

Meanwhile, at the house of their hosts, *mwara* kin attend the mourners. Sisters' sons will groom their mother's brother, shaving him for the first time, dressing him in new clothes they give him, and make him a gift of domestic valuables, such as a new fishing spear. They serve him a meal of sandcrab or chicken, over sago pudding or a bed of rice. The *mwara* kin encircle the head of the mourner with each gift, a gesture of closure and triumph which distinguishes cosmic from domestic exchange as marked and competitive rather than diffuse and unconditional. The mourners sit passively in their new clothes and eat. Their kin, however, proceed to serve a grand, multi-course pig feast to the *mwara* kin.[16] Plates and glasses of liquor are offered courteously but nevertheless remain coded as acts of "murder," or more recently, "incineration."

At a Washing Feast in 1988, the hosts negotiated the crowded floor, carefully monitoring their guests' glasses to keep them full of a beer and wine cocktail. "I am a good man!" effused one senior man as his sons dispensed the grog. "I give 'pigs' [by which he told me he meant the liquor] to everyone! I kill pigs for everyone!" The grandmother of the deceased then encircled the head of her son's *mwara* mother's brother with a bottle of whiskey, and declared, "In the name of my [dead] grandchild – Freddie! Take this!" He and his brothers immediately reciprocated by giving her a case of beer.

In the 1980s, undoubtedly the most locally prestigious and interesting exchange during such events was the number of cartons of South Pacific Lager, the national beer of Papua New Guinea, the commensal kin managed to outgive their feasting partners, the Spirits of Wealth. In 1986, a *sumon goan* in Darapap actually tried to persuade me that beer, wine and whiskey had become more valuable than pigs (see Ogan 1966; Marshall 1983). Whatever the case, lines of women in particular can be seen crossing the village with large wooden or ceramic dishes on their heads, going from their hearths to the house of the mourners. Other families, who may not be related to the chief mourners, may, in this context, decide to take the opportunity to settle their own mortuary debts with *mwara* kin. An ethos of reciprocity, in short, may spread throughout the community. Choruses of songs, lingering from the nighttime festivities, go on. Indeed, sometimes, should the dancers want to continue their performance for a second night and sufficient resources remain to cater for it, they will be invited to resume at dusk. Only when the dancers declare they have been satisfied does the sponsoring insignia-holder present huge baskets of imported foods, topped with strips of raw pork, to the *mwara* kin and then escort them to their canoes, which they paddle homeward, weighted down under a heavy load of gifts.[17] They must leave "defeated" (indebted) by these gifts because, when another occasion arises, their services will be required and they will be obliged to appear, only to depart again indebted, so that the feasting partnership may continue.

Next day, after *sumon* heraldry has been put away, the male cult of the host community celebrates itself by staging a small, recreational banquet called a *baas*. The explicit purpose of a *baas* is to relax the community following the successful completion of any major ritual. But this feast is also coded as a "war." Junior members of the cult prepare *aragen* porridge for their *mwara* mothers' brothers, the cult elders, who then assemble in the hall when a special slit-drum call (*baas debun*) has bidden them to do so. After one of their number spills a spoonful of the porridge through gaps in the floor around the main posts, while calling the personal name of a *brag* spirit each time, "the fight" begins. The *mwara* mothers' brothers "throw spears" at (or "set fire" to) their respective sisters' sons, who retaliate in kind by encircling the heads of their "victims" with a plate of seafood over sago pudding or a bed of rice, before gently depositing it squarely in their laps. Some men stamp the floor and call out the personal name of their war spirit (a spearing cry) to dramatize the gift. If the recipient has made no arrangement "to retaliate" immediately, he must not eat the meal but should send the dish home to his wife to eat, to oblige her to start cooking. After everyone has finished eating, a palaver is convened to air disgruntlements and jealousies which may have arisen during the preparations and management of the recently concluded feast. Candor in this context is believed to resolve them. A solidary ethos is restored between the bereaved and their *mwara* kin and throughout the entire community. Their bodies cleansed of the impurities of the corpse and freed from the menacing ghost, the mourners return to their pure canoe-bodies under the guardianship of the Spirits of Wealth.

Dialogics of womanhood in mortuary rites
Evoking the elusive, lost presence of the mother of early childhood who fed, groomed and defended dependent children from dangers arising from crossing boundaries in moral order, the climax of the reaggregation process, I want to propose, is a grotesque birthing image (cf. Roheim 1943: 110). The conical monster, made of a great mass of leaves, is crammed tight with people (see Plate 22). Might not this image of a human passage through a constricted, medial space – the beach – be an image of birth (cf. Brunton 1980)? Informants, it is true, do not see the image this way. They only call the effigy a *gaingiin*, which is their generic

category of spirit-monster. But several important clues nevertheless point toward this interpretation. Jari, the great culture heroine, removes fire from her genitals to give to her lover (see Meeker, Barlow and Lipset 1986: 39–42). To alleviate the pain of hard labor, said by mothers to feel "hot like fire," midwives will repeatedly douse a woman's legs and stomach with salt water. Infants are literally born splashing through seawater, not to mention the fact that they are immediately washed in it (Barlow, personal communication). Throwing off the "hairy" monster, bathing in the sea, and jumping over fire thus reenacts birth, a moment, I should hasten to add, reached at the behest of the *mwara* siblings, the brothers and sisters who literally *push* their cross-sex partners across the beach and into the water by beating them. From a metaphoric, if not a conscious, standpoint, therefore, washing the body in the sea would seem to culminate a process of pregnancy (cf. Tuzin 1977: 216–21), which is also a period of defilement, vulnerability and seclusion. As a referential device, washing in the sea constitutes a boundary between morally opposed intervals of time and being (Leach 1961a); the cosmic body must pass through a threshold of water in order to change back into its moral form.[18] In Murik, the prototype for this metamorphosis would seem to be a sequence of reproduction. The symbolic equivalent of sexual intercourse would therefore be the moment of death, while mourning would evoke gestation and the Washing Feast would be a form of birth.

Later nineteenth-century views of maternal imagery in mortuary ritual as "primordial," "natural," or "pan-human" were supported by naive romanticism and inadequate evidence (Bachofen 1861: 87–8; see also Frazer 1922). In an attempt to retrieve the symbolism of womanhood in mortuary rites from these essentialist and speculative roots, Bloch and Parry (1982) made a useful, if empirically flawed, functionalist argument (cf. M. Strathern 1987; Damon 1989; and Dureau 1991) in which they claimed that

[f]unerary rituals are an occasion for fertility. This is fertility dispensed by authority, whether it be that of the elders or of the priests, while in the meantime women are left holding the corpse . . . Ideology feeds on the horror of death by first emphasizing it then replacing it by itself. This process is often carried out at the expense of . . . women.   *(Bloch 1982: 227)*

Bloch and Parry assumed that soul stuff is "a limited good" which requires death to renew itself (1982: 9). Birth-mothers must therefore undermine the continuity of the collective body not only through their procreative force, but also in the roles they play during mortuary rites. Legitimate authority – which is male – must come forward to claim birth

and fertility from such women, in order to deny "the irreversible and terminal nature of death and assert . . . [their] own capacity to reproduce
society" (Bloch and Parry 1982: 9). The masters of political order thus
possess a power to convert death into life through magical procreative acts
which cleanse the collective body of defilement caused by women. Culture,
just as Sherry Ortner used to argue, thereby subordinates nature, just as
men subordinate women (1974; see also Lévi-Strauss 1969). Men maintain
political authority during mortuary rites, according to Bloch and Parry,
through the fertility symbolism they claim for themselves and deny in
women.

While it is true that death and mourning are represented through a birth
metaphor in Murik, reproduction is not understood as a "system that is
closed and bounded such that accretion at one node necessarily entails a
corresponding depletion at another" (Kelly 1976: 47–8). There is no sense,
as in the well-known Trobriand case (Malinowski 1954), that a deceased's
spirit need return to society to incarnate a new child. Entirely missing here
is any reciprocal connection between death and fertility. There is no
scarcity of soul stuff. The relationship between the new ghost and procreation is not a contingent one. Nevertheless, the point in the total sequence
of reproduction selected for dramatization during the Washing Feast is
birth, the moment when a newborn leaves the defiled, spirit-ridden, interior body of its mother and enters a moral body of external relationships.
Murik elders do not deny death by making themselves into a source of
fertility that they coopt from women. Their conceit is not so literal: the
relationship of death and mourning to reproduction is analogical, which is
to say, it is yet another rejoinder to the maternal schema. In both processes, the self is withdrawn into an enclosed space within which it is completely dependent upon a "mother." At the end of both, the self is pushed
from this space through a cramped passageway, the threshold of which is
associated with "hair," seawater and fire. In both, the self then enters a
world where its uterine dependency is turned into a social dependency
upon goods and services provided by senior kin who infantilize it. But
there is one important difference between the definitive moments in the
two processes. Death, unlike coitus, is no moment of fertility; it results in
no transmigration or consubstantial exchange. Nor does it promote
growth or health. There is nothing of the corpse to eat; no embodiment of
the soul's vitality from which to take nourishment and agency (cf. Gillison
1983; 1993).[19] On the contrary, the corpse radiates defilement and provides
another opportunity for the Spirits of Wealth to offer their enchanted care
to vulnerable and impure victims in return for gifts of food. At death, the

soul comes into being as a dangerous ghost and those who mourn it elicit image after image of mothering. The cross-sex *mwara* siblings pre-masticate food they feed the mourners. The sister's sons wash and dress the corpse (thereby taking on its impurities just as mothers do with infants' excreta). In return, they are nurtured by the mourners' kin until suddenly their compassion turns grotesque, becoming at once beautiful, comic and aggressive. The appearance of the bejeweled *mwara* sisters' sons, who try to taunt the ghost into joining their performance is both a final attempt to use mischief and trickery to beckon the elder joking partner back to life and agency, and a preliminary maneuver to turn him into a disembodied ghost. No longer their mark, or adversary, the contrast between their vertical mockery and the horizontal passivity of the corpse attests to their final victory and the beginning of his and their new identities. Could any death certificate be more grotesque? The loss of their life-long rival does not find them lost in grief but dancing about the corpse (cf. Barlow 1992). What then happens next? The corpse is placed into a sawed-off lagoon canoe, its coffin, which is one of the final associations between the body and this vehicle in the life-cycle.

Birthing imagery is part of a process through which political status and authority are affirmed (see Chapter 5). Nature, in this sense, is defeated by culture. But in Murik neither nature nor culture is gendered in the way Bloch and Parry imagine them to be. What is defeated by the masters of the political order is death pollution (*mwak*), which defiles both sexes in the same way. It is not caused by women. *Mwak* has no gender. The victory claimed through the Washing Feast, of course, is over the Spirits of Wealth, the grotesque, cosmic "mothers" who have "birthed" and "cared" for the mourners and the ghost of the deceased during the recently concluded mourning period. This victory has no magical effect on fertility. The mourners return to their moral bodies. They become free to marry or pursue prestige and influence within the community which may result in the birth or the decoration of children. The agency for these processes, however, is vested in heraldic images of personhood and their cosmic courtiers, but not in fluids won or lost from the corpse.

The Spirits of Wealth espouse both positive and negative attitudes about the sexuality of young women. They celebrate the autonomy of the phallus, an undomesticated eros that is free of conjugal as well as affinal duty and is independent of the maternal schema. When the *mwara* mother's brother first appears in a child's life, he is not a rival of his sister's son for his mother's sexual attentions and affections, but mocks (albeit somewhat flirtatiously) her sexuality. When he pulls on the penis of his infant nephew, he is not trying to pull it off but "to lengthen it" – for future use. During the firstborn rites of childhood, it is he who introduces his young protégé to the community, he who instructs him how to eat "social" food, and he who shows him how to leave his mother's house. Thereafter, whenever the sister's son engages in activities which reflect further upon the development of his autonomy, the mother's brother may mark the moment by invoking the magical name of his war spirit and by aggressively anointing the child with his spittle; the grotesque blessing places his nephew in his debt and endows the child with cosmic agency of various sorts but does not make his body grow or become fertile. Later, during initiation, when the *mwara* mother's brother removes the youth from his mother's house and carries him off to the male cult house, the uncle reveals axiomatic secrets about his cosmic body (how to play the bamboo flute spirits and how to expel impure feminine blood through the penis). In everyday life, his comedy ridicules and parodies the prudish, self-denying embodiments of the maternal schema, as they appear in the sibling, affinal and ceremonial hierarchies. With grotesque mimicry and sympathy, he patrols public lapses of competence and autonomy by the sister's son, "fighting" expressions of dependency and vulnerability where they do not belong. Finally, when the *mwara* mother's brother dies, the sister's son begins to reverse their relationship. It then becomes his duty to take charge of his rival's soul as it begins to move into a ghostly identity. The Spirits of Wealth create a threshold between the moral and the cosmic bodies. They not only create this threshold but offer themselves as a means of traversing it. Rather than denying women's procreative role in this drama, their dialogue remains explicitly contingent upon the relationship between a man and his classificatory sister, his nephew's mother.

The heraldic double

If the *mwara* kin do not seek to coopt uterine fertility through their role in the Washing Feast, neither do the *sumon*-holders who sponsor it. In addition to squaring their ceremonial debts (see Chapter 5), they want to defeat a second category of hereditary feasting partner, the heraldic double (*wajak*). The rights and duties of the *wajak* are similar to those of the *mwara* mother's brother. Both relationships are adversarial. Horseplay and comic abuses are permitted in both. The partners may arrange trysts for each other. The presence of the *wajak* in a gathering imposes injunctions on everyday exchange through a threat of aggressive generosity. The *wajak* partner will also offer sympathetic mimicry for accidents. The main difference between the two is that while the *mwara* mother's brother serves as his nephew's guardian, the *wajak* is "only" responsible for protecting the heraldry of his partners in the senior sibling group which holds them.

Wajak partners "fight with plates"; they extend sumptuary rights to each other in return for the performance of services which facilitate heraldic exchanges. Should a *wajak* initiate his own or his sister's children, he may also invite the son or daughter of his *wajak* to participate and receive *sumon*. They may exchange housebuilding and consecration services. At the launching ceremony of an outrigger canoe, the *wajak* may be called upon to bring the mast heraldry of the canoe. If a *wajak* dies outside his village, his partner may retrieve the corpse for burial (see Chapter 8). Or, when a *wajak* partner dies, his partner may observe a taboo against eating betel nuts and sago pudding until he goes through a period of mourning. The surviving *wajak* partner should then "guard" his partner's heraldic baskets, by sitting or standing next to them throughout the Washing Feast. This is his last duty to perform for his partner's heraldry. He must protect the *sumon* from an outbreak of violence, which is a chronic possibility during major feasts. Like heraldry, the titleholders and women, the *wajak* is also an agent for social control. He may convene no negotiation. He has no special curses at his disposal. Nevertheless, he himself is a sign of moral intervention: upon his appearance during a fight, a conflict must immediately come to an end. Should a fracas break out when *sumon* insignia are on display, as in the incident I discussed at the end of the last chapter, the *wajak* should remove his partner's heraldry from the scene and quickly take it home for safekeeping. He may then withhold the *sumon* from the sibling group that has been attempting to deploy it until he has been compensated for his service. His "fee" is no bargain: he must be served pig's liver on a plate of sago pudding and receive a basket of imported delicacies topped with a porkbelly. The *wajak*, in sum, is meant to facilitate heraldic exchanges, the shifting pinnacles of prestige in the society. As such, he must be rewarded with only the best foods. Like the *mwara* mother's brother, the *wajak* is also an adversarial guardian, but not, as I say, a repository of fertility.

This feasting partnership is inherited bilaterally. While one sibling group may have inherited *wajak* relations through their mother, another will have inherited theirs through a father. In twenty-four cases of *wajak* inheritance I collected in 1981–2, no rule or pattern of inheritance emerged; nor was one ever explained to me. Many *wajak* partners were residentially dispersed and lived in different villages; but others were coresident in the same village. The most salient principles of the relationship to informants were generational equality, sibling birth order and gender. The *wajak* are sociological doubles. *Wajak* ties pit collateral sibling groups belonging to single *sumon*-holding units against each other in a contest of equals. During major exchanges, the firstborn member of a host sibling group should make a prestation to the firstborn member of their visiting *wajak* sibling group. At the same time, younger sisters and brothers are responsible for feeding and guarding their individual partners: the eldest sister, for example, is said "to fight" with the eldest sister in the guest *wajak* group while younger brother "fights" with younger brother. Each *wajak* attempts "to defeat" his partner with the weapon of hospitality, by matching, or slightly outdoing, the quality and quantity of the previous prestation he received when he visited his guest's village. Characteristically, *wajak* partners *atakasor* when plates of food are presented to them. They may shout somewhat edgy abuse at their host, insulting his generosity and the size of his phallus, particularly if the quantity and quality of the food has been judged wanting.

The symmetrical term of address and reference *wajak* partners use to each other (*jakum*) reflects the heightened level of rivalry in this relationship. It is the case that this term is also used between the *mwara* mother's brother and his nephew. However, while the *mwara* sister's son may use the personal name of his mother's brother in reference, like affines, *wajak* partners may not and do not do so. Whole groups of commensal siblings participating in this relationship thus avoid using the names of their alters by substituting the term *jakum*. The *wajak* also represents a definite moral boundary. The property avoidance they observe is much more severe than that between the *mwara* mother's brother and the sister's son. The *wajak* may not tread upon the land of his partner, enter his house, ride in, or even look at, his canoe. He may not step on his partner's shadow. Trespass, intentional or not, is severely penalized, albeit in a somewhat counterintuitive way. In Western law, a trespasser is punished for his crime and may have to compensate the property owner. In the *wajak* relationship, the property owner *must pay* the trespasser in order to punish him for his crime. Should a *wajak* enter the house of his partner, then the partner should *give* all the goods in it to him. Should he walk on his *wajak*'s land, then he should be given rights to the section of land upon which he has trespassed. Should he even touch the canoe of his feasting partner, for example bumping into it by accident in the dark of night, then he should be given the canoe. The aggressive generosity of the *wajak* is a sanction. Incurring such a major debt is humiliating, not to mention an utter defeat. The only possible redress is a major pig feast, mounted either by the trespasser, or by his sibling group, or his descendants, who inherit the debt.

While the moral body is embedded in a world of generosity, nurture and care, the *wajak* turns this ideal into a threat of total debt, a threat which looms over this relationship like the possibility of a betrayal. The *mwara* mother's brother is a delicious rival who is held in affection. The *wajak* is feared as a more dangerous adversary, the stakes of "battle" being so much higher. The increased level of avoidance in this relationship should not eclipse the fact that this is still a difference of degree, not kind. Although the services provided by the *wajak* do not include instruction or grooming, both feasting partners are included in the same "wealth path": both are Spirits of Wealth. The *wajak* does not attend his partner's heraldry precisely the way the *mwara* mother's brother attends his nephew. His duties do not concern the metamorphosis of insignia in any pedagogical way, only their legitimate bestowal.

The imagery of the *wajak* double further exposes the weakness of Bloch and Parry's ethnocentric assumptions about the gender of nature and culture. The elders who constitute political order in Murik, the insignia-holders and the Spirits of Wealth are no sex-exclusive group seeking to coopt procreative force from women. Their goal is to outgive their feasting partners in order to reproduce a pure, aesthetic body. The deeply heterosexual concept of culture they espouse is under the aegis of an androgynous and gerontocratic body working in concert with their grotesque functionaries to defeat an equally heterosexual concept of nature.

Motherhood, grotesque and moral

Many examples of the grotesque have been reported throughout the Sepik region, irrespective of social structure and linguistic affiliation. The hyperbole of bodily life, debased gestures and bawdy satire have been put to various usages and given different meanings. On Manam Island, for example, it is affines who may have a joking relationship (Wedgwood 1959: 248). In Boroi village, on the mainland coast just opposite Manam, a spouse's classificatory siblings should fall down for their sister's husband, if he does so (S. Josephides 1982). The Boroi do not, however, think that such mimicry is at all funny. On a tributary of the lower Sepik River, the Mundugumor mother's brother may also "fall down" for his sister's son (Mead 1935). But here the mother's brother represents a territorially distinct group, an aggressive, riverine people, while the nephew is a passive, clumsy new immigrant from "the bush" (see also McDowell 1991: 195–6, 200). In Bun village, which is located just downriver from, and is culturally related to, Mundugumor, *kamain* feasting partners also "duplicate misfortunes" for each other in order to restore equality in their relationship

(McDowell 1980: 64). On the middle Sepik River, *wajak*-like feasting partners called *tschambela,* representing rival intermarrying patriclans, "fight" with wealth. Should a sister's son ask his mother's brother for a coconut, the mother's brother "may exclaim at the need of his nephew and go kill a pig for him, for which the nephew would have to repay shell currency" (Bateson 1932: 266, 272–3). Of course, the most well-known burlesque of motherhood in the entire ethnographic record takes place during the Iatmul *naven* ceremony, when a classificatory mother's brother pantomimes debased gestures which are meant to have sexual, reproductive and aggressive references but apparently not comic ones (Bateson 1936: 12; and personal communication: E.K. Silverman). So clearly, this kind of multi-toned genre is not unique to Murik. But besides a few futile efforts to analyze them functionally (e.g., Lindenbaum 1987; Handelman 1979), little cultural sense has been made of these peculiar aspects of personhood since Bateson and Mead's time on the river, in the 1920s and 1930s.

In Murik, the comedy, sympathetic mimicry and aggressive generosity enacted by the Spirits of Wealth do not mediate or mitigate affinal tensions. Neither do they reflect political rivalry between territorially based groups. Nor, for that matter, do they primarily assert equality between participants. In any case, neither affinity, nor property, nor the ideal of equality elucidates why the mimicry of a body falling down is so compelling a message in this culture. Nor do they explain why mimicry is employed at all, or why the heraldic double should *receive* the contents of his partner's house upon trespassing into it. These images, I argue, are rejoinders, which are dialogically related to the moral body. In Murik, this specifically means that they answer a maternal canoe-body. The grotesque symposium of the Spirits of Wealth is not nonsense, but neither is it innocent or utopian. It is rather parodic discourse, a polemical mockery of the hierarchical, but ultimately barren, maternal schema. It does not permanently liberate an aggressive, carnal body, but expresses the contrary side of an ambivalence in which personhood is located in a borderline, carnivalesque state of becoming, not heraldic completion. "Everything in [t]his world lives on the very border of its opposite" (Bakhtin 1965/1984a: 176). Through their nurture, the masters and mistresses of the cultural order – mothers, elder siblings and insignia-holders – tenaciously seek to recreate the maternal schema. Age seeks to maintain chastity and self-denial, to be indulgent and protective, while youth is expected to be wanton, demanding, hungry and helpless. The parody of the *wajak* and the *mwara* mother's brother confounds the serious hierarchical projects of age with laughter. Yet the crooked reflection between the two is evident. The guest who enters a

Murik woman's house prompts her to think of feeding him and to guarantee his safety. The guest who enters the house of his *wajak* prompts a total debt, "a defeat" which may beset generations to come. The mimicry of the *mwara yakabor* turns a domestic mode of influence – exemplifying a desired behavior by performing it oneself – into a sanction. The object of the discipline of the Spirits of Wealth – public incompetence – changes the infantilized image of prostrate helplessness into a misdeed, a wrong. As the person exits the commensal surround populated by a multiplicity of mothers, a debased, nether world of mothers, the Spirits of Wealth challenge him with paradoxical care which remains only one step removed from that in which the domestic group specializes.

The perfection and beauty of the heraldic body, when decorated and completely covered by the ornaments which define it, assert the jural autonomy of the descent group rather than its regional interdependency. However beautiful, such a body remains flawed. Like the economy, it lacks "inner life." Rid of impurities and substances, bereft of passion and agency, its appearance is all facade. The moral body is assisted by a second body, a body which "has no facade, no impenetrable surface . . . It swallows and generates, gives and takes . . . Such a body . . . is never clearly differentiated from the world but is transferred, merged and fused with it. It acquires cosmic dimensions, while the cosmos acquires a bodily nature" (Bakhtin 1965/1984a: 317). In Murik, this is the Spirits of Wealth "who" carry the self across moral boundaries in the community. In addition to their multifaceted evocation of custodial motherhood, the *mwara* kin espouse a more subtle, virtually metacommunicative point (Bateson 1955): they insist that the meaning of life is not entirely serious and that its reality is not entirely fastened down, not entirely single-toned.

The satire, buffoonery and hyperbole that routinely take place upon their stage travesty the repressed, hollow canoe-body of the maternal schema, disarranging it into a thoroughly serio-comical image. These genres temporarily suspend its commonsensical assumptions. Yet the form and meaning of their rigamarole nevertheless respond to its inhibited shape. The Spirits of Wealth do not nakedly reject order. No less than the moral body, what they have to say and do through their great, utterly debased, satire of motherhood remains inextricably embedded in motherhood. While their grotesque evocation of mothering is largely polemical, the moral body they recreate is sheer mimesis. The unmistakable themes of this ambivalent dialogue are thus one and the same.

7

The sexuality and aggression of the cosmic body of man

Glorious heroism, in which soldiers triumph against insurmountable odds, apparently was not the stuff of premodern warfare in the Sepik region (Harrison 1993c). Instead, warriors were collectively empowered by ego-alien spirits and derived aggression from a complex of ritual, ceremonial exchange, and, not least, art (Bowden 1983). The gender of these "cosmic bodies" was exclusively male. But their magico-religious force did not exalt pristine masculinity (see Gewertz 1988). Carved high up the centerpost of Iatmul male cult houses, for example, was an image of a seated, wooden spiritess, her legs spread apart, giving birth, I must suppose. Except to say that her body was a "grandiose female matrix" from which masculine aggression issued forth, Bateson did not specify what her posture meant (1946: 120). Was this an image of womanhood, and "her" procreative capacities, contained within this cosmic body? If so, then a different image of womanhood impelled the Arapesh cult (Fortune 1939), one that was no less fertile, but in a more literal sense. War parties used to go out from Arapesh villages "to pirate" married women, women who were meant "to replenish the land with children" (Fortune 1939: 26). The women who lived outside communal boundaries were foreigners to this masculine body. Not only that, they were already married. These women were approached through an intermediary, magically seduced, and only then collectively apprehended by the warriors. Unlike the inert Iatmul image, such a "woman" was endowed with discretion about, and autonomy over, her love affairs. Arapesh warfare was neither a form of abduction nor bride capture, but was contingent, as Fortune put it, upon obtaining women's consent to have extramarital intercourse and their decision to leave their husbands. The relationship, in other words, of the cosmic body to women in Iatmul may have been more endogamous and reproductively self-sufficient than in Arapesh. Both images nevertheless disclose how bound up collective masculine aggression was with representations of women, as mothers and lovers.

In this chapter, I focus on the position of women in the paramount male cult, the cult that worshiped ancestor spirit-men called *Kakar*. In this extraordinary secret society, living, sentient women, endowed with sexuality and agency, used to enter the men's sanctum, because warfare was held to be contingent upon their committing acts of ritual prostitution to rulers of the cult. The aggression of the cosmic body was not only contingent upon obtaining the consent of a warrior's wife to having extramarital intercourse with her husband's initiator, but it was also held to be contingent upon the warrior-husband's permission to let his wife do it. In order to become a warrior, a man had to be "cured" of sexual dependency upon his "mother-wife." Aggression was explicitly linked to a husband's making emotional sacrifices, not only by renouncing sexual rights in his wife, but by renouncing his emotional attachment to and jealousy of her. He had to cuckold himself. Giving up a wife's sexuality was viewed as the provision of a reciprocal service to a man's cultic partner in return for receiving his rights to cosmic images of aggression. The outstanding figure in the male cult was neither uniquely masculine nor feminine. It was a kind of reversal of the Oedipus complex, compressed onto a single generational plane: a young husband grappling to control his jealous rage and make the decision "to send his mother-wife" to make cosmic love to his "cultic elder-brother." If pre-Oedipal issues, of dependency and differentiation, define the inhibited ethos of the commensal groups in response to the uterine body, the war cult reconsidered the relationship of male aggression in relation to sexual love and loss.

A beautiful warrior

Because it contains defiled, feminine spirits, the birth house is reduced to an outbuilding, a "polluted body" extruded from communal space (see Chapter 3). The cosmic body of men, the male cult house (Murik: *taab*; Tokpisin: *hausboi* or *haus tambaran*), is less toxic. While this building is also a repository of impure spirits, such as flutes, it remains squarely positioned within the villages.[1] Some are larger than the dwellings – the women's houses – by which they are surrounded. But structurally, they are virtually identical. Built of the same bush materials, upon the same frameworks, they have the same rectangular bark floors and stand on the same kind of piling several feet above the ground. They possess the same triangular roofs which rise about twenty or so feet above the floor to a single ridge pole. The crucial difference between the two kinds of building is that the male cult house has twin gables called "noses" (*taab da'ur*). The gables of Murik cult houses neither dominate them like those of the Abelam or Arapesh, nor do they sway up to a peak as in Iatmul.[2] They jut straight out from the roof ridge at both the front and rear ends of the building (see Figure 15). When the building is newly consecrated, coconuts tied to split rattan chains, and other homicide insignia, may dangle from the end of the front gable. Brightly colored geometrical designs, suggesting the huge, intricate frescoes of Abelam and Arapesh cult houses, are sometimes painted on the bark facades of the Murik gables. But they are invisible unless one is standing beneath them and knows to look up. The exterior roof-ridge of certain buildings may also be decorated with a "headdress" of two miniature flotillas of war-canoes, apparently fighting over a small cult house that is positioned between them.

The Iatmul and Abelam cult buildings are classed as feminine "bodies." The Murik *taab* is no woman but a beautiful spirit-man (*brag*). Intimations of womanhood are nevertheless found upon and within "his" body. The cult house has a "penis," but "he" also has "skirts" (*dag*). When consecrated, a bunting of sago fringe is hung along "his" exterior walls. Upon entering the hall, men climb up a ladder and brush through the fringe "like children," they say, "crawling underneath the skirts of their mother."[3] Inside, the walls of the hall are usually lined with several large,

Figure 15: Bungabwar male cult house, Janainamot, Big Murik

beautifully engraved, Gapun/Sanai slit-drums whose "canoe-bodies" are also "dressed in skirts." In the middle of the ceiling is hung a big rectangular decoration (*saidug*) which duplicates the rectangular shape of the hall. Typically about 12 feet long and 3 or 4 feet wide, it is made of sago palm midribs which have been beaten into a 2-foot fringe. The ornament is subdivided into four rectangular sections, each of which stands for one of the age-grades in the cult. Being made of the same sago fringe as women's skirts, the *saidug* evokes them. Hanging above the hall, I believe, is a series of four "skirts" (see Figure 19).

Hidden deep beneath these "skirts" live the "canoe-bodies" of the cult's most sacred spirits. Intricate images may be engraved on the rear post, its "center man" (*wabii nor*). At each end of the roof beam, a serpent's head (*wakun kombatok*) bedaubed in red is carved. Except for clusters of hand-drums hanging on ceiling hooks, a few headdresses, wicker frameworks of effigy figures, or the mast decorations of outrigger canoes, there is no other interior iconography. The floor is not lined with clan platforms like the Iatmul cult houses. Evidence of a different principle of cultic organization is faintly visible, however. Located at diametrically opposite ends of the floor are two square firepits around which senior men regularly sit and light their cigarettes on chilly mornings of casual conversation or during formal cultic assemblies. A second clue lies along the middle of the building, where an inconspicuous horizontal line, formed by the ends of the bark floorboards, divides the floor space in half. For the most part, the interior of the hall appears uncluttered and readily available for the meetings which routinely assemble within it.[4]

The *Kakar* spirits

A small loft is built into the rear, left-hand corner of the *taab*. Sometimes, this loft contains two shelves. The flute spirits, the voices of the *brag*, may be stored on the lower one in plain view. The upper shelf is always closed because the supreme war spirits, the ceremonial spear images called the *Kakar*, are said to "live inside." In Darapap in 1981, about thirty delicately ornamented, anthropomorphic figures were stored in this hiding place. The number of spears varied from one village to the next (for reasons I will go into) but their meaning was uniform. The *Kakar* are individually named, tutelary war spirit-men, "who" possess an unsurpassed power to protect the welfare and prosperity of the community. Organized in a rudimentary hierarchy, their two leaders are the spirits called Sendam and Amumbera. They have a "bodyguard" named Kombek. Although the spear spirits possess male, and patently phallic, canoe-bodies, there is usually a woman or two in their midst (in Darapap, her name is Sangrimanja or Singinasen).

While contact with the canoe-bodies of the *Kakar* spirit-men is deadly to noninitiates – particularly younger siblings, women and children – they endow firstborn, initiated men with invincible aggression. "What a *Kakar* does is fight," an elder told Michael Somare (Beier and Somare 1973: 10). Of all the spirits in Murik religion, Father Schmidt concurred, in "the first place is . . . the sacred spear. The spirit is always in this spear and the spirit gives power in battle" (Schmidt 1926: 76). In order to become possessed by a tutelary spirit who would instruct Murik warriors before they went on a raid, a cult elder absorbed the power of his *Kakar* by eating the "substance" of its "canoe-body," a special *aragen* porridge mixed together with a little soot scraped from one of the spear carvings. "When a *Kakar* enters a person, he begins to shiver and sweat. His ears are blocked. His thinking is changed; he forgets everything. He speaks . . . the words of the spirits" (Beier and Somare 1973: 10). In a possession state, the oracle would suggest tactics and timing for an attack and predict its outcome. The *Kakar* spirit, in other words, was distinguished from any of the masculine "canoe-bodies" in which it traveled. Its sacred potency was held at abeyance from the moral surround. Even after being secluded from the community and ritually sacralized, a man might become its canoe-body only temporarily. He then had "to wash" before rejoining commensal life. The same disjunction between moral order and this cosmic, male body appears in the ethnohistory and legends of the cult. The origin of the *Kakar* spirits, as distinct from the wooden spear imagery in which they are immanent, is not attributed to men, much less to Murik men, but to a woman who was raped near Moim, the village where the Murik ancestor-refugees are believed to have lived before migrating to the lakes.

The Mother of the Kakar[5]

The mother [of the *Kakar*] was named Arake. But their father was not a man. An axe spirit, who was angry at Arake's father, kidnapped and left her to die on an island in the middle of the Sepik. Turtle spirits raped her there and she became pregnant. Arake bore several sea eagles, all but one of whom died.[6] She raised the lone survivor with special care until he grew up and flew his mother back to her father's village, where the Murik ancestors were living. The eagle played badly with other children, scratching out their eyes with his talons. The villagers got angry. Arake herself came forward and killed her son, pouring boiling water over his sleeping basket. She took the eagle's corpse to the edge of the Sepik and buried it. The corpse rotted. Out of the stench of rotting flesh and bones, the *Kakar* spears arose, as did lances (*mansariip*), cassowary bone daggers (*asor*), and fighting magic (*mwar*). Arake saw these things and hurried back to tell her father. Knowing how dangerous they were to women, he took a group of men to help him retrieve the powerful weapons.

While the story makes it clear that their father is not a man, Arake, the mother of the spear spirits, is named and clearly identifiable. The legend stresses that, like all men, their origin lies with a woman, and their powers stem from a mother–child relationship. But what sort of mothering is this? It is an evil vision of motherhood, a vision of motherhood disfigured and rent by conflict. The conception of the *Kakar* is immoral. Arake is kidnapped by a vengeful axe spirit and then sexually violated by turtle spirits. Most of her children are stillborn. Arake nurtures her son, the sea eagle, and he in turn is loyal to her, although he scratches out the eyes of other children. That is, the predator spirit is inimical to the domestic community whose children may not "see" the *Kakar* spirits until initiated. Arake's mothering is ended by murder. She chooses to protect the domestic community from her son, whom she kills and buries at the river's edge, its boundary. Four emblems, each of which represents one of the cult's four age-grades, are then "born." Arake's father and a group of men merely collect them in order to protect, rather than oppress, women. While the sequence of vengeance, kidnap, rape, aggressive play, infanticide and putrefaction, figure reproduction and motherhood in grotesque terms, it also confirms that escaping this feminine mode of agency – namely, gestation, childbirth and mothering – is impossible even for the most potent form of male aggression. Not only does the legend displace the most powerful sign of male agency from men, it displaces them from the source of its creativity, care and protection. The male cosmic body, in other words, has no center. The center is removed, absented: only the feminine periphery remains. The center must therefore deny its absence (Derrida 1970) by dialogically supplementing a presence which displaces woman and "her" domestic order, not to subordinate or subjugate "her" or deny "her" uterine force, but to defend "her." Indeed, the *Kakar* cult does not entirely succeed even in this project. Both the "sons" and "daughters" of Arake share rights, and jointly contribute, to the military powers she bore.

"Two Brothers" redux

Murik villages are divided into ritual matrimoieties (*yarok*) which exchange the *Kakar* spears (cf. Hogbin 1978: 25; Smith 1994: 30). Membership in the *yarok* is restricted to firstborn sisters' sons recruited by their commensal rather than by their *mwara* mother's brothers. The ideal pattern of recruitment is that a firstborn maternal uncle ought to select his elder sister's firstborn son for his moiety, while his sister's husband should recruit the firstborn son of his wife's brother into the opposite one (see Figure 16).[7] While the matrilateral basis of the *Kakar* cult is acknowledged explicitly in the two western Murik villages of Big Murik and Kaup, in the eastern villages of Mendam, Karau and Darapap, it is not. In these latter villages, *Kakar* recruitment is said to be an "open competition." A man may select his own son, his sister's sons, or the sons of any women belonging to the moiety opposite to his own. Michael Somare, whose father was an eastern villager, does mention that should a man want to become a *Kakar* cult elder who has rights to a spear, "he has to have a maternal uncle" who already has the title (Beier and Somare 1973: 8). The forty-five recruitment cases I collected among the eastern Murik, moreover, all follow matrilateral relationships. In these cases, the man's sister, the mother of the boy he took into the cult, was usually two or three degrees removed from the commensal mother's brother.[8]

One moiety is called "big" and "male" while the other is referred to as "small" and "female." These designators do *not* rank them. The *yarok* are only ranked with respect to the spear spirits and the other insignia to which Arake "gave birth." Each moiety (see Figure 17) is subdivided into two age-grades (*orub*): a senior grade (A1), which is called the "father" of the cadet grade, and their "sons" (A3). These two grades, if they are in the

Figure 16: Moiety (*yarok*) recruitment

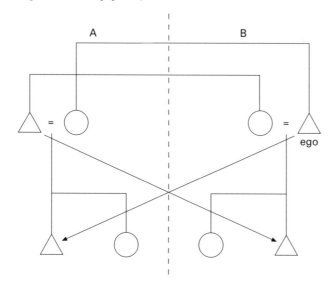

ranking moiety, "hold" the *Kakar* spears and the "dagger" emblems respectively. The opposite moiety is also made up of two age-grades (B2 and B4) which hold the "lances" and "cordyline war magic leaves." While the idiom of relations within each moiety is patrifilial, the two senior grades (A1 and B2) of each are said to be "siblings." The spear-holders are the "elder brother" grade (*tatan orub*). They are senior to, and have authority over, the "younger brother" grade (*dam orub*), the lance-holders. The "elder brother/younger brother" ranks are also used to designate the moieties as whole units since these two grades hold their most important insignia. When rights to the *Kakar* spears are exchanged between them, these ranks are also transferred. "Younger brother" then is promoted to "elder brother." The senior grade in the "elder brother" moiety retires and is replaced by its cadets, who assume the rank of "younger brother" (see Figure 18). In 1981–2, the small, female moiety held the *Kakar* in Darapap and were the "elder brother" moiety. But by 1988, negotiations had nearly concluded to promote the senior members of the "big, male" moiety to the "elder brother" grade which holds the *Kakar* spears.

In addition to emblematic weapons, rank in the *Kakar* cult is made clear whenever men assemble in the hall. In order "to protect" the ceremonial spears, the "elder brother" grade sit about the firepit located beneath the shelf in which they are stored. The senior members of the "younger brother" grade, for their part, will sit at the opposite end of the floor, around the firepit nearest the entrance to the building (see Figure 19). The

Figure 17: *Kakar* cult insignia

two "son" grades may come and light cigarettes in the respective firepits of their "fathers" but should not sit down. They must rather sit along the wall across from the pit. The cult house is divided into "elder brother" and "younger brother" spaces and each side is subdivided again into separate "father" and "son" spaces. Reflected in this metaphorical use of kinship is attention to separation and differentiation as well as dependency and attachment. Inside the sanctum, that is to say, the maternal schema is rearranged, inverted, travestied, but not ignored.

Part of the saliency of the "Two Brothers" story arises from the particular conflict it probes. Domestic and jural morality, to recapitulate its tenets, advocates that elders encourage the dependency of junior kin through acts of maternal indulgence and generosity. Elder brother should give resources to younger brother, but avoid taking anything from him. The younger brother, for his part, is free to take without giving. Their asymmetrical exchange reaches its climax in the strict avoidances between elder brother and younger brother's wife. The two may not even look each other in the eye. Younger brother, by contrast, can freely expect the nurture of elder brother's wife and may even trade ribald abuses with her, so long as they do not regularly share food together. Rarely in Murik kinship is such a boundary so sharply drawn. Younger brother's wife is no resource to elder brother and is taboo to him. Elder brother's wife is an ambiguous "co-wife." The "Two Brothers" story probes the explosive potential for society of the violation of her sexuality by the younger brother, the one "resource" belonging to his elder brother he may not claim. Still, being the subject not only of the most important legend in the culture, but of daily mockery, it would seem that the prospects of such affairs are far from nil.

In premodern times, authority over the *Kakar* spirits used to require a

Figure 18: Moiety succession

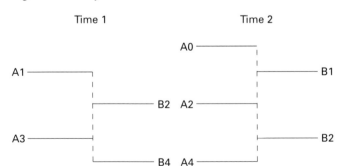

metaphoric violation of this very taboo. The "two brother" age-grades were said to "fight with skirts" for control of the cosmic body. While the "elder brother" grade impersonated the spear spirits, the wives of the men belonging to the "younger brother" grade impersonated Arake, their "mother." Rather than avoiding his younger brother's wife, as affinity would demand, a cultic "elder brother" (A1) allowed himself to be seduced by the junior woman. By the affinal construction of this relationship, such a seduction would have been a double violation. A man would in the first place be having sexual intercourse with his "younger brother's

Figure 19: Firepit rights on the floor of the male cult house

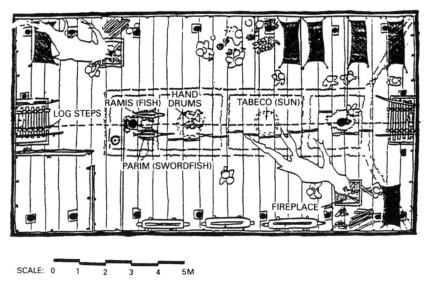

SCALE: 0   1   2   3   4   5M

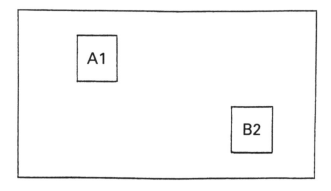

wife." But, as a personification of the *Kakar* spirit, he is also the "son" of Arake, the female spirit who is having intercourse with "the mother who bore him." In the *Kakar* cult, however, this love is not illicit but a means to rank and authority. *Puniim! Puniim! Swaro! Swaro!* (literally, "Vulvas! Vulvas! Copulate! Copulate!") chant the "elder brother" grade during one phase of *Kakar* initiation, when they demand compensation for the display of an esoteric tableau. Their cosmic status was contingent upon the provision of sexual services by the wives of the "younger brother" grade.

Fighting with skirts

Upon entering the "younger brothers" grade, a "son" inherits sex-debts from his "father," the mother's brother who has recruited him (see Figure 17). Throughout his career in the elder brother grade, his "father" received sexual favors from the wife of his opposite number, the junior man who had sought to displace him as a *Kakar*-holder. Not only for control of the *Kakar*, his rival had to "fight with skirts" in many contexts, e.g., before and after battle, when launching a new outrigger, or consecrating a new cult house. The hereditary debt of a junior man was large. Upon entering the "younger brother" grade, he had then to begin to repay his "father's" sex-debts to the man who succeeded his sponsor. In the cultic idiom, he had "to send" his wife "back" to his "elder brother" (B2 > A1) to have sexual intercourse with him. A man, rather than his wife, would count the number of times he had given his wife's sexual services to his *Kakar* initiator by saving coup sticks made of sago palm midribs (*kiinumb*). The debt the new "younger brother" had to repay to become a *Kakar*-holder was equal to the number of *kiinumb* he received from his mother's brother when he was initiated into the cult. "When you give your 'skirt'," the saying went, "you will get the [sacred] things" (*Dago mariabo, mi mwan akorasangetaina*).

Unlike the mundane theory of reproduction, the reproduction of cult membership was closed and bounded. Each time a man was promoted, a new cadet was admitted. Like *sumon* succession, this initiation rite did not require corporeal death, but retirement prompted by exchange. Upon the settlement of sex-debts, the "elder brother" should relinquish his status in the cult and be demoted into the "father" grade (AO). There, no longer in possession of cosmic status, he would continue to maintain a modicum of moral status in the cult. During feasts, he would be consigned to sit at the edge of the hall, where he would only be served "leftovers" to eat but might go on watching and criticizing the way things were being done.

When the last member of the "younger brother" grade has settled his sex-debts, a collective grade-taking ought to take place, which is the *Kakar* initiation. The new ascendant grade should then prepare *aragen* porridge for the retiring "elder brothers." The "younger brother" grade should receive the *Kakar* spears (B2 > B1) and their cadets should receive the daggers (B4 > B3). In the defeated moiety, those who hold the "daggers" are promoted to the grade holding the "lances." That is, they become the new "younger brothers" (A3 > A2) and a new grade of youth enters the cult, putting cordyline war magic leaves into their armbands as they do (A4) (see Figure 18). "Throughout the initiation . . . which may last a week, the initiate's wife will . . . sleep with the man who introduces him into the sacred circle . . . It is the final and greatest gesture of respect" (Beier and Somare 1973: 10).

"Elder brother" and "younger brother" grades are led by the occupants of two offices: the *kombitok* (literally, the "head"), who convenes and closes cult meetings, and the *pokanog* (literally, the "knot" in a tree, or the "knot" tied in a rope or a plaited basket),[9] who may also be called *gapai*, which means the "strong one." This latter man has the responsibility of pre-serving the authority of his grade and moiety. His duties literally include untying the knot in the rope binding the spears after they have been taken down from the loft for display and retying it when putting them away. While his colleague's headship passes matrilaterally from incumbent to heir on the basis of interest expressed by a sister's son in cult affairs, the knot appoints his successor on the basis of a single criterion: he should find the rare man who is "hard" and "strong" enough to control his id. When women enter the cult house bent on seducing their husbands' "elder broth-ers," the knot must remain chaste in the face of their desire. Should he allow himself to be seduced by the wife of his junior partner, who is the wife of the knot in the junior grade, then his moiety will have to give up their rule over the cult. "These junior women [*dam gnasen*, literally 'younger brother daughters'] can [seduce] us as much as they want," a cult elder once told me. "Why? Because our power is protected: we have a 'knot.'"

There was a small, closed door in the rear wall of the male cult house in Darapap in 1981. The door was called the "rear" or "anal penis" (*edug'kaik*) of the building. Women used to enter through it as they came to oblige their husbands' cult partners. The "head" of the "elder brother" grade, in his turn, would enter the building through the front door to convene the sexual license, bid the "younger brothers" to leave their wives and wait together at some opposite corner of the village. Then, summon-ing the knot, who was waiting at home, he would turn the evening over to him: "only the knot," it was said, could carry a torch to illuminate the hall. At his bidding, each woman would then call out the name of her husband's "elder brother." If a single woman was found to be missing, the knot had the right to send all of them back to their husbands to locate her. When everyone was accounted for and had returned to the cult house, the knot might still exercise his prerogative to call off the evening. But should he decide to go on, he would then assign each woman, including his own "younger brother's" wife, to sit with her husband's partner. Inside the cult house, or else outside in some secluded spot on the beach if they preferred, the couples would make cosmic love. The knot would return to the cult house periodically during the night to check whether the "elder brothers" needed anything. But otherwise, he spent the night at home in the company of his wife and the wife of the knot of the rival moiety. At one point, his wife would lead the rest of the women in her grade back to the cult house bearing food. Knowing spells to protect her health from its virulent impurities, as the wife of the knot, she was said to be "the mother" of the whole cult. Of all the women in the senior grade, she alone had the privilege of cooking sago pudding in a big ceremonial pot in front of the building. The rest of the women could cook only inside their houses and then come "to decorate" the pot with delicacies they prepared for their husbands. The knot of the Darapap cult told me about what happened

when everyone . . . finished having intercourse. I return to the *taab* and roll long cigars to give to the women who have successfully seduced their [husband's] "elder brother." [My partner] Kanjo may not smoke though. We count the number of embers glowing from the cigars of the women who are smoking. The women then report to their husbands who await them: "So and so 'shot' me [with his 'spear']. So and so 'shot' me." But Kanjo can only tell her husband, "No. He did not 'shoot' me."

The knot presided over a grotesque, military caricature of marriage and affinal relationships. Among the privileges possessed by the "elder brother" grade was the right to have sexual intercourse with the wives of the junior men (of the "younger brother" grade). Momentarily renouncing their conjugal monopoly of their mother-wives' sexuality, the junior men had to give their *Kakar* initiators the right "to shoot" their vulvas and even encourage the women to seduce these senior men. But while the rest of his grade might "spear" the "younger brothers'" women, the knot alone would remain chaste. Retaining ascendancy in the cult was contingent upon his resistance to the sexual advances of his partner's mother-wife, which rejection was assisted by the presence of his own mother-wife.[10] When he was finally seduced, however, succession and initiation would both immediately take place. The senior grade promotes the "younger brothers," who displace them from sovereignty. Father Schmidt apparently observed this rite: "The spears are taken from the little shelf in the *taab* and stood against the wall. The initiates must sit before their . . . [*Kakar* which] make the boys strong and courageous so that they will not run away in battle" (1923/4: 33). The spear spirit-men were displayed and named for each heir. With the retirement of the old senior grade, a new cadet grade was admitted into the defeated moiety. The reigning knot in the Darapap cult recalled the trickery that was involved in the seduction of his predecessor and the initiation which ensued.

The ex-knot in the retired grade was dying in the early 1950s. He made a death-bed request to his sister's sons, Simbua and Murakau, to find a way to settle his sex-debt with Bwata'u, the reigning knot, in order to reclaim the *Kakar* for their moiety. When he died, his sister's sons tied the *dua'takiin* headband (made of women's hair), which stood for his knotship, around his forehead. Next day, Simbua gave Tutup, a young woman he had just married, a bundle of tobacco leaves and asked her to take the headband from the corpse. Tutup flirted with Bwata'u, the knot, and told him to look for her later on one of the paths to the beach. After the first time they had intercourse, Bwata'u wanted to send her away. But Tutup insisted that they have intercourse again. Her desire flattered him, so he agreed. After the second time, Tutup showed him the tobacco and the headband.

Bwata'u was horrified. Removing her skirt, Tutup put flower petals in her pubic hair [to signal her success] and returned naked to the cult house. A few men in the "younger brother's grade" saw her. Immediately appreciating her nudity, they beat the slit-drums all night in celebration.

When a woman successfully seduces her husband's "elder brother," she should not wash immediately. She is escorted home by her husband to prepare a dish of sago pudding, or a sago pancake for him, a courtesy which renews their relationship. "As the two walk home," a senior woman named Kiso explained to me,

the wife wonders, "Will he be angry with me?" Meantime, the husband's . . . [*mwara* mother's brother or cultic partner] will secretly follow them at a distance and stand nearby the house to overhear if [the husband] gets mad . . . If he becomes upset, and the two fight, then another time, for another reason, when [the husband] brings his wife, to his [cultic partner], he will refuse [to have sex] with her . . . [The husband] will remain just like a woman! Because of this "law," men fear getting angry with their wives after they [have had sex] with their [cult partners].

Recall the dish of sago pudding women gave Sapendo, the spirit-man. In this context, such a meal does not compensate a man for being a lover, as it did in the story, but compensates a husband for having permitted his wife to take his cult partner as her lover. By giving a dish of sago pudding to her husband, the wife signals that her dalliance has not been adulterous but in his service. By eating the food, the husband necessarily ingests the mystical traces of her intercourse (*son menumb*) that still defile her. By giving and partaking of the sexual fluid-spirits which have penetrated the food, the commensality of husband and wife indicates that the husband is willing to accept his wife. Her marriage renewed, only then does the woman wash herself. Her cosmic adultery has not cuckolded her husband. Her gift repairs what was a morally difficult moment in their relationship. The question is: morally difficult for whom, the husband, the wife or the couple?

Ginau once told me how he understood the relationship between wife-lending and the power of the *Kakar* (voicing a widespread view among senior informants of both sexes). "When a man is born," he supposed, "he is afraid. He is afraid of *dying and losing his wife*, so he is useless and weak in battle. But having given away his woman, he has already 'extracted his liver,' so this fear is lessened" (italics mine). By his reasoning, controlling the aggressive agency of these spirits required a training in emotional disengagement from erotic attachment and dependency. "So long as she feeds me properly," the "head" of the "elder brother" grade used to tell me, "I do not care if my wife takes lovers. Why? Because I am a *gapa'nor* [literally, 'strong man']. I can send her to meet lovers. I will not be angry. The man who allows himself to become angry, who talks hither and yon, is merely bullshit. It is no good if I get mad and someone dies from [my spells]." But overcoming erotic attachment was difficult to do. "For . . . white men," the knot in Darapap said, "it is easy to give up your women [sexually]. You receive money. But with us, it is too hard. *We cry when we send our women to the cult house*" (italics mine). Loosening the jealous grip of sexual attachment was barely possible. Instead of the rage a cuckolded husband would feel in the event of his wife's "secular" adultery, a man was expected to feel gratitude to her for succeeding at being unfaithful to him in the service of his cult status. The *Kakar* cult turned the jealousy of husbands into an Oedipal source for masculine agency. It was the women who entered the highly toxic, masculine body. It was the women who seduced the senior men representing the spirits that otherwise made them sick and die. It was the women who had to give themselves to a grade of men, otherwise taboo, the "husband's elder brothers." This is why a Mendam man once remarked to me, "the ancestors used to say that all power came from women."

Women, however, felt challenged, rather than coerced, by the opportunity to seduce their husbands' cult partners "to make the name of [their] husband into that of a big man" (*main ya'ut nor apo totimeri*). Wives did not fear the duty, both Murakau and Minjamok, his wife, insisted to me, because it was a secret held strictly by the woman, her husband and his cultic partner. No one else knew anything about it. It was the husband, rather than the wife, they both agreed, who "might have the fears" about "another man . . . 'holding' something belonging to him."[11] It was the husband's deep attachment to his mother-wife's sexuality which might prevent her from going to the cult house to seduce his *Kakar* initiator and not misgivings of her own. Minjamok added that even the very first time she had intercourse there she felt no qualms. "It is our way," she demurred to my insistent questions. Kiso, another senior woman, was more emphatic: "Fear? What is there to fear?! This is not something to fear! It is not warfare! . . . You [usually make love] with your husband! A man! Only a man will come! . . . It is not many times. [With your husband or] in affairs, you are always [making love]! This is different. This is limited to the men's house, to the *Kakar*." By contrast, the relationship between husbands' sexual attachment to their mother-wives and this cosmic agency touches the heart of the Mangrove Men. The prominence of this problem is confirmed by the legend of Sendam, the spirit-man some credit with the invention of organized warfare, the moiety system and wife-lending in the cult. The first episodes of his story render the issue of masculine sexual attachment even more starkly, if this is possible, than the sexual exchanges which took place over the *Kakar* spirits.

Sendam, father of warfare

Like the "Two Brothers," Sendam hails from an area outside the Murik Lakes (the village today called Samap). And, like the *Kakar*, his adventures begin with a grotesque evocation of motherhood. The mother of Sendam was a wild pig; his father a villager. His mother, that is to say, was not only a woman, but also an image of value which men can obtain independently of women's labor. Not being raised domestically, she becomes the object of male violence: the men of the father's village hunt her down. In retaliation, Sendam destroys everything and everyone in it (save his sister and her children) and then leaves for the Murik coast where he finds men fighting each other without form or even weapons. He introduces spears and spearthrowers to them and teaches them how to fight each other in opposed pairs, a principle of dual organization some say he also applied to the male cult moieties. Before "his arrival," in other words, warriors fought chaotically. A symmetrical form of a contest between individuals, or groups of men, they had not invented. An image of a self diametrically or antagonistically opposed to the other is not, according to the legend, intrinsic or natural to men, but had to be learned or imported from outside the boundaries of culture. The story goes on to take up the great dilemma of the Mangrove Men, confirming the conclusion that, in order for the self to oppose the other, men must somehow free themselves of emotional attachment to mother-wives.

After making his gifts of weapons and dualism to Murik warriors, Sendam went on toward the Sepik River, where the village today called Kopar was under attack by two sea-eagle spirits, who were husband and wife. Climbing the tree in which they nested for the night, Sendam killed the couple just at the very moment they were making love in their mosquito basket. Saving a community from a predatory image of conjugal intercourse, he "remove[d] the curse of men's sexual dependency upon women" (Meeker, Barlow and Lipset 1986: 62). In return for rescuing them, the village men wanted to celebrate and honor Sendam with a great feast. But the hero refused their offer. He only wanted women. Each man brought his wife to the cult house and the spirit-man had intercourse with them, one by one. Having liberated the community from a predatory image of conjugal intercourse, Sendam then became its greatest husband, monopolizing the sexuality of all its mother-wives to himself. The next episodes were recited to me in Tokpisin by Nangumwa of Mendam.

But the one woman he desired most, was refused him by her husband. Sendam [turned himself into] a tree. The woman's husband went out with his dogs to hunt wild pigs. One of them smelled Sendam in the tree and began to bark. The man admired the tree and decided to use it as the centerpost in the cult house he was building. He brought men who tried to chop it down, but Sendam held it up with a creeper vine. The man sent his son climbing up the tree to cut the vine. Sendam saw the boy. Still infatuated by his mother, he asked him, "Does your mother not have pubic hair on her genitals?" The question shocked the boy, who scurried back down in a panic to tell his father what happened. Exasperated, the man ordered his son to ignore the spirit-man's foolishness and just cut the vine. The boy did and the tree fell to the ground.

The men trimmed off its branches and tried to carry it back to the village. The tree trunk became heavy. They returned to the village to fetch their wives for Sendam. He . . . [turned into a] man again and had intercourse with each of the women. Only then did the tree trunk become light enough for the men to carry. They moved on a bit further until it got heavy again and once more they had to satisfy Sendam. This happened several times until the men finally reached the cult house. But before they could erect the post, their women once more had to lie down and have intercourse with him.

The grotesque relief of the spirit-man's body is not confined by a closed, impenetrable surface that is complete and separate from nature. The body of Sendam is one of exaggerated proportion and huge protrusions. He can change into a tree trunk. His genital passion is virtually insatiable. Limitless virility does not, however, cause Sendam to become, like every-man, conjugally attached to, and emotionally dependent upon, women. Instead, his virility frightens sons and dominates husbands. A woman need only be pubescent – rather than his wife – to become an object of his desire. But to what extent can we say that the hero successfully resolves the dilemma that sexual attachment and emotional dependency pose for the cosmic body of men when the absence of just one woman provokes him so vehemently? Along the same lines, the legend also presents an image which attests to the husband's discomfort when permitting cosmic adultery. The lone woman is refused to Sendam by whom? Her husband. She herself does not refuse. And then, in a subsequent episode, Sendam becomes agitated by another image of conjugal intercourse in which he does not participate.

Sendam crossed the Sepik River. He came upon a childless couple making love in their garden. After they finished and left, he rolled in the damp ground where they had been lying and turned himself into a crying baby. The woman heard his cries and came back to the garden. She adopted Sendam as her foundling child and took him home. The infant Sendam cried all night. She turned him from one side to the other but could not settle him down until she laid him down upon her breasts and let him nurse. He immediately turned back into a man and made love to his mother. Then he slept [as a baby].

Confronted by a scene of another man making conjugal love with a woman, Sendam must devise other strategies. He becomes the woman's baby who can monopolize her nurture by day. But at night, the claims of infancy fail. Dissatisfied with his mother's breast, Sendam resumes a sexually potent body which enables him to assert the phallic claims of manhood. He thereby resolves a problem faced by ordinary husbands who are tabooed from the sexual attentions of their wives during the early childhood of their offspring. Transforming his body back and forth from nursing infant to virile man, he is able to monopolize both the nurture and the sexuality of a wife-mother simultaneously. The hero becomes an extraordinary, albeit grotesque, exemplar of a warrior-husband. In 1982, no less than three cult houses in Murik villages bore his name (see Plate 23). The irony of his exploits and eponomy, as well as the very format of exchange he may have introduced to the cultic system, is how forcibly they disclose the impossibility of his solution and, by implication, the irretriev-

ability of the masculine condition (see also Allen 1988: 80). Such, as
Gluckman used to say, is the appeal of the sacred.

Despite the contemporary presence of Sendam cult houses, the protocol
and beliefs in terms of which the *Kakar* spirits of the 1980s were understood and attended were quite different from what they had been in 1918.
The directions of change – toward both persistence and emasculation –
taken by the cult in response to colonial oppression were complicated. The
events described in the following four sections of this chapter are meant to
illustrate and assess these changes and the ambivalent dialogue between
the *Kakar* spirits and the twentieth-century forces which have sought to
bar their autonomous agency.

The end of headhunting
At the request of several, culturally "progressive," young men, Father
Schmidt established a permanent Catholic outstation on the Murik coast
in 1913. Schmidt took a dim view, as might well be imagined, of *Kakar*
spirits, "fornication" and headhunting. But from what I was able to piece
together, the missionary did not exert violent pressure on cult or culture.
Such cruelty was reserved for "a renegade Catholic priest named
Heinriech Luttmer" who apparently went from village to village in the
Lower Sepik during the mid-1930s demanding "that . . . men reveal their

Plate 23: Sendam male cult house in Mendam

sacr[a] . . . to women and children" (Kulick 1992: 163; see also Huber 1988). In one village (Kaup), he took it upon himself to break the *Kakar* spears over his knee (see Plate 24). Father Schmidt dutifully stored the pieces in the attic of his house until 1936 when he gave them to a young American ethnographer, Pierre Ledoux, who was doing research in the Murik Lakes. Obviously, early colonial institutions had but limited cross-cultural sensitivities. But I want to make it clear that even before this foreign repression had begun, which is to say prior to the settlement of Father Schmidt on the Murik coast, young Murik men were eagerly seeking jobs in the German police force (Somare 1975: 17). The protocol for consecration and initiation rites, which called for ritual homicide, had already begun to be repudiated or, at least, to be neglected. Colonial state and mission were only the most obvious forces of change (see also Hauser-Schaublin 1993). Endogenous impulses were also at work before the initial penetration of modernity.

Plate 24: Broken *Kakar* spears, collected in 1936 by L.P. Ledoux and donated to the American Museum of Natural History

Headhunting seems to have been limited in frequency and scope to annual or biannual forays, not in each of the Murik villages, but among all of them. Murik warriors typically ambushed lone individuals with whom no one maintained trade relations. Such forays might temporarily ally multiple cultic groups in the Lower Sepik region, one of which might have been retaliating against an enemy.[12] The victims, moreover, were not chosen at random. There was always a direct relationship between local politics and regional warfare. Two elderly men whom I knew in 1982 recalled a mock-dirge they learned as children which had been composed by Ariapan villagers upon seeing the image of two widows keening, following the murder of their husband.

> O o o o your husband's penis is gone!
> Now you don't have anything,
> they have cut it.
> Now you go have nothing.
> ai ai ai ai o o o o

The song was meant to taunt the widows whose husband was widely feared and disliked. My informants were quick to deny that Murik warriors, who had been involved sixty-four years earlier, had cut off the corpse's penis. They wanted to emphasize, even so many years later, that the war party had been solicited by the people living in the victim's own village who wanted him dead because he was believed to be a powerful sorcerer who allegedly killed many women and children. Of this side of the story – and its implicit claim, if a pun is permissible, that headhunting was an acephalous form of capital punishment – Father Schmidt apparently had no knowledge or belief. He only knew that

the men of Janain were opening a new spirit house, named Bungabwar. They went out and got a man from Ariapan and brought back his head.

Brag masks came down from the houses. They were richly decorated and shook as they surrounded the head. The spirit [mask] slurped at the blood about the head and then shoved it to the next mask. Blood dripped from the mouths of the masks.

On 22 November, in the evening, the men entered the *taab* in procession with torches and noise, the procession was even frolicsome. [Inside] men danced before the masks and the head and stained the snake head motifs carved at the ends of the ridgeposts with blood. All night in the *taab*, men sang the spells of the *brag* spirit. The next day, they heated up a big pot of water at the beach and boiled loose the flesh of the head. The *wajak* [of the man who threw the first spear] cleaned the flesh from the skull and set it out in the sun to dry. Afterwards, the skull was hung in the cult house.   (*Schmidt 1923–4: 700*)

Suffice it to say, the head of the serpent condensed an image of aggression, a site of agency, and, iconically, a sign distinctive of manhood.

Painted red with the blood of a victim of the *Kakar* spirits, the image synthesizes the gender, fluid and capacity of their grotesque bodies. But this was the last time the serpent image was so anointed, and it spelled the end of the premodern *Kakar* cult.

Father Schmidt quickly sent news of the murder to the colonial administration, then located at Aitape, about 200 miles along the coast to the west. The District Officer and native police arrived in due time, burned down the new cult house (see Kulick 1992: 276 n. 4) whose consecration Schmidt had observed, and destroyed its *Kakar* spears. The *Kakar* cults in Darapap and Karau were also accused of participating in the war party, but their spears escaped a similar fate due to a longstanding patron–client relationship between Sagiwa, Ginau's father, and Father Kirschbaum, the great ethnographer-priest who headed the Catholic Mission in the Lower Sepik prior to World War II. Upon hearing of the murder, Kirschbaum met with the colonial officials and arranged a deal in which Sagiwa confessed to the limited complicity of the two cults in it and made an offer to pay indemnity. In return, the authorities would (and did) pardon the *Kakar* cults in Darapap and Karau. Kirschbaum failed to intervene on behalf of the village of Mendam, however, and their *Kakar* spears were destroyed, even though the cult there had not been part of the war party. In the event, headhunting ended: the *Kakar* spirit-men ceded their right to force to the colonial state (cf. Rodman and Cooper 1979). In Big Murik, where the spears had been burned, no new *Kakar* were carved or imported. The moieties stopped exchanging cultic authority until 1981, when Kanari rebuilt the Bungabwar cult house in conjunction with his succession to his elder sister's *sumon*ship (see Lipset 1990 and Chapter 4). The building was meant to house newly carved *Kakar* spears which he had commissioned a trading partner in a Lower Sepik village to carve.

Bungabwar rebuilt

A male cult house is constructed in a sequence of ritual workdays during which the "younger brother" moiety provides support for the "elder brother" moiety in return for pig feasts. The junior group erects the center-post, raises the ridgepost, thatches the roof, hangs its bunting, and, finally, consecrates the new building.[13] But the cultic rendition of sibling piety, being overlaid with metaphors of military organization and violence, is ambivalent. Services exchanged within and between the moieties are both spoken of as "warfare." The cadet grade of the junior moiety serve their cultic "fathers" as "soldiers" staging "an attack," in return for which they receive a meal. The rival moieties also "fight with food" during these occasions.

I went along to Big Murik with a group of senior *Kakar*-holders from Darapap in 1981 and had an opportunity to experience this military metaphor at first hand. Cult recruitment had continued but there were no members of the reigning age-grade alive in the community. Darapap elders were invited to stand in as surrogate "elder brothers" on this particular day, while the junior moiety attached the "skirts" to the outside walls of the building and hung up the interior sago-fringe decoration (*saidug*) from its ceiling. After completing their work, which went uneventfully, all the men repaired to opposite corners of the hall. The "elder brother" and "younger brother" grades sat around their respective firepits, with members of their respective cadet grades seated nearby. Two meals were served. The "elder brothers' sons" (A3) served the "younger brothers" on behalf of their cultic "fathers." And the "younger brothers" (B2) served their cultic "sons" (B4) (see Figure 18). Just before everyone began to eat, plates of food were subtracted from those which had been given to the "younger brothers" and were sent back to their wives who had cooked.

One of the "elder brothers" (A1) took spoonfuls of liquid from a plate of sago pudding and poured them through the gaps in the floor around several of the houseposts, calling out the names of *Kakar* spirits as he did. Stragglers continued to enter the cavernous, yet crowded, hall. As each man looked for an appropriate place to sit down, several youths would call out the name of the latecomer to invite him to come sit down and dine with them.

A squabble broke out over me. Kanari, the sponsor of the event, called across the hall to invite me to come sit down and eat with the "younger brother" moiety of which he was a member. With uncharacteristic forcefulness, an elder next to me told me not to budge. Kanari insisted with what appeared to me to be real belligerence, and I felt unsure and confused about what to do since I understood we were his guests.

"Elder brother" had fed "younger brother" for "his work." Likewise, "younger brother" fed "his son." Being the "elder brother," the guests did not receive food but tendered it. However, the nurture was simultaneously destructive, yet dutiful and generative. Kanari offered me something to eat, I only very much later came to understand, not simply as a good Murik host, but to claim me from, and score a defeat over, the "elder brother" moiety to which I had been assigned. Typically, inexperienced youths can be tricked into eating with the incorrect moiety, which, should they do so, results in their being "captured" by that moiety. When the moieties "fight with food" to compensate for services they have received, in other words, they also "fight" to reproduce themselves.

When the serpent head motifs adorning the ridgeposts inside the cult house were anointed in 1981, no trophy head was paraded through the village (see Plate 25). No *brag* masks slurped human blood. The serpents were stained with red betel nut spittle (a substitution anticipated in the "Two Brothers" legend, see Chapter 2). The sacred spears possessed no one in this cult house: there were no ceremonial spears for which the moiety women competed. The phallic desire of the cosmic body, and its associated system of sexual exchanges, had been cut off. The moiety system persisted largely in the context of cult house construction ritual. The "Two Brothers" who built the big hall sat in their respective corners, taunting each other with invitations to eat. The cosmic body had been castrated, or at least domesticated, by the bureaucratic body of the colonial state.

The photographed *Kakar*

In villages where the *Kakar* spears had not been destroyed, grade-taking had also come to a halt. The last transfer of authority in Darapap took place in the early 1950s (upon the seduction of the knot described above). Recruitment continued, but no alternative token of compensation was

Plate 25: The folk opera, *Aimaru*, was performed at the consecration rite for Bungabwar cult house

devised to substitute for "skirts." Rival moieties remained locked in position until 1973, when Michael Somare, as a young leader of the emerging state, committed a ritual blunder which started a renegotiation of the value of cosmic authority. Together with his friend, Uli Beier, the art impresario, Somare sought to protect the remaining *Kakar* spears in Darapap and Karau villages from theft and resale in the international primitive art market. He therefore asked elders in the cults in these two villages for permission to photograph the carvings. Such a display, said outraged elders, would tip off a transfer of rank between the moieties. Somare, then thirty-five, was too young. The knot of the grade had not been seduced. The head of the elder brother grade demanded "skirts." Eventually, after some negotiation, the elder brothers agreed to an alternative "payment." The younger brothers would grant provisional status in their moiety to Beier, a man in his fifties, who would then be promoted along with the rest of them, when the spears were displayed by the senior grade, who would step aside in order to give them the opportunity to photograph the tableau. Somare would only be allowed to serve as Beier's interpreter.

Beier, and the younger brothers grade, were taken to a secret spot in a casuarina stand near the beach. Upon presenting a basket of *canarium* almonds and a meal to the elder brothers, twenty-nine *Kakar* spears held erect by a small scaffolding were revealed to them (see Plate 24).

Wino and Karok [two senior grade-holders] both wore . . . headbands and heavy shell decorations and squatted on either side of the line . . . We remained in complete silence, and were finally told we could take photographs. Flutes were blown . . . The total impact of the scene in the grove was quite overpowering . . . I was aware of the seriousness of the occasion . . . I could see that Wino and Karok were deeply moved by the event, but I did not, at that time, realize what a *huge sacrifice* the old men were making by allowing me to see the objects.   *(Somare 1975: 32–3, italics mine)*

The "huge sacrifice," of course, was giving up their cosmic offices and rights. Somare was genuinely taken aback by this outcome. "The day Wino and Karok showed us the *kakar* images on the beach," he lamented, "was the last day on which they handled" them (Somare 1975: 33). Because they forswore the sexual services of young women, the elders had begun to revalue cult initiation. Somare condemned their compromise as a pragmatic betrayal. In the atmosphere of cultural revival that was going on in the run-up to the independence of the state (see Somare 1975; Lipset 1989), perhaps this move did seem nihilistic. Some twenty years later, however, the image of elder brothers abdicating power to the younger brother grade on such a basis seems more ambiguous, a compromise to be sure, but also a "maternal" sacrifice of no little authenticity.

The stolen *Kakar*

We returned once again to Darapap in 1988. People immediately cautioned us not to let our two little boys play or swim along the lake shore because a large crocodile, which had attacked one youth, was lurking in the shallows. But the full meaning of their warning was not clear to us until we began to hear about E, a mother of five, who recently committed suicide. J, her husband, had been involved in several love affairs, it was said, and had wanted to take a second wife. The couple argued. J beat her repeatedly until she overdosed herself with chloroquine, the antimalarial drug. Both J and his father then fled to the provincial capital, "shamelessly," many people felt, while E's aging parents were left behind pitifully observing mourning taboos for their daughter. The husband, a fervent Seventh Day Adventist who did not believe in the *Kakar*, allegedly had stolen two of the oldest spears, with the idea of selling them for several outboard motors and a dinghy.

In the male cult house in Darapap, I found a freshly cut coconut frond covering the portal of their loft and began to look into the incident in earnest. People recalled how, the year before, a young man had brazenly entered the male cult house in neighboring Karau and taken one of the ceremonial spears, claiming that it belonged to him to do with as he pleased. Soon afterwards, one of the senior members of the elder brothers grade in the Darapap *Kakar* cult had had a dream which made his skin tingle. Given what had just happened, the elders interpreted his dream as a message to make the preparations necessary before taking out their own *Kakar* spears so they might count them.

They "stilled" the village. Women and children were required to stop working and hide in their houses so as to protect them from contact with the *Kakar* spirits. The wives of the younger brothers grade cooked a meal for the ranking moiety to compensate them for counting the spears.

The moieties met in the cult house in the middle of the day. The elder brothers grade sent the younger brothers grade from the hall and erected *kakramenung* spears decorated with rings of cassowary feathers outside the door to taboo entry into it. Four members of the elder brothers grade then took down the ceremonial spears. The sister's son of the knot untied the bundle into which they had been bound. The rest of the grade sat still, bowing their heads. The tense silence in the cult house was unbroken by laughter. The count began and two spears were found to be missing. The elders spent the rest of the night in seclusion inside the hall.

Next morning, after both elder brothers and the younger brothers grades reassembled in the cult hall and had eaten their separate meals of fish and sago pudding, the *Kakar* were returned to the loft. The men immediately went to wash in the sea. Elders of both moieties harangued the youth but leveled no accusations against any specific individual. A few people noticed that one teenager, who had also failed to appear the night before, did not attend the meeting. Later that afternoon, while he was spearfishing in the lakes, the young man was attacked by a large crocodile, the very crocodile "who" was the "canoe-body" of one of the two missing *Kakar*. Robbed of the spear "canoe-body" in which his spirit otherwise traveled, it was thought, his spirit had been forced into his crocodile "canoe-body" which then went after the youth, apparently because he was an accomplice to the theft.

Meanwhile, the marriage of J and E was rapidly deteriorating. J threw his wife down their house ladder several times. And she then overdosed herself. While her kin were keening over her corpse, its skin turned yellow and dark black pouches appeared under her eyes. The changes, which seemed to resemble the eyes of the missing *Kakar* spears, incriminated her husband. Having stolen the spears, it was alleged, he had hidden them *inside his house* near the mosquito net where his wife slept. People assumed that E's suicide was the sanction of an angry spirit. Upon returning in 1993, however, I learned that the woman's suicide was then thought to have been unrelated to the *Kakar* theft of which her husband had been falsely accused. E had "really" killed herself because of rivalry with her elder sister who had been having an ongoing affair with the husband. E, it was said, felt such intense shame over the tension in her sororal relationship that the prospect of having to go on eating together with her elder sister had been intolerable. The alleged thief of the *Kakar*, the philandering husband, threatened to sue anyone who leveled further accusations against him. "We just don't know what happened to those two spears," shrugged Marabo, who told me the story.

Culture and historical practice

These four incidents – the burning and then the rebuilding of the cult house, the photography and the theft of the *Kakar* – raise questions about the relationship of culture to historical practice. During the well-known events leading up to his murder in 1779 (Sahlins 1981; 1985; 1995), being wrapped in priestly red cloth and named for a god or a chief (Obeyesekere 1992), the local meaning of Captain Cook was determined by Hawaiian cosmology. In the *Kakar* cult, although a long sequence of acculturation is being played out rather than a moment of first contact (see Mosko 1992), Sahlins' general point remains valid. Commonsensical assumptions applied in response to novel events can be seen to go on creating unforeseen changes in those assumptions. In Murik, the intruders were not uniformly novel, much less wholly foreign. Nevertheless, practices that derived from the enduring cultural schema continued to assimilate the external forces they represented, which altered this schema.

To wit: as a result of the display forced upon the Darapap *Kakar* cult by the alleged theft, a grade-taking was negotiated. A new group of men were promoted to the elder brothers grade and a group of junior men were initiated as cadets of the new younger brothers grade. The new younger brothers, it was agreed, would "pay" the retiring incumbents K100 (K1 = US$1.25) in return for relinquishing their authority in the cult. The money, they said, was "a substitute for skirts." The "price" was set by two senior men who had been the last members in their grade to send women to their cult partners. Gifts of money, food and liquor had displaced "skirts," or rather had become the semantic equivalent of "skirts," not only in Darapap but throughout the Murik Lakes. The masculine attempt to worship a deliberate sublimation of sexual jealousy was abandoned. The moiety system still operated: the "Two Brothers" continued to compete to recruit sisters' sons and continued "to fight" with food and services. More importantly, the position of the "canoe-bodies" of the *Kakar* – their ceremonial spears – continued to be worshiped as paramount forces in Murik religion, possessing the greatest potency within and without the community. The reproduction of the cosmic body, once set in the difficult emotions aroused by ritual wife-swapping, had been moved toward an exchange of disembodied commodities (see Mauss 1925; Gregory 1980; 1982; M. Strathern 1988a), but syncretically: for "culture is precisely the organization of the current situation in terms of the past" (Sahlins 1985: 155) and the current situation permitted neither a total repudiation of the past nor the exact duplication of the foreign body. The relationship of one to the other, in my terms, was not unilateral but dialogical.

Despite the gradual emasculation of the *Kakar* cult during the twentieth century, mothers' brothers continued to build new cult houses and recruit their sisters' sons on into the 1990s. The moieties, however, did not regularly transfer authority either in the Murik villages where the spears had been destroyed by colonial officials earlier in the century, or in the villages where they were still stored in the men's cult houses. Many questions about the cult must therefore remain unanswered. I do not know, for example, how frequently the moieties exchanged rights in the *Kakar* spears during premodern days. I do not know how old a man might have been when he "retired." I do not know very much about the subjective experience of participating in the sexual exchanges, either for men or, in particular, for young women. The ethnographic gaps aside, the central image of the *Kakar* cult remains unmistakably Oedipal: a "younger brother" had "to give" sexual privileges in his mother-wife, a bowl of *aragen* porridge and a pig feast to his initiator in order to obtain that man's supernatural status, prestige and agency, in order, that is to say, to obtain his *Kakar* spear. Throughout the cult, elders holding rights to the spear spirits did so until the seduction of their whole age-grade was completed by the wives of their rivals, the junior age-grade leading the opposite moiety. And everywhere, the ascendancy of the senior grade was protected by the chastity of the knot, whose power "bound" them to authority over this cosmic body. The adage that "a single woman takes down the *Kakar* from their loft" is known throughout the Murik Lakes. And the crucial woman upon whom grade-taking was contingent is always said to be the wife of the heir of the ranking knot who aspires to succeed him. In each of the *Kakar* cults, her sexuality is not linked to procreative force but to the cosmic power an individual man gains by renouncing his attachment to it.

According to my argument, the maternal schema is a heterosexual rejoinder to uterine motherhood, while the grotesque body espoused by the Spirits of Wealth and the *Kakar* cult is, in turn, an ambivalent, military rejoinder to men's sexual dependency upon the maternal body. Now we can see that still, even at the end of the twentieth century, the male cult was continuing to answer the emasculating forces of modernity in terms of this dialogue. The unmerged yet mutually interrelated logic of these contradictory creeds is audible enough. But sociologically, we might well ask, what is happening to the voice of local-level political authority in so far as it no longer derives ultimate power from men sublimating Oedipal jealousy?

Religion and society

In Melanesian ethnology, very few societies have been reported in which the jural equality of the genders has reached the level of the Murik ceremonial system (cf. Landtman 1927; Lepowsky 1994). While there are many cultural differences between male and female – in the division of labor, affinal and cosmic duties – the gender of the jural body, as defined by who may possess rights to citizenship, rank and authority, excludes neither male nor female. The polity consists of multiple, coresident sibling groups, each of which is internally stratified by principles of seniority and primogeniture. Under the ritual leadership of male and female firstborn elders, these sibling groups maintain ties to multiple, cognatic, insignia-holding groups whose membership is dispersed throughout the Lower Sepik region. Committees of senior men and women, that is to say, direct heraldic bodies without regard to sex or residence. The groups they manage overlap ambiguously. They are so amorphous that marriages between them tend to be both endogamous and exogamous at once. In any case, neither the sibling groups nor the insignia-holding groups prescribe categories of marriageable spouses, and Murik affinal groups are not strongly opposed to each other either sociologically or metaphorically. Affines were not enemies at war; they are only "flying foxes," or thieves. The Murik do not legitimize marriage through exchanges of bridewealth between jurally distinct groups. Instead, male and female titlebearers conduct ceremonial exchange *within* the ceremonial groups they lead, with the goal of differentiating them from each other rather than linking them together. The fluid boundaries of the cognatic ceremonial groups, the principle of gender equality, and the low level of affinal tension, are all reflected in the *Kakar*.

Both men and women are vulnerable to supernatural impurities which are held to derive, on the male side, from mystical residues of women's procreative or sexual fluids (as well as the "canoe-bodies" of female spirits), and, on the female side, from the substances of male spirits. The cosmic bodies of men and women, in short, defile each other. The ambivalence with which husbands, at least, enter into the practice of ritual wife-lending as they seek the authority and agency of the *Kakar* spirits, far from being disarticulated with, or dividing, this thoroughly heterosexual culture against itself, must be seen as its greatest exemplar. Of all the secrets a husband learns in this cult, the one *he and his wife* are taught and expected to guard most absolutely is about the sexual exchanges among himself, his wife and his *Kakar* initiator. Through her act of seduction, not only would the husband become promoted in cult rank and privilege but so would she become a "big woman" in the cult. The critical difference between the cosmic status of men and women, of course, is that the *Kakar* are vested exclusively in the "canoe-bodies" of senior men. How is the meaning of this privilege organized?

First of all, the lockout of noninitiates – youth and women – from the cult house is semipermeable rather than total. From time to time, not only women, but uninitiated children, enter the "mouth" of the cosmic body. Throughout childhood, mothers' brothers may invite their charges to come inside to nourish and instruct them. During the *brag* flute initiation, the performance of the flutes was stressed, not the secret revealed. Senior men recalled how initiated women used to dress up in men's rags and dance lewdly in their own cult house during this *rite de passage*, mocking with their laughter the incompetent attempts of novices to play the flutes. The cult was little concerned with controlling knowledge of its artifice (cf. Hays 1988). Now it is true that, during this initiation, the youths were sequestered from the domestic group and beaten with eels. But to what ends? They were isolated from women for up to six months during two phases of the initiatory cycle when they were taught how to cleanse themselves of masculine pollution contracted from ritual activity and feminine pollution contracted from sexual intercourse. They were given love magic to make them virile. Lastly, they received hereditary regalia which differentiated them as jural persons. The abuse they were made to suffer was a gauntlet linked to praise and renewal. "The one who is thrashed," as Bakhtin observes in one of his many exegeses of medieval carnival, "is decorated" (1965/1984a: 206). The ritual potency of the youths as warriors was also enhanced. But the whole emphasis of the rite was to provide magico-religious operations to engage with, and immunize the domestic community from, that potency, and to teach youths operations to protect their power from the community. The point was to endow young men with cosmic bodies, and carefully to circumscribe those bodies from women and children in order to protect the noninitiates from power, rather than denude or deprive them of it.

The *brag* initiation is contradictory. A man is entitled to fight but also to make love and marry. The metamorphoses the initiates underwent in the "womb" of the cosmic body render them warrior/lovers (cf. Hiatt 1971). Advancing through the *Kakar* cult, the masculine body becomes less phallic, or phallocentric. Men had to become emotionally distanced from both their own sexuality and that of their wives in order to take cultic rank. Men had to learn to give up, rather than take, sexual privileges from women in order to acquire cosmic agency. Ritual heterosexuality was a lengthy process of socialization for the sublimation of desire for and erotic attachment to women. Should a man want to accept a lover in the cult, he had to yield cosmic rank and authority in it. Should he want to become and remain a man armed with the power of the *Kakar*, he had to "fight" desire, or "fight" by manipulating other men's desire for his "mother-wife." Masculine agency reveals another dimension of the relationship of men and women, the intensity of emotional attachment husbands felt for wives. Winning control of the *Kakar* spirits was the great reward for the painful achievement of learning emotional detachment from a feminine image of fleshly temptation. The source of a man's agency and downfall, the sensual bodies of women, was regarded with ambivalence by the cosmic, masculine body. The grade of the "younger brothers" fought the "elder brothers," coercing them with the allure of their wives to "shoot their spears" (cf. Roscoe 1994b: 59–60). In the moiety rivalry, women presented themselves as "victims" to their husbands' *Kakar* initiators, "victims" whom the elders could not restrain their desire "to shoot." But each cult elder only possessed a limited number of "spears," and once he had "shot" them all, his rank in the cult expired. The strategy of the *pokanog*, the knot charged with defending the authority of the ascendant moiety, was to stay home. While sexual license was going on within the cosmic body, he would take his "younger brother's wife" to spend the evening there *in the company of his wife*, where he, and his moiety's authority, were safe from her seductive allure. This secret image of the dependency of the cosmic body upon a married couple, rather than upon manhood isolated, suggests that Durkheim will not be mocked by the *Kakar* cult. The premise of his sociology of religion, when applied to Melanesian male cults, is that a sacred body will express and confirm – mirror – the gender inequalities constituted by the dominant political body (Allen 1967). The image of the knot sitting at home with his wife and his partner's wife in order to defend his moiety's ascendency *from his desire for the latter woman*, and the adage that "a single woman takes down the *Kakar* from their loft," are therefore faithful renderings of a sociology whose fundamental differences, both among its constituent groups and between the sexes, are cast in terms of gender equality and interdependency, instead of misogyny and androcentrism (cf. Keesing 1982).

It was a woman's sexuality which most disturbed the cosmic body. At one time, the divorce of procreation beliefs from sexual intercourse was one of the most hotly debated ethnographic conundrums that Melanesian ethnography managed to produce (see Malinowski 1927; 1929; Leach 1954; 1966; and Spiro 1982). But parthenogenesis is not the major analytical issue at stake in Murik religion. The meaning of sexual intercourse in the *Kakar* cult had nothing to do with the transfer or collection of semen as life-force.[14] In the cosmic context, sexual intercourse was not an exchange of body fluids, but a reciprocal service that was counted, like trophy heads, with coup sticks. The provision of such an intimate service was understood as a kind of "currency" that repaid a debt and as an irresistible "weapon" that was meant to arouse rival men "to shoot their spears." The powers of the spirits were not contained in sexual fluids.[15] The cult was reproduced – authority over sacred aggression was won and lost – through images of reciprocal services, rendered in terms of sexual desire, renunciation and warfare, but not uterine reproduction.[16]

If the *Kakar* cult does not contest the sociology of religion, it does raise another, no less difficult, exegetical problem. Several students of Melanesian male cults (Allen 1967; J. F. Wiener 1988b; and Harrison 1990a) have agreed that controlling fertility – the manipulation, or differentiation, of procreative force – is the *sine qua non* of these cosmic bodies (cf. Langness 1967; 1974). But in Murik, neither the heraldic system of status attribution nor the male cult makes such claims. At best, fertility was negatively vested in the cult. The spear spirits had to be secluded from the community, not because they contained crucial procreative substances, but, as we saw in the case of the stolen *Kakar*, because of their extreme toxicity to the "reproductive force" of society, namely mothers and children. Moreover, the cosmic love which couples used to make yielded no sort of biogenetic substance. In a word, the *Kakar* spirits were the subject of a phallic, rather than a fertility, cult (cf. Whitehead 1986; 1987).[17]

A sibling Oedipus

From a psychodynamic point of view, the greatest force in Murik culture – the *Kakar* spirits – must surely "represent" the mastery of a primary loss, a mastery of a displacement from a primary source of gratification, the maternal object of desire (Roheim 1942; Mahler et al. 1975). Like Spiro's useful notion of a culturally constituted defense mechanism (1965), the cult would seem literally to have rechanneled and concentrated rage felt over the lost mother into the society's most potent phallus (or agency). Thus Sendam, the warrior-hero, is expert not in impossible acts of aggression, but in mastering the Oedipal predicament imposed upon men by reproduction and generational succession. In his story, as in his cult, such mastery is demonstrated by the rehearsal of loss and the possession of other men's "wife-mothers." Terminologically, we know that Murik wives are their "husbands' mothers," but they are also "mothers" in another, esoteric sense. In the triangulated "fight" between the cult partners who "spear" their "younger brothers' wives," the younger women sent by their husbands to seduce these cult partners personify Arake, the legendary mother of the spear spirits whom the senior men personified. The junior men "send their mothers" when they send their wives, to seduce their cultic "elder brothers" in order to retire these men from cultic status, sexual rights and authority. In other words, they "send their mothers" to displace these men's claim to personify the spirits of the phallic *Kakar* spears. The "elder brothers" could take the love of these double "mothers," but had to sacrifice cultic status when they did so. The dense metaphoric reversal of domestic taboos in this secret exchange was not in the least immoral or unconscious, however. It was explicitly acknowledged by both men and women, if not as the "royal road," at least as "the main path," to sacred authority. The cosmic body of the *Kakar* did not possess or seek to remove procreative force from women. At stake – for men – through the process of learning to renounce their sexual attachment to their "wife-mothers" was succession to cosmic ascendancy. Such a prize, at such an intimate cost, behooves any functional analyst of this institution to rethink it in view of Oedipal theory and early childhood development.

In Oedipal theory, the successful resolution of sexual rivalry for the love of the "wife-mother" is accomplished through the imagined threat of castration, and results in the development of conscience, the son's superego, the moral component of the self, which later yields in adult men the capacity to marry women and father children. In the *Kakar* cult, the successful resolution of sexual conflict between "younger brother" and "elder brother" over the "mother" yielded rights to a power so destructive that it had to be isolated from the community in order to protect that group's reproductive capacity, e.g., women and children. According to Oedipal theory, in other words, paternal force, or rather an image of paternal force, masters incestuous jealousy over the mother to create a husband and father out of a son. In the Murik cult, an enchanted image of matrilateral force masters incestuous and conjugal jealousies over the "wife-mother" to create warriors out of "younger brothers."

Clearly, the Murik family complex and ritual heterosexuality in *Kakar* initiation, not to mention the tale of Sendam, all suggest that this institution will not challenge the putative universality of the Oedipus Complex in the same way Trobriand matriliny was once claimed to have done (Malinowski 1927). But it does touch on the issue of cross-cultural variability, as that case did; i.e., that systems of erotic and bodily meanings, and the family organizations upon which they draw, need not be confined either within the unconscious or by a single intergenerational, nucleated model of desire (see Obeyesekera 1990). From an ethnographic point of view, the value of the questions raised by the Oedipal theory and its varieties must be the better sense they can help to make of particular cases (Epstein 1979: 181; Paul 1989: 189; see also P. Riesman 1986). Of the many possible questions that might be asked about the *Kakar* cult, one of the most obvious is why the cosmic rival turns out to be a metaphoric "elder brother" instead of a "father" (see Herskovits and Herskovits 1958).

Recall the matricentric Murik family in which even the dwelling house itself is mystically identified with the mother. And recall how the image of an indulgent, superabundant, nurturant mother, with a gaggle of dependent children clinging to her skirt, towers symbolically over the moral order, like a lighthouse. Now I would add the following premodern data: fathers avoided their firstborn sons and were meant to observe a postpartum sex taboo for more or less two years following each birth (until the child could walk and begin to talk) and mother and infant would sleep together during this time. At the birth of the second child, the father became no less of a maternal figure of abundant nurture. The event, in short, which spelled the end of a child's monopoly on his mother's love was not observing the primal scene, as Oedipal orthodoxy would insist; for in Murik, little sexuality takes place at home. (Love is not made in the dwelling, but usually seeks the privacy of a secluded spot near, or on, the beach.) What therefore dethroned a firstborn son from his mother's attentions was not the return of his father but the arrival of a younger sibling. The firstborn was displaced from the love and care of his mother just about the moment when his attachment to her was becoming charged with his phallic identity, that is, around the age of three. Worse: with the appearance of a younger brother or sister, the firstborn had to begin the extremely difficult process of learning the sacrificial, maternal role expected of him as elder brother (see Barlow 1990 and Chapter 3). The shifting, or compressing, of intergenerational conflict over the love of the mother-wife onto an intragenerational axis in the *Kakar* cult makes eminent developmental sense. In Murik, it was and remains the cradle of siblingship, rather than the return and intervention of the father, which disrupts maternal attachment, and creates conflict as well as a sense of vulnerability in the firstborn.

According to psychoanalytic theory, the unconscious is a repository for volatile desire and jealous rage, emotions which are governed by conscience and repressive institutions, like male initiation rites. Alternatively, Bettelheim (1954) argued that the unconscious is an agency seeking a moral position for the self in society. In particular, young men attempt to control gender differences through symbols of fertility and other ritual acts which afford them mastery of their ambivalence about and envy of women's bodies. A second pschodynamic question to ask about the *Kakar* cult is then to what extent it repressed desire or imparted sex-role mastery? As a process which retrains and rechannels male sexual attachment through a painful socialization to achieve emotional autonomy from women, it would seem to have served both punitive and instructive ends. While being beaten with the tail of an eel and learning urethral bleeding were painful experiences, as a whole, initiation was not meant to be punitive but instructive and empowering. The *Kakar* cult sought to instill emotional mastery over sexual dependency upon the "wife-mother."

But by the 1980s, if not earlier, ritual heterosexuality – as symbol and in practice – had fallen into disfavor and disuse. Men continued to recruit their sisters' sons into the rival moieties. The "two brothers fought with plates" as they built new cult houses or outrigger canoes and they continued to sit around their designated firepits in the halls. But money replaced sexual intercourse "to pay for" the transfer of cosmic authority.[18] Men no longer learned to master Oedipal jealousy but continued to view acts of cultic nurture as metaphoric "weapons." Why had food persisted as an instrument of cosmic violence, but not sexuality? Did this result from sociological change? Or just lack of colonial oppression? Certainly, it had not been caused by any reduction, much less the elimination, of sexual jealousy in the society at large. With the end of sister-exchange, rights to a sister's body – in her sexuality – were no longer vested in the sibling group, but, meanwhile, the role of the "wife-mother" remained unaltered. A woman's nurture was still construed to be available to her children, her husband and her brother on demand. The meaning of her nurture was still organized in terms of the maternal schema. No brother would ever think to dispute a husband's right to beat his sister for failing to feed him properly. Had early childhood changed? Would any self-respecting Murik mother think to style her nurture in any other but indulgent terms? Would any mother expect children to practice anything but the "giving in," or sacrificial, model, of sibling authority? No (see Barlow 1985b). Pre-Oedipal issues, the initial differentiation of self from other, the denial of attachment and dependency, oral satisfaction, the very salience of food in the maternal and sibling domains, remained enormously important in the cultural construction of early childhood. By contrast, the significance of Oedipal issues, establishing the boundaries of the self, libidinous claims in feminine sexuality and sex-role identification, had become, at least in a social structural sense, less significant. However severed from ritual love and the difficult emotions it aroused, and however removed the *Kakar* spirits had become from their former station as an extrasocial conduit for these emotions, Murik society remained rent by the strange, unlimited range of motives upon which sexual jealousy may feed (Simmel 1955). Cultic seductions might have ended; the problems they sought to distill and resolve – namely, Oedipal desire and jealousy – certainly had not. It should therefore come as no surprise that the final two chapters of this book will concern the lived relationship of sexual jealousy and the male cult to the reproduction of collective order.

PART III

DIALOGICS OF THE MATERNAL SCHEMA IN SOCIAL CONTROL

Plate 26: Bearing heraldry, Mendam youths make peace with Darapap villagers (1982)

8

Conflict and the reproduction of society

While Barlow and I were stuck in Wewak town, waiting for the monsoon winds to subside during the first months of 1981, the kind of escalating pattern of conflict which Bateson called "symmetrical schismogenesis" (1936: 177) was taking place between the villages of Darapap and Mendam (see Chapters 1 and 2). By the time we finally managed to arrange for transport, the first indication that there was a moral crisis going on among these two eastern Murik communities presented itself the moment we reached Darapap late in the afternoon of March 2.

A long train of youths led us across a slick, muddy tide-flat to the empty house where we were to stay. On the way, a middle-aged man rushed up and enthusiastically introduced himself as Kaibong. We could not fail to notice the bloody, still coagulating, gash squarely in the middle of his forehead, just above his eyes.

A few hours later, the story of this man's injury began to come out. Luke, then a man of about thirty, came and told us about what had happened. There had been four fistfights. The first had broken out during a soccer match between the respective Sports Clubs of Darapap and Mendam. Kaibong had been hit from behind with a spark plug. But, Luke assured Barlow and me, Darapap people were not dangerous "like Highlanders." We should not worry about our personal safety as we came and went in the village.

Five or six teenage boys, all of whom had participated in the brawls, came to greet us a few days afterwards. Since the fourth fracas, they had been staying in the *haus mangro*, the bivouacs dotting the inland edges of the lakes, to which families repair for weeks or months at a time while fishing. Village life was too quiet and dull, they complained, so they had gone off to be by themselves. But as their conversation turned to the events of recent weeks, the youths became animated. Summarizing key episodes, they ventured nothing about their contribution to its causes, except to say that after the first fight, the Mendam had taunted them about "having rounded up the pigs in Darapap," and they had still been smarting over this insult, when, a week or so later, two Mendam youths came to Darapap to deliver a ceremonial invitation to Ginau, the *wajak* partner of the couriers' grandfather. In retaliation, our extremely pleasant and poised guests had beat them up. The following Sunday, a big group of Mendam men then came to Darapap and a brutal, hand-to-hand fight broke out between nearly everyone in the two villages.

While the youths were still feeling very angry, village elders were assembling day and night in the male cult houses to debate a proposal advanced by Marabo Game, one of their number, that they meet with their Mendam counterparts in Karau, the intermediate village (see Map 2), to negotiate a truce. The location made sense to them. Not having been party to the conflict, the male house in Karau was thought to be an appropriate site to begin talking about a settlement process. What was more, according to the refugee ethnohistory to which both feuding villages adhered, Karau was their original settlement, and evidence of this former unity – *Kakar* spears – were still kept in storage in the cult house there. The venue apparently appealed to the Mendam as well and I was given to understand that the meeting was imminent. I asked to be permitted to come and watch it. Joel Gobare, a middle-aged man who was then the local government councilor representing the two villages of Darapap and Karau,[1] conveyed my request to the elders, who raised no objection but seemed flattered.

This chapter is an account of the debate into which we were plunged, a debate which was about how, during an early moment in the postcolonial history of Papua New Guinea, the relationship between conflict and moral order might be understood and represented. It analyzes an intricate process of events, during which negotiations for the reinvention of a rite of reconciliation between these two feuding villages were set in train and brought to a conclusion. The process itself, as well as the semiotics of its conclusion, I shall argue, recreated the dialogue between the maternal schema and its interior nemesis, the uterine and sexual body of woman.

"One law, not two"

In the early morning of our second day in Darapap village, Barlow and I sat talking about the brawls with one of Luke's wives. Marabo Game suddenly burst into the house. "Never mind your smalltalk!" he ordered me. "Stop what you are doing! We are going to Karau!" A few minutes later I found myself among a group of about twelve men crossing the broad channel between Darapap and Karau in a small flotilla of canoes. After paddling about fifteen minutes, me clutching desperately at the gunwales, we pulled up our vessels onto the shore, and walked down the beach for another half-mile into Karau. Once there, each man immediately repaired to kin for a morning meal of sago breads and smoked fish before finding his way to the small male cult hall located more or less in the center of the community. The men of Darapap and Karau then proceeded to lie about for several hours waiting for the arrival of their Mendam counterparts. The Spirits of Wealth, mothers' brothers and sisters' sons, abused and mocked each other (the scene is described at the beginning of Chapter 1) until, much later in the morning, the Mendam finally arrived. In all, about twenty-five men, coming from each of the three eastern Murik villages, had assembled.

Standing underneath the loft in which the *Kakar* spears were stored, Marabo convened the meeting. Proposing to institute a settlement process which, he said, was commonly used during the "era of our fathers," he spoke alternately in Murik and Tokpisin. Marabo recalled how peace used to be made between enemies who had been at war. A taboo would be established which would be ended by reciprocal pig feasts, the first of which would be sponsored by the assailant. The embargo phase of the process, which he called *a'iinaro* (literally, "there is talk"), meant that kin or affines were not to visit [to trade with] each other, but might go on trading sago, betel nuts or coconuts, meeting in neutral Karau village, or in canoes anywhere else in the lakes.

The audience clucked and shouted even as he spoke: "*Ariito!* (Murik: Good!); *Stret!* (Tokpisin: Correct!); *Ehhh!*; *Gutpela!* (Tokpisin: Right!). Marabo invited the Mendam to come to Darapap village for the first feast bearing their heraldry as they came ashore, a display which would conclude the state of war between them. The Mendam would be served a pig feast and individuals might compensate one another.

But as Marabo went on, his invitation, and the customary remedy he sought to define, seemed to take on the trappings of an opportunistic strategy rather than an admission of culpability or an unbiased attempt to restore order. Attention to him became divided. Four men sat off to one side of the rectangular hall, facing each other, talking quietly among themselves, while another pair sat opposite the first group conducting their own separate conversation. There was an outburst of slapstick among the Spirits of Wealth. Having made (and seemingly won) his main point, Marabo broached a different topic of mutual concern.

"Karau must stand between Mendam and Darapap," he yelled out, "by offering something everyone might share and which the fathers used to distribute to make peace: tobacco!" As the hosts left the hall to retrieve it, one of them mumbled that he wanted payment for such a gift.

During speeches which followed, the volume and tempo of mirth continued to escalate. At one point, Saub Sana, one of the most senior men in Karau, stood up and began quite histrionically to speak in the Manam Island vernacular, in which only he was fluent. Sitting behind him, Marabo jabbed rolled-up newspaper into the crotch of his *mwara* sister's son. Caught off guard, the man lurched forward and jumped up. Suddenly, several big, conical sheaths of tobacco thudded onto the floor in the middle of the hall. Marabo quickly picked up the bound tobacco. Thanking the Karau elders for hosting the meeting, he handed it on to Tamau, his younger brother, to divide up and distribute. Marabo then sat down, boasting to me that because he was a firstborn *sumon goan*, his younger brother was obligated to help him.

One by one, elders stood to talk about past incidents of the settlement process Marabo had proposed. I had the impression that only the most senior of them had actually observed such an event at first hand. A leading senior man from Karau summed up their consensus: "Now in Papua New Guinea," he said, "there are not two types of law, of the state and of custom, just one!" The remark, which hailed custom, was spontaneously applauded by everyone in the hall. Darai, another Karau elder, quickly picked up the point: "In the old days, the 'money' of the ancestors would have been exchanged to make peace: boars' tusks and dogs' teeth ornaments. Now we can still do something like this. We do not need the [village] councilor or the ADO [Administrative District Officer]. We can settle it ourselves!" But then another man, who was from Mendam, stood up, smiled in a peculiar way, and demanded that the proposed settlement feasts take priority over every other one that was pending between the two villages. He was referring to Ginau's succession feast (see Chapter 5) which then seemed imminent and was supposed to involve exchanges between Spirits of Wealth from both villages.

The speechmaking subsided. Then Kombek, a very elderly Karau man, picked himself up off the floor and demanded that the drunkards who had started the fight be thrown out of Mendam. He ambled across the hall in search of a piece of string with which to tie up the leaves of tobacco he had received. Passing by a *mwara* sister's son seated below him, Kombek lifted his leg over the younger man's head. The man winced. A wave of laughter broke over the hall as plates of seafood and sago jelly began to arrive carried by junior men. Little groups formed around the plates and everyone began to eat. As another one of Marabo's younger brothers bent over to pick up a ceramic bowl of flower cakes to pass out, his sister's son goosed him in the buttocks.

About three hours had passed. The Darapap got ready to walk back to their canoes before they drifted off with the rising tides. But another course of food suddenly appeared and everyone reseated themselves for a final repast of various kinds of seafood over sago jelly, or a bed of rice.

What face did mediation take during this meeting? In whom was it located? What was its ethos? Set in a sacralized, distinctively masculine space, in which the outbreak of conflict was supernaturally sanctioned and penalized by expensive fines, the meeting was convened by the gracious nurture of a neutral third party, the Karau male cult, whose strategy of conciliation was that of a host, dispensing seductive signs (tobacco and food) before and during it, but otherwise taking little active part. The remark made by one Karau elder about the autonomy of customary law was no more than a summary comment. Of the two other hosts who spoke, one did so in a foreign language and the other's comment was taken with bemusement. The sequence of discussion was punctuated not by the persuasions, or exhortation, of any individual but was dispersed by the comedy of the Spirits of Wealth, whose horseplay began and concluded it. The settlement was negotiated bilaterally by the visiting dignitaries, who were themselves the disputing parties. Responsibility for the meeting, in this sense, belonged not to the Karau male cult, who hosted it, but directly to Darapap and Mendam leaders, the antagonists.

Thus Marabo Game, the initial mediator, was no discrete, neutral authority, who was authorized to compel a settlement by withholding or controlling resources. He was rather responding to the willingness of both disputing parties to negotiate. Since no adjudicatory third party, either legal or divinatory, had appeared, questions of cause or liability were hardly broached (Koch 1979). Instead, the goal of the meeting was to create a settlement procedure. The folk model of conflict and conflict resolution upon which it drew seemed tripartite: (1) the state of war was to be (2) suspended in favor of a rule of avoidance between the two villages that would be (3) concluded by public ceremonial exchanges which would immediately normalize relations between them. Support for this understanding was couched in a rhetoric of local autonomy, historical precedent, the memory culture of both communities, as well as in the gifts provided by the host Karau, who took no sort of superordinate position. At best, the consensus seemed unstable. Demanding that no exceptions to the intervillage taboo be tolerated, a single, contrary voice was heard, the implication being that the settlement procedure would be no less subject to competitive pressures than any other ceremonial process. If the agenda was not unassailable, however many precedents for it might be recalled, then might not every step of the process have to be renegotiated, including its sponsorship?

Liability, the self and money lost

Marabo Game's plan immediately began to unravel. Since neither Darapap nor Mendam was willing to accept primary liability, the idea was floated that both villages compete to invite the other to the first feast. Alternatively, went a second suggestion, the male cults of the two villages might meet at a neutral site, play their flute spirits, and share a feast. Within Darapap village, meanwhile, Marabo and Tamau, his younger brother, convened a meeting in the male cult house to announce that they would collect K90 (US$120) and use the money to purchase a pig from one of their trading partners whom they had not yet selected or contacted. Tamau took it upon himself to take responsibility for this work on behalf of his elder brother. He had just returned from meeting an inland trading partner in Boig and had betel nuts to distribute. Like tobacco, giving betel nuts is understood as a moral gesture which draws people together to air and resolve grievances.

Tamau placed a thin sheet of bark on the floor. The bark was to be dried, beaten soft and worn as a new loincloth during the settlement feast by Malai, the Darapap youth who had been involved in the dispute which had originally caused the brawls.

Young men began to laugh and tease each other until Mange, a bony, very slight, elder, yelled to them to come to attention since they, and not the senior men, were at fault. James Kaparo, a leading middle-aged man in the community, immediately put K10 down on the bark and commented that everyone ought to take responsibility for the fighting. Bate gave each of his two sons K5 (US$7.50) to donate. A younger man, who had been writing down the donors, read off the names on his list, and the men talked about how the brawling had escalated. Tamau wondered to which village he might go for the pig and mentioned trading partners in Singarin, on the lower Sepik River, and in Boig, across the lakes. The mood of the meeting lightened up considerably after James Kaparo, who had charge of the money, folded it up, and reminded everyone that many young men had not yet contributed. By its end, however, Marabo was complaining that while village elders were "working hard" for their children, Ginau was doing little in town but sending messages behind their backs about his own feast (see Chapter 5).

In this and subsequent meetings, two related, but contrary, notions of agency were being contested (see White and Watson-Gegeo 1990: 7). The first was centripetal, Marabo and Tamau's claim upon the sponsorship of the reconciliation rite. The second, by contrast, dispersed the obligation to make redress for the injuries caused during the conflict throughout the community (Allott, Epstein and Gluckman 1969). Even though pigs were largely bought for money during the 1980s, rights to buy them from specific trading partners remain a hereditary privilege. The identity of the donor/vendor was therefore critical in determining heraldic sponsorship of the feast. But while Marabo and Tamau wanted to take the credit, they were not prepared to foot the entire bill. The two men therefore faced the problem of casting the net of liability more widely than might otherwise have been expected. There is no village-wide concept of liability in Murik culture. Liability is vested in the self, the sibling group and its heraldic extensions, one's age-grade, to a certain extent, and finally in the cosmic body of the male cult, but not one of these categories is coterminous with an entire community. No individual person is obligated to help compensate a victim on the basis of simply being coresident in a culprit's village. The campaign led by the two brothers, Marabo and Tamau, to solicit money to buy a pig had to tread a line between attributing individual liability, e.g., the list they kept of contributors' names and their claim to sponsorship of the reconciliation rite, and collective liability, in which the inclusiveness of the conflict was emphasized and responsibility for it was not located in any one man. While the image of the bark lying on the floor iconically referenced a specific youth as having been party to the initial dispute, Tamau's speech made no mention of the youth's specific role in starting the conflict.

A second meeting was convened in the Darapap cult house one week later. The previously donated money was placed in the middle of the assembly. Tamau made a long speech summarizing the four brawls.

Discussion turned to other subjects: a photograph of a boat circulated. The village fishing cooperative was hoping to buy the vessel and there was talk about how much money had been collected for this boat fund. Daniel Wambu, a middle-aged man who had been involved in the fighting, complained that giving money to buy the boat would bring future happiness to the village, while the money given to pay for the Mendam pig was a waste. Other senior men were also bothered by this.

Two dishes of fish and sago pudding, brought by the President of the

Sports Club, Reuben Wapo, were set down next to the pig money. Daniel Wambu told me that Reuben's mother, a Mendam woman, had sent the food because she felt shame about the last brawl. Her husband had actually fought one of her brothers. She wanted the settlement process to go forward on account of this and because the taboo was restricting access to her family's sago stands adjacent to Mendam village.

Only a few old men moved to the middle of the hall to have something to eat: most went on sleeping. The young talked among themselves. Despite the annoyance of some at wasting money, consensus remained strong about proceeding with the settlement rite. Several middle-aged men were secretly planning to start a fund to purchase a second pig and sponsor a feast for the elders in gratitude for their "help." At that time, I assumed that the "help" for which they felt obliged was in organizing the reconciliation rite.

The elders' attempt to define "the facts" of the case and create a sentiment of collective liability had two effects. On the one hand, it was successful. The money (K90) was collected and a second fund was established. But, on the other, elders were blaming youths for having caused the conflict which was diverting resources from the village fishing cooperative. Evidently, the reconciliation process would not only require a narrowing of differences between the feuding parties, but a renewal of solidary relations within the village. The latter tensions aside, relations cross-cutting the two communities (Colson 1953), their common ethnohistoric ties, and the network of affinal relations motivated countervailing, integrative voices and actions. One meeting was catered by a Mendam woman married and living virilocally in Darapap. And even during the brawls themselves, there had been an attempt to organize the fighting into individual grudge matches (like Sendam, the "father" of warfare). The cultural "distance" between the two villages was minimal. Their differences were nonetheless not so tractable.

The whole village of Darapap slowly assembled, roughly by age and sex, around the dirt surface of the basketball court, for its routine Monday morning councilor's meeting four days after the meeting described above. Word went around the gathering that, the night before, Lester Murakau had been heckled by Mendam youths as he paddled by on his way home from Kopar, a village located at the mouth of the Sepik, where he had gone to trade for sago. "Woman!" they shouted. As Lester told his story to his peers and his father recounted it to adult men, the women of Darapap, who normally keep a modest distance from men in public, lest one of them pollute an elder with menstrual blood, crossed the basketball court to listen. The story visibly agitated Lester Murakau's generation, the adolescent bachelors and newly married men. While waiting for the councilor to start the meeting, the young men passed time in rowdy horseplay, slapping, shoving and jostling each other, the restlessness of which was virtually angry.

Once begun, the day's speeches focused on chronic problems, paying school fees and organizing the repair of an outrigger canoe for the Darapap Fishing Company. Only toward the end of the meeting did the councilor return to the Mendam situation to say that he felt it was too soon to respond to the attack on Lester. Yawi, a senior woman, then strode into the middle of the court. Picking up this point, she angrily denounced the youths who wanted to resume the fight. As she went on, men began to drift off to help move the canoe hull to a shady place where its leaks could be caulked.

At a subsequent meeting two weeks later, the councilor, who had returned from a district-wide meeting, reported about various national and provincial matters. The most ludicrous plan, which he credited to the Lower Sepik MP, was to construct a vehicular bridge over the Sepik River at Angoram even though there were no roads on the other side. Turning to more local issues, he criticized the youths for playing guitars and cards, rather than cooperating with their elders on a housebuilding project.

Marabo then worked himself into an absolute frenzy. He had just spent an entire week sitting in the cult house, he screamed, waiting to hear from the young men. "But no one showed up. I settled the fighting with Mendam because I felt disgraced. We agreed to taboo going back and forth to Mendam. Now the young men must meet and decide what they are going to do."

Holding a toddler in his arms, Jakai Smith stepped forward to respond to the charges made by his "father" (father's elder brother). The young men had met secretly, he announced, and had agreed to buy a second pig for the elders to thank them for their "help." The money for the first pig, which James Kaparo was holding, had been collected and donations for the second one were now being accepted. "The talk [among us] is not dead," Jakai Smith assured the elders. "We are just speaking quietly."

Although I have documented that women and men possess equal ceremonial rights in Murik, no doubt hearing the strident voice, indeed any voice, of a woman speaking in public presents a striking contrast to the greater part of village Melanesia (cf. Kulick 1992: 104; Alder 1922: 129, 148ff). However, the political status of Murik women is only of passing importance to my present concerns. What I want to point out is a contrast between the ethos permitted in meetings held inside houses and that in those held outside. The issues broached – organizing public works, condemning the immorality of youth, getting on with the reconciliation feast – were often the same. But in a total of twenty-one months spent in Murik villages, I rarely saw comparable outbursts either inside a dwelling or in the male cult house. Being noncommensal, as well as being set in a neutered space, meetings convened by the state are denuded of compelling signs of "maternal" morality and tend to resort to genres of coercive oratory, which is to say, to aggressive displays of self-assertion. The interior of architectural space is thus meant to remain pure: violence is tabooed from its heraldic premises. The state is confined to amoral spaces in the village, where its law must rely upon raw force rather than cooked, or catered, symbols of morality. Of course, during this early postcolonial period, when state power was being dispersed, and grandiose development projects captivated national leadership, no squad of police was readily available to impose order on two remote, rural villages.

The truce between Darapap and Mendam sucessfully forestalled further outbreaks of violence. Adversaries averted eyes while paddling past each other in the lakes; or turned away as they passed by the rival village. The first six weeks of claims to sponsor the settlement feast had yielded nothing, however; then the following two incidents took place.

James Kaparo (see Plate 27) lent his outboard motor to Tamau, his mother's brother, who wanted to go visit one of his sons, a schoolteacher in Bogia, to get money for the settlement feast. But while he was crossing the mouth of the Sepik, the motor exploded. Tamau sent a message back to the village that he was going to have to wait to have it fixed while his son saved the money to pay for the repairs.

James Kaparo decided to take the pig money to town to put it in the bank for safekeeping. While waiting for a bus he put his netbag and the bank passbook with the money in it down on the bench next to where he sat. Although he picked up the net bag when he boarded the bus, he somehow missed the passbook and the money. By the time he realized what had happened, and had frantically returned to the bus stop, of course the money had disappeared. Kaparo returned to Darapap openly penitent, promising to pay for the pig himself.

The two brothers who initially advocated instituting the ritual procedure and collected money for its requisite pig had now both quit the village. Tamau, the younger brother, was away on what became an extended visit with kin. Marabo, the elder brother, also left to care for a tubercular daughter who needed treatment at Boram Hospital in Wewak town.

Plate 27: James Kaparo of Darapap

Worse, the pig fund had now vanished. The loss of the outboard motor and the money had also brought to a halt a collection for a second pig feast which the two younger leaders, James Kaparo and Jakai Smith, who were both firstborn heirs to two important *sumon*ships, had wanted to offer their "fathers." The violence between Mendam and Darapap, in short, had provoked a host of ritual ambitions. No one seriously suggested "lumping it" (Nader 1980). No one argued that no action be taken. No one denied that some form of ceremonial exchange should be mounted to resolve the conflict. No such challenge was raised, for example, by the Seventh Day Adventists in Darapap who regularly objected to such performances as "Satanic." The negotiation of the reconciliation process did not concern the value of reinventing tradition at this point. There was consensus that a ritual mode of redress should serve as a remedy. The negotiations in Darapap were political: who would get the pig to underwrite the rite?

A narrowing of differences
Several months followed during which no further attempts were made to take control of the settlement process. This lapse eventuated in fear. Murik pessimists will always insist that the inevitable outcome of neglecting to fulfill redressive obligations is not simply "getting away with" or "lumping" it, but accident, illness or death. Grievances harbored in secret and left unsettled they expect will fester into ailments or misfortune because antagonisms must give rise to mystical revenge (that is, through sorcery). Any number of victims in the households of the collateral siblings of disputants may suddenly become ill. Not until the indemnity, the provision of a feast, or an exchange of betel nuts or tobacco, has taken place, can the parties and their commensal kin reliably stop worrying about magical retaliation.

Late one night, about six months after our arrival in Darapap, we heard a faint slit-drum beat (*debun*) coming from the eastern lakes. It was a *pre-debun*, people told us next day, the *debun* signaling the death of a Murik warrior who had fallen in battle. News trickled into the village that Debwase, the local government councilor in Mendam, had been killed by a wild pig he was hunting.

After seeing his wife off to market in town, it was said, Debwase had escorted two Kaduar Island trading partners to the beach, where he helped them launch their outriggers. Accompanied by two classificatory "sons," he then set out to cut down some betel nuts. The three of them walked in the bush for a few minutes and came to a clearing where they heard the loud grunts of a pig. The two boys fled to a safe distance, but

Debwase, armed only with his hatchet, stood his ground, as if immobilized. The pig charged him, and gouged out his groin and stomach. Debwase did not try to protect himself with his hatchet. The two youths returned to the clearing and saw the animal standing over their "father." Before disappearing into the bush, the pig grunted again and dragged a hoof across the ground. The boys pulled Debwase, still alive, back to the beach. He looked up at them, laughed in a peculiar way, and just before he died declared, "Now I am leaving you: I grieve for my 'mother' and her children."

The circumstances of the death raised many questions in Darapap: why had Debwase failed to fight back? What was the significance of the way the pig dragged his foot over the ground before trotting off? What was the meaning of Debwase's strange laugh as he died? Even though he was a well-liked man and a state official, no one doubted that the oddness of the death could only be explained by a single cause: retributive sorcery (*timiit*). If the overt purpose of the review of "the evidence" which went on in Darapap for days after the incident was to establish a motive and identify not only the specific spell, but the sorcerer and the victim's adversary, the man who had "hired" the sorcerer, an entailed purpose was to deflect culpability from Darapap villagers. If no one of them was guilty of murdering the Mendam councilor, then differences between the two feuding villages might be narrowing. People talked about where the attack took place, recent tensions in the dead man's relationships, the dishes from which he had recently eaten and, of course, the women in his life.

Some senior men immediately realized that the locale was the very area where Sendam, the spirit-man, who was the son of a wild pig mother, had killed the predatory sea eagle spirits in their treetop nest (see Chapter 7). No doubt, ran one train of speculation, some food scrap stolen from Debwase's plate had been bespelled while in the mouth of a wooden pig figurine, representing Sendam's mother. Others added that this land was also the subject of an old land dispute between Mendam and Singarin, a sago-supplying village on the lower river. Just a few days before his death, the two councilors of Darapap and Mendam had stopped off in Singarin on their way downriver from a district-wide meeting in Angoram. They had a meal there, Joel Gobare, the Darapap councilor reported, of which Debwase was openly suspicious.

After "the weapon" had been identified, the first allegation situated the sorcerer as hailing from outside the Murik Lakes. But neither the motive nor the assailant was immediately clear. One train of speculation was that a Mendam man, a man, that is to say, from the victim's home village, was the culprit. Debwase was not killed by a cuckolded husband, in this view, but nearly so. He had recently accused and humiliated the man before a big, village-wide meeting for seducing another man's wife and committing perversions upon her. The councilor had denounced the man for exposing his lover's genitals during their lovemaking. The accused had been reduced to speechlessness. Debwase, in other words, had been mystically killed for trying, as a representative of the state, to intervene in a case of adultery. The strange death, in short, was the result of a problem for which no one in Darapap was responsible.

Asking no one's permission, five Mendam women who had married and come to live in Darapap, wrapped towels turban-style about their heads to show their defilement (*mwak*) and paddled off to mourn for their deceased "brother." The next day, their husbands, despite having fought in each of the brawls, followed suit. For the first time in six months, the truce between the two feuding villages had been breached.

Nangumwa, a senior Mendam *sumon*-bearer, watched the arrival of the women from Darapap through one of the wind-doors in the male cult house. Seeing the towels around their heads, he told me later, made him feel remorse about the unresolved situation. "'These were our sisters,' I realized when I saw these women come ashore. I felt that we [in Darapap and Mendam] were all 'brothers.' The fight had to end."

Next day, he sent a gift of betel nuts back with the returning husbands to give to Sauma, his *wajak* feasting partner in Darapap, to renew the settlement process. Amid a village-wide outpouring of sympathy, Darapap elders debated what would be the most appropriate response to Nangumwa's gift and the councilor's death. Several men vowed to sponsor the reconciliation feast quickly.

"Opposition," according to William Blake, "is true friendship" (1975: 20). With quite a bit less parsimony and elegance, but perhaps more theoretical precision, functionalist sociologists (e.g., Simmel 1955; Coser 1956), have also argued that opposition, or just the evocation of opposition, can intensify political cohesion of a potentially cohesive group by concentrating authority within it, the result of which may yield concerted action. With the appearance of an outside threat, the solidarity of the two feuding Murik communities did indeed increase. The image of the "sister-wives" returning home created a moral consensus between them. The women had stepped into a conflict, ongoing as it was. Functionalist explanation obviously cannot explicate the logic of cultural meanings in terms of which particular peoples constitute their moral orders in the face of outside threats. It cannot explain why this image of women was effective. Why, to ask the more complete question, had the appearance of the five women invigorated a moral ethos between Darapap and Mendam while the image of the exposed genitals of a woman had subverted it? "Women are like *sumon*," a man once commented to me. "They are our 'queens.'" He was expressing the great value Murik men afford women. The major point in my analysis, however, has been to pursue just the reverse direction of this analogy. The *sumon* insignia, I have argued, are like women. In their association with the quietism of the maternal schema, the insignia and their titleholders are understood to possess the moral authority to intervene and stop conflicts which Oedipal jealousy over actual women creates in society. The repressed, hollow quality of the body in the maternal schema is dialogically opposed to the internal passions it seeks to suppress.

Conflict remembered

Consensus enlarged between the two villages. Within each of them, differences about the meaning of the brawls – the question of liability – narrowed. Amid the ethos of fear (cf. Schwartz 1973) which followed the councilor's death, every generation of Darapap men suddenly became more than willing to detail the sordid roles they had played in the brawls to me.

Malai and his age-mate, Beldon Gabir, who were then about twenty years old, agreed that the original cause of the conflict was a torn photograph. Malai explained that just after he married Lady, a woman from Karau, her ex-husband, William Aumbe, demanded that she return the picture of himself. She did so, but one corner of it was slightly torn. Malai then began to hear rumors that William Aumbe wanted to fight him for having deliberately damaged the photograph. Malai said, "I went to [him] . . . and told him that I had not done anything to his photograph and that I was not angry at him. But he was angry because I had taken Lady [from him] and married her." My two young informants also agreed that when the Sports Clubs of Mendam and Darapap met in Karau to play soccer, William Aumbe had tried to inflame tempers between the players to get revenge.

The Murik ethnotheory of conflict (Avruch and Black 1993: 132) is understood in terms of Oedipal jealousies. From the perspectives of these two youths, indeed, the trouble began in a dispute between a cuckolded husband, his wife and her second husband. But instead of Oedipus, I want to attend to the torn photograph for a moment. A design, an image of contested desire, was here said to have ruptured the moral body. Recall the legend of the "Two Brothers" who came to blows over a magico-religious design tattooed across the vulva and thighs of the elder brother's wife. Their siblingship was subverted by the sexual jealousy that an aesthetic, seductive image had provoked. Although they had updated it so well, when talking about how the brawls started, my two informants never explicitly compared the photograph to this legendary prototype. They were concerned with the cosmic powers which were implemented on their behalf during the fighting.

A series of soccer matches, sponsored by the Sports Clubs of Darapap and Mendam were played on a pitch at Karau, the village located halfway between them. The first one went off without incident. During the second game, which was held between younger adolescent boys, Barake, one of William Aumbe's friends from Mendam, got quite drunk on methylated spirits and began pulling up the goal posts to make a nuisance of himself. Komet, a young man from Darapap who had been refereeing the game, told Barake to stop interfering with play. Barake insulted him and a fist-fight broke out between the two. The head of the Darapap Sports Club quickly pulled his side off the field and led everyone walking down the beach toward their canoes. The Mendam gave chase and beat up several youths. "Not many of our youths," Malai swore to me in utter seriousness, "had fought before."

Meanwhile, Joel Gobare, the councilor of Darapap and Karau, was sitting in the cult house in Karau, talking about tobacco. When word reached him of the fighting, he and Debwase, the Mendam councilor who was later killed, immediately made their way out to the beach to try to settle things down and get an explanation of what had happened. After the story came out, Joel Gobare fined Barake K4 (US$5), which he apparently paid on the spot. Barake and Komet shook hands and everyone went home.

The Darapap youths crossed the channel in their canoes. As they reached the foreshore of the village, slit-drums resounding from the cult house called them to a meeting in the hall. "The elders," Malai remembered, "were ashamed of us for running away from an attack. Marabo, Tamau and Murakau explained that in the past, no one could possibly have defeated Darapap. We had acted like cowards, they said. They were deeply let down and we felt ashamed."

Malai's point about the youth of Darapap not having "fought before" was not lost on the senior generation of men in the village (the same elders who would later take charge of organizing the settlement rite). At this moment in the conflict-process, however, their disappointment with the youths was not for having broken the peace but for having lost the fight. The youths' "cowardice" had exposed the elders' negligence in not having initiated them into the war cult and thus exposed them to the mystical sanctions of the *brag* spirit-men for having failed to do so. From the elders' point of view, not only were the honor and the reproduction of the cult at stake, but so was their very well-being. The problem created for them by the first brawl did not have its roots only in sexual jealousy, but in gaps in the sequence of initiated generations. The compensation exacted by the local representatives of the state, the two village councilors, could do nothing to remedy this condition. James Kaparo, a middle-aged man, discussed further aspects of the background of the conflict with me, which I summarize below.

A few weeks before the soccer-game melee, two groups of men from Darapap and Karau had gone to Mendam to attend a launching rite for a new outrigger canoe. Ginau had been invited to serve as *wajak* partner and several of his younger brothers, sons and daughters went along to help him. They personified Kaideban, the war spirit of their descent group, whose flutes they played before fitting the *sumon* heraldry to the top of the mast pole of the canoe which was being consecrated. Ginau himself put his Yamdar *sumon* basket on the canoe platform. For this "work," the *wajak* group received a pig feast and a big basket of foods with a porkbelly set atop it to take home.

One week later, the men of Mendam held a *baas* (banquet) in their cult house so their war spirits could celebrate the conclusion of another phase of outrigger consecration. Apparently, heavy drinking went on and several fistfights broke out while heraldry was *still in place* on the mast pole.

The Mendam visited Darapap on board the new outrigger a few weeks later as part of a final phase of the launching rite in which the canoe is taken on a first tour of Murik villages, and were invited to sit down to a meal in the cult house.

Ginau had heard about the fighting in Mendam. When debate began, he asked the guests whether they had repaired their *sumon* emblem. The Mendam at first denied that the *sumon* had been on display during the fracas. But Ginau insisted that "the fathers" had never put away their heraldry until a new outrigger had completed its tour of the Murik villages in which the group who owned the canoe resided. That tour was obviously still going on. The Mendam left Darapap in disgrace, acknowledging that they would have to mend the broken *sumon* by giving a pig feast to Ginau and the rest of the senior insignia-holders in the descent group in order to complete their launching rite.

In a matter of weeks, Ginau received a message inviting him to come to the *sumon* restoration feast. The invitation was brought to Darapap by Barake and Bakur, two young sons of the canoe owner. Still incensed about the first brawl, and the indignities they had suffered, Malai and several friends beat up the two messengers. Ginau refused to attend the restoration feast because of this attack. He sent his heraldic basket with his younger brothers to receive the prestation from his *wajak* partner. He himself left for town to stay with his sister's daughter, which was where we met him in late January 1981.

In addition to Oedipal conflict between youth and the matter of cultic negligence, a ceremonial dispute had been dividing leadership in the two villages. Ginau's demand that the *sumon* insignia be repaired before the end of the canoe rite effectively obstructed the intentions of his rival kinsmen. His subsequent exit from the village arena was motivated by a concern for his emblems. The situation was getting out of control, he felt the violence enclosing him, and did not want to defile the purity of his *sumon* insignia any more than they already were. Such partisan concerns did not impede others, however.

The two Mendam youths made their way home a few days later with the story of the assault. On a Friday, the assistant councilor in Karau, who frequently visited his wife's kin in Mendam, brought word to Darapap that Mendam was planning to come to fight on Sunday, two days hence. On Saturday night, senior men in both villages sequestered the youths for the whole night in the male cult houses. For the first time in their lives, Malai and Beldon Gabir participated in ritual ablutions for warfare.

MALAI: We were not allowed to eat [food prepared] by our wives [or sisters]. Our mothers brought food to our fathers, who brought it up to us in the cult house. We could not eat coconuts or *nimbon* [sago pudding]. We were only allowed to eat fried sago bread and [smoked] fish. Marabo told us about the *Kakar* [spear spirits], and asked them to help us fight. "You great grandfathers cannot sleep," he called out to them. "For tomorrow you must watch over these children while they make a big fight." The elders talked all night but let us go to sleep.

At dawn, they woke us up and told us to go wash in the sea but not to stray in the village on our way to the beach. The cult house "spit" us out. Our wives were forbidden to see us. Only our mothers were allowed to cook the food we ate that morning. Then we waited for the arrival of the Mendam.

BELDON: One thing. That night some fathers individually "gave" their sons [that is, bespelled them with] war magic. They gave them barks to chew and magical lime gourds.

MALAI: Just before we left the cult house, we rubbed ashes from the firepit over our bodies. This [substance] is the power of the *Kakar*. The "fathers" held up bunches of ginger leaves by the ladder which we brushed past as we climbed down. The Mendam expected the worst that day because on the way to Darapap they saw an omen, a crocodile eating a fish, which they understood meant that their defeat was certain.

With the evocation of outside threat, the male cult elders segregated the youths from women (i.e., not only mothers, but also wives and sisters) by secluding them in their sanctum. According to my informants, only post-menopausal mothers were allowed to cook their food. The foods they were permitted to eat – fried sago breads and smoked fish – were "dry" not "wet." Coconuts, betel nuts or sago jelly were forbidden them. The implication would be that their aggression and agency were vulnerable to mystical contact with sexual and menstrual fluids. The youths were there-fore allowed to eat only "dry" foods, either fried or smoked, but not boiled or full of fluid, or foods prepared by "wet," young women in their sexually active, childbearing years (cf. Lewis 1980: 138–40; see also Smith 1994: 103).[2] Only "dry," senior women went on cooking for them. The "dry" bodies of postreproductive women could not contaminate the food they prepared for the men to eat. But even so, the youths were ordered to avoid all women as they walked through the village at dawn to go secure their souls in their bodies by washing in the sea. The "fathers" fed them, bespelled them and gave them magical barks to chew to safeguard them in battle. The warriors rubbed ashes from the *Kakar*-holders' firepit, which was said to contain their power, upon themselves.[3] The councilor of Darapap, Joel Gobare, recalled the fighting.

When "the enemy" arrived, I went to meet them at the lake shore to ask whether they had come under the "rule of law or war." When warriors went on a raid in the era of headhunting, the usual strategy was to travel at night, surround the village and wait until dawn to attack, just as the unsuspecting enemy arose from sleep. The warriors would kill everyone and then stick a palm frond [Murik: *parub*; Tokpisin: *tanget*] into the ground to proclaim the victory. Just as I reached the lakefront, the Mendam threw a palm frond and I understood their intent. I returned to the cult house, called to our youths to come down and march through the village. Fighting broke out immediately and spread to the edges of the mangrove swamps. The Mendam finally retreated to their canoes. In Darapap, we felt that we had knocked them down, kicked them hard and forced them to flee.

The representatives of the state played a moral role during the first inci-dent, arranging as they did a compensation exchange. In this episode, the councilor became a partisan actor who led the youths into "battle." Local divisions, in other words, had completely enveloped the state. Why had an image of a crocodile eating a fish presaged defeat to the Mendam? The "canoe-body" of Amumbera, the paramount *Kakar* spirit, is a crocodile (see Chapter 7). The Mendam cult house had burned down long ago. No new *Kakar* spears had been carved or imported to replace them. But in 1982, this *Kakar* had not yet been stolen, and remained an active presence in Darapap. More: the fish "eaten" by the crocodile was recognized as the "canoe-body" of a war spirit owned by a Mendam woman. The discursive context of intervillage warfare, at this moment, was couched entirely in customary signs and portents: the force of the state, as a discursive genre, had vanished.

MALAI: Just before the fighting ended, the elders left us and returned to the cult house. When we went there [later], we were very surprised. They had not told us what they were thinking or what we should do. They had taken down the *kakar-menung*, the [ceremonial] spears decorated with cassowary feathers they keep hidden in the rafters of the cult house. Each of them had spread his legs apart and was holding two spears across them. We were told to crawl through a tunnel of legs and spears the men had made. As we did, the men bespelled us. Simbua, the last man in the line, beat us and encircled our heads with a *tanget* [cluster of ginger leaves]. The elders puckered their lips and made smacking noises. Simbua held coconut leaves over our heads and only then were we allowed to go back to eating, smoking and drinking in a regular way.

"The spear spirits are meant for fighting and do not come down without reason," the elders told us. "Before you [children] were in your mothers' wombs/stomachs. Now you have been in our wombs/stomachs. You have seen [the sacred power of] these things. In the future, should something like this happen, you will see them again."

The process of turning the warriors' cosmic bodies back into moral bodies was represented by a synthesis of aggression and birthing. The youths left the "wombs/stomachs" (*sar*) of their "fathers" by crawling through men's legs and spears. I am not certain about the meaning of the smacking noises the cult elders made after their "children" crawled through their legs, but perhaps these evoke the suckling sounds of babies. Expressly, they are associated with the spells which mystically turn the "hot," aggressive cosmic bodies of the warriors "cold." Certainly, the ginger leaves (*gorongol* or *kakir*) with which an elder encircled the heads of the youths were meant to do this: they "closed" the "hot" state of aggression, I was told, the condition which had been "opened" as the youths brushed through these leaves as they left the cult house to go to war. Lastly, the coconut leaves, which are otherwise used as property markers, were held over the heads of the youth to conclude the interval of the cult's authority over their bodies. They could then return to their quotidian diets and defiled relationships with the young, fertile mothers who fed them.

The passage back and forth between their cosmic and moral embodiments was made through a grotesque birth image. Momentarily, therefore, the male cult house became the functional equivalent of a birth house. This metamorphosis was rendered in images of digestion/gestation and expectoration/birth that drew life and death together into a single, destructive yet fertile, body. The elders became guardians of their "children." Having been "mothered" so attentively, the youths went on to say, they now felt morally obligated to these men. At least, this was the way they spoke about it to me eight months later. Having been initiated into the war cult (see Chapter 6), they were changed from disrespectful, overindulged youths into powerful warriors. In return for this consideration, the youths spoke of a readiness to be obedient to, and cooperative with, their "fathers/mothers." The maternal schema had been recreated in this grotesque birthing sequence. But not for long. A letter was brought to Darapap: the Mendam wanted to return to the beach at Karau, the site of the original conflict, one week hence and settle things once and for all.

The night before the last brawl, the youths were again secluded in the cult house. At dawn, they paddled across the channel and walked the beach to Karau, where the senior men had them sit down in a row near the breakers. (Women followed later in the morning.) As they waited, Malai and Beldon observed several elders blessing the beach with their spittle. The whole morning passed but no one was permitted to wander off to visit kin and get something to eat. Around midday, the Mendam finally appeared, marching in seven lines along the beach from the east. They were led by their ex-councilor who tried to organize fistfights between single pairs of men with specific grudges to settle. The first man he called was Sol, his own "brother" in Darapap, whom he himself wanted to fight. A riot quickly erupted that was easily the most inclusive and brutal of the conflicts which preceded it.

BELDON: Unlike the other two brawls, this one lasted a long time. Boys, old men and women [fought] too. The power of the cult house was so strong that even our youngest boys felt unafraid. They fought just like big men.

MALAI: Some of our elders fought. But others did not. Mwaima fought. Somebody hit him in the neck with a screwdriver. And Barake cut his ear with a ring from behind. Then [Mwaima's son] James Kaparo took a club and beat up Barake very badly. Kaibong got hit between the eyes with a sparkplug. One of Daniel's son-in-laws hit him in the stomach and he fell down. We carried him to the side where he regained consciousness. Daniel asked us what had happened and when we told him he threw himself back into the fray. Smith, James Kaparo and Kaibong surrounded Daniel's son-in-law and beat him up. Nobody came to help him. He crawled away [on his hands and knees] like a pig! Benjamin fought with the first husband of his wife, the man from whom he had seduced her.

BELDON: When people were fighting during this last round, they were thinking about their past quarrels!

MALAI: If two people had argued before, they looked to fight each other.

BELDON: Us? We were all alright. But them! Plenty of them – blood. They did not use their strongest [magical] power . . . Anenbi [of Darapap] is a very powerful magician! Whenever we go and play soccer or compete in something against another village, we win because of him, because of the power of his magical barks.

There was no residual rage or embarrassment; people spoke euphorically about this last brawl many months later. Mwaima, the arthritic old *sumon goan*, who otherwise rarely left his verandah, told me how he had felt no pain when the riot broke out. The stories of gang attacks on old men, sucker punches with spark plugs, which were recounted with great satisfaction, suggest that, in spite of the best efforts of Sendam, the culture-hero, notions of a self opposed to others, e.g., of a "fair fight" or "individual heroism," were far from the forefront of anyone's idea of bravery. Indeed, the quality most notable to the two twenty-year-olds, Malai and Beldon, was ego-alien, magically induced durability in battle. Except for a handful of men, such as Ginau, the very elderly, and the two village councilors, Debwase and Joel Gobare, all the men in both villages fought. Youth and age joined forces to attack affines. Reports were contradictory about how many women became involved. Malai and Beldon thought that the women from the two villages had fought, while others saw them as merely taunting each other.

But, for my purposes, the critical point was not who fought or how. The point was that the quietist tenets of the maternal schema associated with age, siblingship, affinity and even womanhood were all violated. Instead of becoming increasingly chaotic or random, however, the violence apparently became informed by longstanding Oedipal conflicts. Antagonists who sought each other out tended to be first and second husbands of the same woman with specific grudges to settle. In other words, as the fracas escalated, sexual jealousy appeared to motivate it. Malai went on to describe its end.

If the fight had not been stopped somebody would have been killed. Finally, Joseph Kabong, a warder [working at] the jail in Popondetta, and Joel Gobare, our councilor, yelled out: "The fight is over!" We divided up on the beach at Karau where we had fought before. We both lined up and shook hands . . . We returned to Darapap and they returned to Mendam.

An image of the state, which had been totally enveloped by events, ultimately reemerged to end the conflict. But was its salience due to the legitimacy of the civil statuses the two men held or did it result from a willingness on the part of the combatants to quit brawling? I will return to this question below. At this point in the story, the problem for the Murik cults was again to return the cosmic bodies of their warriors to their moral forms.

MALAI: When we got back to Darapap, we did not go to our houses. We went back to the cult house. The elders stood up the crossed spears and opened their legs and we crawled through. They beat our chests and then encircled a *tanget* [magical cluster] of ginger leaves about our heads. We sat down along the sides of the cult house. They hung coconut frond necklaces about our necks and put the spears away.[4] We washed, collected our bed things and went back home.

The many threads of this conflict – the initial sexual jealousy between the two youths as well as that between many of the men who fought, the rivalry between *sumon*-holders, the atrophy of the male cult, the withering of civil authority, the intrinsic allure of fighting – had amounted to nothing less than a comprehensive challenge to the entire normative code of the culture, that is to say, to the maternal schema. In the aftermath of the councilor's death, a willingness to reconsider the conflict without shame had developed. The feuding parties had moved into a new phase of the reconciliation process. But still no formal mechanism, no remedy, had been agreed upon to take it to a conclusion. And still, no sponsor had stepped forward to lead it.

Pigs, knots and sorcery

Barlow and I had also been observing the truce; we had done no research in Mendam village during our first eight months on the Murik coast. After the councilor was killed, and the five women had disregarded it, we too began to visit that village on a weekly basis. Being morally identified with the men of the "enemy village," I was evidently looked upon with suspicion in Mendam and got only the most reluctant sort of coopera- tion from the men there. Barlow was more warmly received. Dakuk, one of her two adopted mothers, was one of the Mendam women married and living virilocally in Darapap (see Plate 28). Dakuk had plenty of eager kin there who wanted to begin to know their new, white daughter. Ginau came along with us on our second visit to Mendam in the hopes of opening up discussion about settling the conflict. Several elders immediately offered to suspend the taboo so that his *wajak* partner might invite him to attend a feast the Mendam youths were giving that same night in compensation for the ritual "work" the elders had done to prepare them to fight. The youths had purchased two pigs from Koil Island.[5] In return, the Mendam elders were to teach them the spells of their war spirits. Ginau accepted; so he and I spent the night in the male cult house there. The rhythms of the long evening, needless to say, were

Plate 28: Dakuk of Darapap

interspersed by the grotesque antics of the Spirits of Wealth and several courses of food.

Next morning, we returned to Darapap with the departure gift of Ginau's *wajak* partner, a basketful of pork and other ceremonial delicacies. Ginau quickly distributed the food among his junior siblings and their firstborn children to oblige them to help prepare a feast. He wanted to broach the proposal put forward by Nangumwa and Sanimba, two Mendam elders, that Ginau and his *wajak* partner reciprocally exchange baskets of food, like the one he had received, display their heraldry and end the conflict between the two villages.

The exception made for Ginau, not to mention his continued desire to resolve the conflict, did indicate that differences between the villages were narrowing. At the same time, within Mendam itself, age and youth were also showing renewed consensus. But in Darapap, as months passed following the councilor's death, the willingness to go ahead with the settlement rite once again fell into the cracks of political division in the community, dispersing commitment to follow through with it into concerns which seemed more pressing. In the event, Ginau could muster only the barest minimum of support for the Mendam *wajak* initiative. Moreover, as the following incident revealed, relations between himself and his *wajak* partners proved too unstable to serve as a pretext for the settlement rite.

Some coconuts piled up next to Ginau's cult house were stolen and he accused one of his grandsons of theft. It was alleged that the youth lunged and pushed Ginau down. Ginau insisted that he had only slipped, having become badly overexcited. Soon afterward, I went to Mendam and foolishly mentioned what happened to Nangumwa and other elders there.[6] Nangumwa immediately started planning "to stand up" Ginau by giving him a pig feast to restore his broken heraldry. He wanted to retaliate for Ginau's insistence that the Mendam repair their emblem after it had been damaged because a brawl had broken out during their canoe launching rite.

Upon hearing of these plans, Ginau chided me for spreading rumors and failing to appreciate the gravity of *wajak* duties. We went back to Mendam and he deposited a folded ginger leaf there, a sign which authoritatively falsified the story about his fall. Ginau prevented the threatened *wajak* prestation. But neither he nor anyone else was able to come up with an acceptable settlement procedure during the next few months.

The urgency of reaching a settlement faded as many elders became preoccupied by the imminent arrival of a ship, the *World Discoverer*, which was supposed to bring hundreds of tourists to a "great market" of carvings and baskets in Darapap. One day, as I sat in the cult house with men who were carving masks and miniature slit-drums by the score, one of the most senior of them leaned over to me and muttered that the longer the taboo went on, and people ignored it, the more likely someone would suffer.

In addition to resentments against Ginau and the anticipated windfall from tourists, lingering ceremonial rivalries in the two feuding villages still interfered with the reconciliation process. The limits of the dispute had not been established by the death of the Mendam councilor. A second evocation of an external threat, my informant seemed to imply, would be necessary. The spectre of mystical sanction to which he referred was the sorcery of ghosts (*nabran timiit*) or ancestor spirits (*brag timiit*) who tend to punish senior men or their kin for neglecting to fulfill ritual obligations. Only a few elders were left discussing possible remedies in desultory conversations. They recalled stories to each other about peacemaking rites staged during precolonial days until the details of one particular incident stirred up a great deal of excitement among them.

An armed riot between two Murik villages had been unilaterally settled by a lone man who secretly imported a pig, gave it to his adversary, who then cooked and distributed it among the rest of his village. When the peacemaker arrived with his pig, the recipients tied ginger leaves (*gorongol*) over the prow of his canoe which "closed" the warfare which had been "opened" when warriors brushed through clusters of ginger leaves as they left the cult house to go and fight. When the Mendam came to Darapap to end the taboo, the elders agreed, ginger leaves should also be tied across the prows of their canoes just at the moment they reached the foreshore. The visitors would then go on to tie ginger leaves to the pilings of every building in Darapap.

Having flagged the cultural prominence of "tether" imagery in the Sepik region in Chapter 3, I here want to collate the great metaphoric significance of acts of binding, tying knots, folding and so forth in the poetics of Murik sociomoral processes. A knot may be tied in front of a newborn's eyes to secure his soul in his body (see Chapter 3). Male initiates are bound in new loincloths on entering jural adulthood. A knot is tied in a vine (*nog*) round the forehead of a mourner to protect him from soul-loss (see Chapter 6). Space is bounded by knots. The leaves of a coconut frond may be tied together as a property marker to prevent trespassers from entering one's gardens or coconut groves.[7] Time may also be marked (bounded) by tying knots. Pairs of trading partners tie knots in leaf *tangets* on a daily basis to serve as mnemonic devices that synchronize a future meeting to exchange goods at some designated spot. By remaining chaste, the man serving as knot in the *Kakar* cult does nothing less than protect the authority of his age-grade. The truth of a statement may be asserted by displaying a knot. When Ginau sought to staunch rumors about an alleged injury, he presented a folded ginger leaf to his feasting partner to prevent the pig exchange which threatened him. Lastly, in the context of Oedipal rage, knots come apart. The lashings of the elder brother spirit-man's outrigger canoe break up, having been deliberately lashed in a weak way by his younger brother. Tying knots is a metaphoric gesture in Murik culture, signifying as it does a transition from a state of moral unbounded-ness to one of moral boundedness, or solidarity.[8] Elders in Darapap now felt ready to proceed with the reconciliation rite. The moral canoe-body would be relashed with creeper vines, vines, needless to say, that were not understood as umbilical. The image of moral reproduction which so appealed to them was embedded in a metaphor of canoe building rather than parturition.

A death, a stolen corpse and a suicide attempt

Fourteen months after the instigation of the whole peacemaking process, one of Ginau's most senior *wajak* partners suddenly died while attending a consecration feast for an urban male cult house that had been erected in Wewak town. Ginau immediately began to insist upon his right to go to Wewak and retrieve the body for burial. Evidently, the deceased had per-formed the same service for Ginau in the late 1970s, after his wife died at the Aid Post in Mendam, where she had been taken for treatment. Ginau had then compensated his *wajak* by giving him one pig, for which he paid K90 (US$120), a great amount of garden produce and several big sacks of rice. Now he wanted to square his debt and claim the feast owed him. In the cult halls, elders were accusing youths of repeatedly bungling the settlement. Word went around the village that a meeting would be held to discuss the situation. As it opened, one of Ginau's younger brothers summarized the emerging opportunity and challenged the youths to support "combining the fulfillment of Ginau's [*wajak* duty] . . . with that [reconciliation] thing!" Daniel Wambu, another of Ginau's "younger brothers," then charged Reuben Mamar, the President of the Sports Club, to make a decision and act on it.

REUBEN: Now you are talking behind our backs, accusing us of laziness. But we have never made a feast before. You should "steer" us. You are our "fathers." But all you do is gossip.

JAMES KAPARO (see Plate 27): You [Reuben] and Luke are already fathers. This is your village. We [your elder brothers] sit here [in the cult house] in front of your "fathers." We are not talking behind your backs. If you are angry, be angry! That is just a feeling! We are tired of you people and your [sports] whistles. We sit here trying to find a way to settle this problem for you, the youth.

Luke parried James Kaparo's attack upon his generation by quickly reminding everyone present who had lost the money which they had collected for the Mendam pig.

JAMES KAPARO: How many times have I told you? I don't have "a path." I did not eat the money [e.g., spend it on food]. I lost it. If I had an outboard, I would go first thing . . . I think you people must have fought over my wife.[9]

Ginau announced that his son and grandsons, who lived in town, had already taken the body of his *wajak* to the town morgue, had bought decorations and had built a coffin for it. After they brought the body back to Darapap, he would return it to Mendam for burial where ginger leaves would be tied upon the prow of his canoe at the moment his canoe reached the foreshore.

Next day, Ginau went to Karau, whose copra boat he wanted to use to transport the body from town, to explain his plan. His itinerary and intentions were received quite positively. Nor were any objections raised by four senior Mendam men who were there waiting for a ride to Wewak. Two days later, a few other Mendam men paddled to Darapap collecting membership dues for the "Pangu Pati" in connection with the 1982 national elections and reconfirmed the favorable response to Ginau's plan in their village. Ginger leaves, they promised, would be tied on the prow of the canoe bearing the corpse. Ginau would return to Darapap until the initial grieving was over and the body had been buried. The Mendam would then send for the Darapap to come for a small pig feast. Other details remained to be worked out, they said, but no one in Mendam questioned Ginau's claim to settle his *wajak* debt and thereby set the reconciliation rite in train.

The use of mortuary services as a pretext for conflict resolution is another example of the ambivalent relationship between interior and exterior images of the body in this culture. The death encompassed both poles of becoming. It was at once a terrifying murder and grounds for moral renewal. In the wake of the *wajak*'s death, the settlement process between the villages resumed. This was not a uterine or transmigratory renewal. It was an opportunity for the Spirits of Wealth to right the canoe-body. Wanting guidance, indeed, the "sons" literally asked their "fathers" to "steer" them. Agreement had now been reached about how to represent the peace. Appropriate symbols were selected and leadership had come forward to sponsor the rite. All that remained was the performance itself.

But two younger brothers of the deceased *wajak*, Dana and Sarewa, bore a grudge against Ginau. When their two "sons" had gone to invite him to a feast, and were assaulted, why had Ginau failed to intervene and protect them under the auspices of their common heraldry? The two men, who were staying in town, stole the *wajak*'s corpse from the morgue and buried it themselves in the Catholic cemetery at Wirui under the cover of darkness. The next day, when Ginau's grandsons went to collect the body, they found it gone.

There was a great outpouring of pity for Ginau in Darapap and Karau. Absolutely furious, he vowed never to participate in any feast mounted in Mendam until he received a pig from his surviving *wajak*s there. He then left for town. Dana and Sarewa, the two bodysnatchers, immediately came to meet him. Dana explained that they had stolen and buried the body because they believed that no ritual exchanges should precede lifting the taboo between the two villages. But he also admitted that, had Ginau protected his two boys, he would have had no qualms about accepting his burial claim. A knowledgeable senior man from Karau also told me that Dana was still angry because Ginau had refused to attend the feast to restore the "broken" *sumon* heraldry after he himself had insisted that the feast be mounted. None of this was said directly to Ginau, only to his kin. Ginau, for his part, was sulking when the bodysnatchers had come to talk, and refused to meet them.

The agreement to combine the mortuary service with the reconciliation feast had not been sufficiently comprehensive. The process of negotiating a suitable settlement procedure was going to require a prior agreement upon an acceptable range of unresolved differences between the parties. And Ginau, despite having taken a lead in the final moments of the drama, was clearly too controversial a figure to command its final act. Not only was someone else going to have to do it, but the moral context between the communities that would make such leadership possible still seemed incomplete, or immature, and seemed to require a renewed sense of urgency, perhaps yet another evocation of mystical force.

An all-night mortuary feast was staged in Wewak for kin of Ginau's deceased *wajak*. The next morning, one of the most senior insignia-holders in Mendam was found, after having wandered the streets of Wewak all night, following omens, he said. A few days later, the elder, a man named Kubisa, returned to Mendam where a small feast was sponsored by his family to air the causes of what people were calling his *baba'yagot* (literally, "crazy sickness") in the hopes of curing it through open discussion. The madness, it was agreed, was a ghost's sorcery (*nabran timiit*), the result of failing to allow Ginau the opportunity to settle his burial debt.

In the predawn of the following day, Kubisa paddled alone into the lakes. When the sun came up, his wife awoke and alerted neighbors. A search party went after him: they found him perched on a branch high up a mangrove tree. He tied a singlet over his eyes as they watched and jumped into the muck exposed by the low tide. Kubisa badly fractured a leg and had to be brought to hospital in Angoram (the Lower Sepik administrative center).

Rumors reaching Darapap greatly exaggerated the extent of the injury and created no little apprehension in the village. People spoke of little else and developed a complicated interpretation of the injured man's motives. Being one of the most senior *sumon*-holders in his descent group in Mendam, Kubisa clearly understood how heraldry ought to be handled and was extremely sensitive to sorcery which sanctions ritual negligence. He was disturbed about the burial of Ginau's *wajak*, because he believed that repayment of debts in that relationship was an indisputable right and because, as the heir of the deceased feasting partner, he had inherited the obligation to settle his elder brother's outstanding *wajak* debts. Privately, it was also said that Kubisa had been involved in a longstanding love affair with his "father's wife" (FBW). The night of the urban mortuary feast, Kubisa's son had discovered them *in flagrante delicto* and had told his mother about it. The son's betrayal was unbearable to the father, it was said in Darapap, as was the shame of going on living with his son who knew about his affair.

One month later, in June 1982, Tamoane of Darapap went to visit Kubisa, his "elder brother" (MBS), then receiving care in Boram Hospital in Wewak. Kubisa took Tamoane into his confidence and confessed that the son's discovery of his affair had indeed driven him to attempt suicide. Even then the thought that his own son had seen him *with his own eyes* having intercourse was still upsetting him. "This is the 'truth' (*nogo'iin*) of the story," declared Tamoane's younger brother, my informant. "He would not deceive his own brother."

Profoundly unstable emotional charges, seeming inevitably to implicate Oedipal jealousies, adhere to the *sumon* heraldry. But the effect of the suicide attempt was nevertheless to further the restoration of moral relations between the two villages. Following the councilor's death, Mendam women, married and living virilocally in Darapap, had first broken the taboo in order to mourn for their "brother." My own fieldwork and the national election campaign further weakened it. The death of Ginau's *wajak* partner and the attempted suicide had impelled its renegotiation. The disposition to settle had not been coerced by any third party as much as it had emerged from the cultural interpretation of events. The taboo had come apart because of the collective fiction that it was becoming increasingly dangerous to maintain. If Kubisa's madness, at least to *cognoscenti*, was not "really" caused by the mystical sanction against heraldic or ritual negligence, but by another "family romance" that was completely unrelated to the various ceremonial disputes, then, diagnosing his suicide attempt as being caused by tutelary spirits gave both communities a credible and compelling rationalization for going on with their negotiations. With motivation to proceed and a script in place, all that remained to determine was the sponsor: who would import the pig?

The outcome
Two weeks after the suicide attempt, the Karau copra boat (see Plate 29) made a round-trip to the two Schouten Islands, Kaduar and Blupblup, for which no fare was charged (its captain was trying to mollify widespread suspicion that he and his wife were extorting profits from the cooperatively

Plate 29: The *Bar Nor* ("Mangrove Man"), the copra boat on which Sauma traveled to trade for the pig he used to underwrite the reconciliation rite with the Mendam

owned vessel). A few senior people from the two feuding villages were invited to go along by their *wajak* partners. Among them was Sauma of Darapap.

Partly at the urging of Dakuk (see Plate 28), his co-wife from Mendam, Sauma agreed to make the trip to obtain a pig for the settlement rite. He took about ten "trade baskets" (*save suun*) and two other baskets of fresh seafood and various kinds of sago delicacies that Dakuk and her daughters prepared to give to her trading partner in Kaduar Island. He also took five more baskets of shellfish collected by Kiso, his first wife, to give to her trading partner in Blupblup Island. She was expecting an equal number of buckets of *canarium* almonds in return to use in a firstborn rite she was planning for an infant granddaughter. Sauma carried nothing earmarked for his own trade "brother" in Blupblup Island. Neither did he take any cash. His plan was to rely upon his good standing with this "brother" and bespell him with his pig trading magic (*nembren timiit*) to charm a pig from him, as he told me, "on credit."

Why did Sauma agree to try to sponsor the reconciliation rite? He wanted to renew his wife's access to her family's sago resources near Mendam. He wanted to appear virtuous in Mendam because he feared that the younger brothers of Dakuk's first husband living there were still plotting against him for having seduced her away from their elder brother some twenty years earlier. He felt responsible for Malai, his sister's son, the youth who had been involved in the original dispute. He also sought to settle the conflict in deference to Ginau, his collateral elder brother. With the taboo ended, Ginau's *wajak* partners from Mendam would be permitted to attend his mortuary feast. Sauma wanted to facilitate the latter ceremony because he sought to succeed Jong, his elder brother, as *sumon*-holder in their sibling group. He considered his succession ancillary to staging Ginau's feast. Ginau was genealogically senior to Sauma's elder brother; and Kiso, Sauma's first wife, had been married to Ginau when she took up with him during the 1950s. Sauma was not some sort of Nuer leopard-skin chief (Evans-Pritchard 1940), or Ifugao go-between (Barton 1919). He was no impartial, structurally neutral, third party. Despite his partisan agenda, neither was he a self-seeking subversive, planning devious strategies to exploit or exacerbate social division. To put it in the Murik idiom, Sauma wanted to win for himself and his elder brother the honor of displaying their heraldry during the reconciliation rite.

As the Karau boat returned from the islands, deep blasts from a conch shell quickly drew a crowd of children to the beach to watch it slowly chug back through the Darapap channel into the lakes. Its mast pole was decorated with green bunches of betel nuts. Its hold was choked with garden produce. Eight pigs lay trussed on deck. The boat received, quite simply, an ecstatic reception. Every ambulatory person in the village turned out to watch and help unload its savory cargo. A senior man hobbled up to a heap of buckets of almonds, which his eldest son had brought back from Kaduar Island, and beamed. Typically, when an outrigger canoe loaded with tradestore goods arrives from town – a regular, almost biweekly, event during the dry season – a raucous scene ensues of people screaming for kin to come help unload their things. On this occasion, however, there was no shortage of eager, young volunteers ready to assist. Older male youths sprang forward to carry the trussed pigs. Younger boys and girls hefted large bunches of bananas up onto their heads and paraded across the tidal flats. Suddenly, in the midst of this process, an elderly man from Karau, standing on the deck of the boat, began waving an enormous bunch of betel nuts in the air. "Who will take this?! Damnit! Who will take this?" he shouted. "Don't I have a single *wajak* in this lousy village?!"

When everything was carried off, and the boat had departed for Karau, the crowd dissipated, leaving Sauma's first wife, Kiso, alone squatting by the lakefront shore. She began to shriek that her co-wife and husband had divided their goods before she had arrived on the scene and had stolen the greater part of her share. For several hours afterwards, Sauma avoided the woman, who kept up a loud barrage of invective as she sat in her doorway (cf. Kulick 1992: 104f). Only late that night did he go to her to try to explain that he and Dakuk had not deprived her of anything. Her trading partner had simply been away in town, and not been around to organize, borrow, etc. a number of buckets of almonds equal to the number of Murik baskets she had sent him.

Next day, Sauma came and, calling me *wandiik*, the affinal term of address which he rarely used with me, proceeded to ask to be taken to Mendam. From his reference to my status as his son-in-law (I had married Barlow, his adopted daughter), I inferred that I was meant to honor his request as a form of brideservice.

Although it seemed doubtful that any sound could be heard above the din of my 15-hp outboard motor, Sauma insisted on blowing his conch shell to herald the passage of his pig several times along the way. As the canoe slid ashore in Mendam, youths bound its prow, not with ginger leaves, which were unavailable for some reason, but with coconut leaves. Elders then invited us into the cult house to set up a time and place for the settlement feast. In two days' time, it was quickly agreed, the Mendam men would come to Darapap bearing their heraldry and carrying ginger leaves to tie on the pilings of all the buildings in the village. When one Mendam man proposed that the feast be delayed for a few days so that the several youths who had been involved in the original violence might be able to return from Wewak, Sauma gently resisted his suggestion, recommending instead that kin of these absentees be found to serve as their surrogates. Mendam elders assured him that his pig would be cooked immediately and the meat would be carefully divided up and distributed so that every man in the village who had fought might taste it. As Sauma was honored with a meal of crab and sago pudding, at the opposite end of the hall the grotesque slapstick and phallic abuse broke out among the Spirits of Wealth.

The arrival of foodstuffs from the Schouten Islands inevitably exacerbates domestic tensions in Murik villages. Imported goods are usually precisely targeted for use in the prestige economy, not to mention being regarded as great and relatively rare delicacies. Expectations among those who stay behind are always frustrated because the landing of a canoe rarely realizes anyone's hopes. Kiso's dispute with her husband, however, did not obstruct the settlement process. Sauma's pig-trading magic had been effective: he had obtained a pig, the rule of recognition for the settlement rite which both communities felt compelled to accept. No one complained that the pig had been detached from the feast which the Mendam had then to prepare for themselves. Tying knots over his canoe prow, neither did Sauma quibble about the kind of leaves that were used. Differences over sequence were resolved without rancor. In spite of the bitterness of Sauma's first wife, that is to say, final negotiations were carried out amiably.

Douglas Newton (1987) has interpreted the premodern poetics of a reconciliation rite among the Manambu of the middle Sepik River (cf. Harrison 1990b: 592) in which an assailant force-fed chunks of sago impaled on arrows to an enemy he had captured and the men then broke their shields. The act of nurture evidently turned the assailant into a grotesque mother. By receiving food, the victim, who would otherwise have been expected to retaliate, became a mother's child, who would not knowingly spill the blood of the "woman" who fed him. While the conciliatory gesture cast the assailant as a "male mother," he was not just metamorphosed from his cosmic body. Motifs engraved upon his shield also evoke the image of "a mother [who] guards a child from the father who strikes it" (Newton 1987: 251). The act of nurture was superimposed upon a weapon which made the figure of motherhood, the shield, unnecessary in the combatants' relationship. The logic of these images is not derived from the conversion of blows to gift exchange, what Modjeska once called an "exchange-value cycle" (1982: 54), or from a renewal of political equality, but, again, from an ambivalent schema of motherhood. No less conspicuous was this schema in the outcome of the present case.

In the two days which followed Sauma's return from Mendam, the community of Darapap showed its apprehension in several ways. One couple got into such a loud argument about firewood that a neighbor, who was one of the husband's *mwara* sisters' sons, overheard and quickly tipped half-dozen of the man's nephews, each of whom turned up to give him a piece of wood. Onlookers greeted their queue with whoops and cheers. Near the men's cult house, meanwhile, a senior man badly cut his foot while he was making ready a *sumon* ornament a mother wanted her son to display during the settlement ritual. The elder, it was assumed, did not possess the credentials to do such work. Refusing to permit its illegitimate display, the ancestors of this ornament had clouded his attention. At an afternoon meeting in the cult house, the anxieties of the youths came out. Understanding that they were to bind ginger leaves over the Mendam canoe prows, they wanted to know where precisely on the lakefront they were supposed to meet their enemies' vessels. The instructions elders issued did not require the youths to spend the night in seclusion from women. Their bodies evidently did not need to be purified from women for the purpose of making peace.

Approximately seventy-five men left Mendam aboard two vessels, a long motorized outrigger canoe and an old flat-bottomed boat. Nangumwa, a leading Mendam elder, followed at a distance, paddling his own canoe. He did not want "to hear the smell" of feminine sexual impurities that young men possess in the morning, being too lazy or ignorant to know "to wash" after getting up.

Upon reaching the foreshore at Darapap, a few hours after sunrise on 17 May 1982, the Mendam were met by several firstborn youths who tied ginger leaves over the prows of their vessels (see Plate 30). Then, Darai, a senior insignia bearer from Karau, waded into the shallows to welcome the Mendam as they began to climb down into the shallows. He held up Magoj, a boar's tusk *sumon* belonging to the Mongaren *poang*, on behalf of Aprawa, the father of Malai, the youth around whom the fighting had begun. Following him was Sauma, who stood next to his sister, Makembo, the mother of Malai. Makembo wore a *sumon* on her chest, called *Usiig Murup*, which was a string bag studded with half a dozen boars' tusks. This insignia belonged to the Kaun *poang* which Sauma and his elder brother represented. She wore it on behalf of her son because, as he told me, he had not yet been initiated and was therefore not yet privileged to wear heraldry. Last in the line, wearing no *sumon* at all, stood Jakai Smith, who represented both his mother, who was Malai's birth mother, and the ties of Malai's deceased father to the Sait descent group (see Plate 26).

In a moment of obvious duress, a polity which had been opposed to, and then had separated from, an adversary, became transformed into an "emphatically perceivable" filial image (Simmel 1964: 340). The moral order, ruptured by Oedipal desire and rage, represented its reproduction as detachable, aesthetic images upheld or worn upon the exterior body. The heraldry Malai might have worn did not in fact decorate him; only his bilateral genitors were adorned. But even if he had worn them, the meaning of the ornaments would not have been cosmetic. They would not have been meant, as Simmel suggested, to beautify the body for egocentric or narcissistic purposes. The display of heraldry signaled the pure being-for-the-other of the young man and the being-for-him of his kin. Recreated and asserted in the pageant were elements of the maternal schema: peace, filiation and the non-uterine reproduction of personhood, enclosed, stilled and rid of inner life.

The Mendam waded ashore bearing masses of 4-foot-long ginger leaves.

Plate 30: Upon the arrival of the Mendam, Malai tied a ginger leaf around the prow of their vessel

Upon reaching the tidal flat, they followed a path of coconut midribs which had been laid out to look like canoe rollers. The visitors were led by William Barake and Bakur, the two youths who had been beaten up in Darapap during the second episode of the fighting. Bakur held the Kombek *sumon*, a cluster of dogs' teeth ornaments, bird-of-paradise plumage and a possum skin headband. William Barake held the Mindamot heraldry, which comprised a pair of boars' tusks, bird-of-paradise plumage and the big *sumon* basket named Wankau (see Plate 26).

As the guests filed past their hosts shaking hands, three groups of Seventh Day Adventists from Darapap, the "Women's Welfare Association," "Good Samaritans" and "Pathfinders Youth," suddenly marched up to the lakefront, smartly dressed in neat green shorts and yellow T-shirts. With flags of their Mission and the state held aloft, they burst into a hymn before taking their places in the receiving line. The beat of a slit-drum, bidding the men to come to the cult house for a *baas* feast, became audible.

The Mendam made their way from one end of Darapap to the other, fanning out to tie ginger leaves round the piles of every building (see Plate 31). The leaves stood out like bright green ribbons against the weathered-down, drab brown sago thatch. Still leading the march were the two *sumon*-bearers. Several times along the way, individuals belonging to the descent groups the insignias represented stopped the procession to split open a mature coconut, take a sip and anoint them by spraying the milk. Upon reaching the edge of the community, the men turned back toward its center, where the cult house stood. Both groups of men then crowded into the hall. *Aragen*, the sweet, milky, ceremonial porridge immediately appeared for the guests and a breakfast of freshly smoked fish and sago breads soon followed.

In overseas relations, setting out canoe-rollers for the arrival of trade partners signals that docking privileges have been extended and that the guests' safety will be guaranteed during their stay. This convention was originated by Andena, the elder-brother spirit. Although symbols of Western morality and personhood, the Adventist villagers in Mission uniforms, holding flags, were no less a part of the scene than the bearers of heraldry moving through the community. As the rite developed, only heraldry and the male cult continued to play significant roles. Upheld by the individuals who had caused the conflict, the sweep of the *sumon* through the village "closed the war," it was said. The Mendam trailed after their insignia, tying knots upon all the houses, as if the community was a newborn. Their insignia received the blessing of local kin, who swigged juice from coconuts, evoking the breast, and sprayed it over the passing emblems. The men withdrew beneath "the skirts" of the cosmic, male body, where the guests received the milky porridge of the *brag* spirits in honor of their assembly. The recreation of the maternal schema could not have been made more plain in the grotesque banquet imagery which ensued. Recall that the Spirits of Wealth, the classificatory mothers' brothers and their sisters' sons, metaphorically "kill" each other with "spears" during a *baas* feast by exchanging dishes of food. Utensils of motherhood – plates – become ordinance. Right to a certain oratorical frankness about issues is then granted. The male cult cleansed the heraldic, maternal body beset as "she" was by the defilement of sexuality and conflict with his "spears" and ceremonial porridge. A palaver was convened.

Speechmaking, compensation payments, and more food took up the rest of the day. Premodern indemnity consisted of boars' tusks or dogs' teeth headbands. The privilege of transacting this kind of wealth in this or any context was exclusively vested in firstborn insignia-holders in a senior sibling group, their sole "owners." But anyone can use money. One-half of

Plate 31: Led by the instigators of the conflict from both villages, the Mendam walked through Darapap carrying heraldry. Sauma is wearing the singlet which reads, "Mi tasol." Walking immediately to his right, wearing a vest of boars' tusks, is Makembo, the mother of Malai. Note the young man in the white shirt tying a ginger leaf to the housepost in the center of the background

the ten exchanges made that day were made in teeth or shell ornaments; the rest were paid in cash. All of the ornaments were given by firstborn sons; the money was not. The "wealth slit-drumbeat" (*mwara debun*) was sounded after each exchange to signal to the local community and perhaps to other Murik villages within earshot. When a boar's tusk was exchanged, the *usiig* or *sumon debun* was sounded, which distinguished the transaction of this prestigious valuable.

In spite of the great rhetorical effort which had been put into creating a collective sense of liability during the first weeks of the reconciliation process, beginning with the presentation of the pig, each of the ten payments was made by individuals and sublocal groups, and each pertained to specific acts of violence. Yanga'ii, the elderly man whose front teeth had been knocked out, received an outfit of new clothes from his *mwara* sister's sons and a compensation payment of K5 (US$6.00) from one of their sons, while the young man continued to deny having kicked him even as he gave him the money. The final exchange was presented to the Mindamot, but not the Kombek, heraldic group, even though the Mendam displayed both of these emblems upon arrival. Nangumwa, the recipient, then redistributed two separate gifts of food, in his dual statuses as the senior Mindamot *sumon goan* and the *wajak* feasting partner of the three brothers who were sponsoring the event, Jong, Sauma and Daniel Wambu. The latter man, being last born, presented Nangumwa K10 for insulting him during the fourth fight.

According to Marilyn Strathern (1985b), an ethos of masculine bravado accompanies indemnification in the western highlands. There, acts of compensation evidence and boost the power of individual men either to dominate others through violence or to mediate social relations symbolically by exchanging material goods. In this case, indemnity redressed specific acts of violence, but the exchanges differentiated injuries done to the heraldic titleholders and hereditary feasting partners from injuries done to youths. Moreover, instead of being given by aggressive men, Murik elders served as the "canoe-bodies" for insignia and spirits. Handshakes were limp and explanations of the injury were offered in tones of hushed embarrassment rather than boldness. As the donor encircled the head of the recipient with his gift (like a victorious war spirit), the percussion of a slit-drum rumbled out from the cult house to signal the collective agency of the male cult to women and children listening outside. Inside, the exchanges signaled the renewal of specific relationships. Symbolically, the individual and collective ability to turn blows into gifts was not problematic for Murik men. What was problematic was coordinating the fractious relationships that the heraldry represented. Masculine agency was vested in ensembles of collectively held ornaments. Such a unitary representation of legitimacy as detachable from the self is absent in western highlands poetics (M. Strathern 1985b: 124). In Murik, relationships are understood in terms of metaphors of unities which can indeed be "broken," then reassembled, not directly through the fleshly "strength" of individual men, but through the heterosexual agency of heraldic and cultic actors. The admonitions that went on after breakfast were voiced without swagger or conceit.

"We are all Iji'gob," declared Nangumwa, the Mendam elder to whom Sauma had given his pig. The common ancestors (*nogam*) of the two villages had fled the middle Sepik to drift about the North Coast before resettling the village of Iji'gob on the Murik coast. The village quickly fell apart and the *nogam* moved to the site which became known as Karau. The villages of Mendam and Darapap that followed later were, as he put it, "the head and tail of one serpent" whose midsection was the village of Karau.

Others followed to profess the great "shame" (*yanga'nabiin*) they felt about the brawling. One man spoke of the many marriages between the communities and cited the many namesakes living in the two villages as evidence of these ties. Several men went on to reveal personal names of mutual ancestors.

Darai of Karau pursued the common genealogy of the two villages heraldically. Ordering Malai and Bakur, two young men who had been involved in the second incident, to stand up before the hall, he held an emblem made of bird-of-paradise plumage attached to a boar's tusk between them. The youths, he pointed out, were "brothers" who shared a relationship (*yakabor*, literally, a "path") to the *sumon*, which he called Magoj, and to a common ancestor whom he also named. And being siblings, they should have known of their mutual claim to the insignia. They should have known of their common genealogy (*dekara'iin*, literally, "stand up talk"). They should have known better than to fight each other. The two youths shifted their weight from one side to the other and studied their feet closely.

A course of seafood and sago pudding was served, while Joel Gobare, the Darapap councilor, rehearsed the sequence of the conflict, episode by episode, and concluded that sports and alcohol did not mix. "In the future, before you start another fight, you youths must remember how easy it is break [relationships], but how hard it is to put them back together."

The newly elected Mendam councilor quickly dissented from this moral, demanding, "Where were my four *poang* 'brothers' when Bakur, my son, was attacked by Malai and the others?"

"I," snapped one of them, "was in Kis [village] meeting my trading partner."

Discussion turned from residual differences to issues on which the two sides might agree. A young man from Karau who had spread rumors about Malai and the photograph, was really to blame. He was the one who had caused the whole conflict (and was not present in the hall).

A final meal was presented to one of the *wajak* feasting partners in attendence. Then Yamuna, the catechist from Mendam, concluded the speeches with the observation that perhaps the brawls had a good effect on the two villages. Many kin had renewed relationships which had been permitted to lapse in recent years. When the Darapap came to Mendam for the reciprocal feast, he went on, they must plan to spend the whole night because so many more (genealogical) stories were left to tell and so many more compensation exchanges were left to make. The catechist then issued a blanket invitation to the entire village of Darapap to come to Mendam to attend a great Washing Feast which was nearly ready for the widow and other kin of Debwase, the dead councilor.

It was getting to be late afternoon. The northwest winds were picking up. The men of Mendam and Darapap climbed down from the cult house and were joined by the women and children of the village to escort their guests to the lakefront. The Mendam were bidden *awo*, the ubiquitous Murik salutation, and set off home. None of the hosts dispersed until their three vessels had disappeared completely from sight.

Political talk, as Myers and Brenneis have observed, may assert "public understandings, of both specific events and more general assumptions about the social world" (1984: 28). Indeed. The confessions and harangues delivered during the day had set out a remarkably comprehensive, if somewhat sanctimonious, summary of Murik morality: the common ethnohistory, notions of affinal duty, bilateral kinship, and cognatic descent, of heraldry, siblingship, as well as the basis of the whole symposium, cultic commensality. Except during some sacralized exchanges between *mwara* kin, nothing whatsoever was said about the war spirits, or the venue within which the event took place. Nothing was said about the *Kakar* spirits, whose cosmic powers both communities shared. Perhaps the implicit justification for this collective silence might have related to the role the male cults had played to foment the conflict. The ethos of the day had been tense and serious: there had been no laughter. Differences lingered. The vilification of Karau was unrestrained. The invitation extended by the Mendam catechist was not only to attend what he touted as a bigger and better phase of the reconciliation process but also to attend an even grander celebration, the Washing Feast for the kin of the dead councilor. These were not unmotivated invitations. Currently, there is considerable theoretical acquiescence to the view that talk (and action) constitute moral order while remaining contingent upon preexisting sociopolitical forms (see, e.g., Keesing 1990: 497). Yet there is also an older, less optimistic view of order in society, which is that no act of redress works. When does the feud ever die?

Darapap men scattered to bush toilets after the departure of the Mendam and then returned to their houses to get something to eat before reassembling in the cult house. A lengthy silence enclosed their subsequent gathering. Kangai, an insignia-holder who was also the reigning head of the *Kakar* cult and a member of the most senior generation in the village, finally murmured, *Moan apo dewan!* (literally, "That was something big!") and went on to express his relief that everything had gone off without a hitch. Others agreed about the great good they had done. Karau village caused the conflict. Darapap had settled it. Two visitors from another Murik village, who had come to observe the proceedings, promised to carry home the news of what had happened and departed. Sauma, whose insignia had won the day, then stood up to speak.

Stepping gingerly through the middle of a floor lined with attentive men watching him and waiting for his words, he paced, evidently thinking. Ever so quietly, he began to discuss one of the *sumon* of his group – a net bag with six boars' tusks attached to it (*Usiig Murup* of the Kaun people) – which his sister had worn that morning on behalf of Malai, her son. The pig had been his and the feast had been sponsored in the name of the Kaun heraldry held by his elder brother and himself. He sat down, but quickly sprang back to his feet, as if he had remembered something else he wanted to say. Suddenly, he began to weep. "Jakai!" he called out to a young man. "Your mother was also crying when the Mendam came ashore. Your mother cried today because you did not wear your *sumon* when you went to meet them. But who was here in the village to assemble it for you?" Sauma paused, as if he would go on speaking. Instead, he wiped his eyes, and reseated himself.

Heated debate ensued. Men detailed failed plans to sponsor the feast. One man actually rebuked two "sons" for disobeying their elder brother, when he proposed to go ask one of their trading partners for a pig. "Our *sumon*, Gimoro-Matui, could have been displayed!" the elder yelled. "Our *sumon* could have had coconuts broken for it today! But never mind! The *sumon* of Jong and Sauma has taken the honor." Another man reminded everyone that other feasts still remained. This point provoked James Kaparo, who had lost the original money and had been conspicuously absent all day, to stand and speak.

I spent the day in the lakes fishing. I was so ashamed that I had no cash to pay the men I had fought. I will get a pig "to stand up" the fathers who fell during the fighting. I wanted to bring the pig [for the settlement feast]. But I still have no "path" [e.g., outboard motor]. We are a rubbish village. We are poor. How can I get to market? Parts of my outboard have been stolen. Now one of the senior men has cut me off. There is nothing more to say – except that this [pig feast] for the elders is still left!

Another middle-aged man, who had also failed to get a pig, demanded that the "taboo go on until the Mendam reciprocate our pig. No Darapap ought to go to the Washing Feast there!" But his mandate sounded no less empty than Kaparo's vow. Senior men returned to discuss the genealogies disclosed during the day and then another, unrelated, *sumon* dispute came up (see Lipset 1985). It was dusk. Shadows were lengthening. Ghosts and spirits were afoot. Men began to slip off to the seashore to wash.

The victor's reticence during this *post mortem* raises several questions, perhaps the most elementary of which is why Sauma's voice had been so small. Although his pig had precipitated the settlement, Sauma was not the firstborn in his sibling group. He had not yet succeeded his elder brother as its insignia-holder. He had spoken but little during the day and afterwards, apart from this momentary outburst, his performance had been strangely subdued. Rather than crushing the floorboards as a *sumon goan* might have done, he had trodden lightly across them. But for the shy personality of his elder brother, a man who rarely figured in collective life, Sauma should not have been speaking for the *sumon* of his group at all. The very next day after the rite of reconciliation, Sauma and his sister met in the house of their elder brother, Jong, and had a meal together before Jong carefully detached the six boars' tusks from the net bag and put the valuables away in a *sumon* basket which he hung on a central housepost.

Why had Sauma wept? His tears were a sort of mimicry in sympathy for the woman who had cried during the rite because her firstborn son had worn no heraldry to greet the Mendam. "I am not guilty of this oversight: it is not my fault" was one explicit message of Sauma's tears. They were also a plea of innocence to dissuade any tutelary ghosts who might ensorcel him for the ritual negligence of the absent *sumon*-holders. Perhaps his tears were also a plea that they not kill him for illegitimate sponsorship of the whole event. A word about why this woman had been brought to tears. She had sponsored her son's initiation herself but was also upset because

Malai, the youth for whom the rite was staged, remained her son, although he had been raised by her elder sister. Her descent group had thus gone unrepresented at this critical juncture. The great assembly legitimated by the display of *sumon* had been accompanied by a mother's tears, in other words, by a mother's expression of loss. Sauma's tears and the tears of Jakai's mother again expose the depth of the relationship between heraldry and the referential object to which they are a rejoinder. Attachment, or commitment, to Murik *sumon*, is embedded in strongly felt sentiments, but, more specifically, these sentiments adhere to a maternal figure. Could they be otherwise and remain culturally so significant?

In what senses did the rite of reconciliation resolve the feud? Even though the reciprocal feast was not offered immediately, relations between the two villages did return to normal. The intervillage taboo was lifted (see Lindstrom 1990: 386). A few days after the settlement rite, some Darapap people did attend the Washing Feast for the councilor's widow, and kin resumed visiting trade on a regular basis. Individual fears persisted, however. On the eve of the Washing Feast, for example, Sauma decided not to go. We had agreed to attend it together. But just as we were ready to leave, after having told me how proud he was of his Mendam heritage, how he was "really a Mendam," he conceded that he remained unwilling to spend a whole night there. Even after the settlement feast, the thought of eating in close quarters with the brothers of his second wife's first husband still seemed to him to be more like a trap than an entertainment. The display of heraldry, in other words, had concluded the state of enmity between the two villages, as was its purpose. But the peace of the *sumon* did not settle grievances which predated the conflict it was meant to settle.

A reciprocal metaphor

Following Victor Turner (1957), P.H. Gulliver (1979: 121–78; see also Swartz et al. 1966) has also argued that, as moral crises escalate, leaders begin to bring adjustive or redressive mechanisms to bear upon them. Beginning with a search for an arena in which literally to sit down and meet, they initiate negotiations between the disputing parties. The resolution process seeks first to define its agenda and the meaning of the conflict, in short, to pinpoint outstanding differences, before preliminary and final bargaining take place. A ritual outcome is then formulated and executed. The phases of such sequences, Gulliver allowed, are variable. They may overlap, recur, or be abandoned. The negotiation process I have tracked in this chapter no doubt documents Gulliver's view, however loosely or strictly one wants to hold to it.

1 *A moral crisis, defined as a disagreement which cannot be settled dyadically, enters public discourse.* The fighting escalated from sexual jealousy between two young men until two entire villages were pitted against each other.

2 *Negotiations begin when a mutual desire to stop the dispute and settle it exists: the parties must agree to contain differences (either voluntarily or through coercion).* The two villages made several agreements to stop fighting, just as they brawled several times.

3 *An arena is sought by leaders, who are not party to the dispute, in which to meet and begin to define an agenda for the settlement process.* This was precisely when Barlow and I appeared on the scene. A meeting took place in a neutral space, the male cult house of an intermediate village. Elders who attended the meeting and made attempts to define a settlement agenda had participated directly or mystically in the brawls. A taboo was instituted which was meant to be observed until the two villages exchanged reciprocal pig feasts.

4 *An exploration of the meaning of the dispute follows; key differences between the parties, the causes of, and events in, their dispute are defined.* In many meetings, elders rehearsed the meaning of the conflict again and again as they tried to persuade youths of their collective liability and motivate them to contribute money to pay for the settlement ritual. The negotiation of norms of relationship between generations of men emerged as a related dynamic in the reconciliation process.

5 *In the course of the negotiation, the meaning of the dispute may change, becoming more obfuscated or more specific; but differences between the parties ultimately narrow, and areas of tolerable agreement emerge. The "leeways" of morality* (Llewellyn and Hoebel 1941), *which consist of a tolerable range of behavior that both parties can accept, are drawn out.* With the death of the village councilor, debate about the conflict was renewed. Both internal and intervillage differences narrowed. The latter were effected by a group of initiated Mendam women, married and living in Darapap, who broke the taboo to go to mourn for the deceased. A consensus began to develop, in part because the appearance of these women signified peace – the cessation of conflict – in Murik culture. But the negotiation process remained stalled. Residual indignation and the development of new disputes among ritual leaders thwarted several efforts to formulate an acceptable

settlement procedure. Its sponsorship was still being negotiated by rival leaders.

6   *The process of bargaining about core differences results in the gradual enlargement of consensus about what they mean. Final bargaining begins: a ritual procedure is selected.* The widening consensus was finally impelled by the threat of sanction. This force derived from meanings imparted to three events: a death, the deliberate defiance of a ritual protocol and an attempted suicide. The suicide was understood as the mystical consequence of neglecting to settle the conflict formally. The sponsor had not instigated earlier phases of the negotiation. The ritual procedure, its symbols, location and timing, were quickly affirmed.

7   *The outcome is executed and formal recognition of the settlement goes into effect immediately.* As the Mendam arrived in Darapap for the first time in eighteen months, their canoe prows were wrapped with ginger leaves. They were formally greeted like trading partners. Both villages displayed sacred emblems. The visitors tied ginger leaves on the piles of all the buildings in Darapap. They received a day-long feast. Individual hosts made compensation payments. Speeches were made revealing common ancestry, heraldry and other cross-cutting relationships. Peace resumed immediately.

But however accurate a framework for the descriptive analysis of negotiation processes it offers, Gulliver's model fails to account either for the strategies and motives of the parties involved or for the effect of any particular culture upon them (shortcomings which he acknowledges). I have dealt at length with strategy and motive. Now I want to conclude this chapter by turning once more to the cultural construction, or what I prefer to call the dialogics, of conflict resolution processes in Murik.

Not only does a metaphor of reproduction inform the symbolism and meaning of conflict resolution in Murik, but a metaphor of conflict resolution informs the symbolism and meaning of reproduction. In both, cause is often, if not exclusively, located in triangulated sexual rivalry. Following both intercourse and violence, the "body" becomes "pregnant." It is beset by mystical impurities and vulnerable to sorcery attack. A woman or a victim must then be quarantined from moral and cosmic bodies, which she endangers and which may endanger "her." The culmination of both processes – birthing and *sumon* display – at once threatens and reproduces the moral and the cosmic bodies. During both, knots are tied upon or over the body. In the final phases of both reproduction and conflict resolution, while women cook for and nurture children, the cosmic body prepares "his" sacred milky porridge for the pleasure of the community. The two processes, shall we say, enact a contrary, reciprocal image of each other. Conflict and conflict resolution rites elicit a metaphor of procreative process and, vice versa, procreative process elicits conflict and rites of conflict resolution. They constitute a single, equivocal metaphor.

Despite the beautiful array of sacred emblems which complete the moral body, it remains little more than an inhibited facade. The defiled, uterine forces, sexual desires and Oedipal rages within it, not to mention the myriad services rendered by the cult, are indissociable from and indispensable to it. Rather than being an empty vehicle, this grotesque interior brims over with impurities, arousing the passions which inevitably lead to jealousy and violence. This latter body is not just impure, defiant or subversive, it is the immoral interior of an exterior form "who" raises and decorates children. The reproduction of Murik society, given this kind of a contradictory inner and outer body, must remain a malady which flares up from time to time into full-scale rioting. No complete cure or recovery from such an affliction is, or could be, possible. In the course of all of the scandals, moral crises and expiatory experiments that I have chronicled in this chapter a tense dialogue is lived, a tense dialogue about the reproduction of this disjunctive body. The last chapter of this book will confirm that what I have documented at such length is no aberration but an ongoing problem in Murik communities as they coexist within the postcolonial state.

9

Social control and law

Sir Michael Somare, a firstborn son and heir of a prominent Murik *sumon*-holder, became the first prime minister of Papua New Guinea in 1975. In order to placate earlier challenges leveled by several secessionist movements, under Somare's stewardship (1975–81) the state embarked on a course not of consolidating its powers and resources but of "giving in" – dispersing them to the provinces (see Lipset 1986).[1] One unintended consequence of his policy of decentralization was to quicken a renewal of intertribal violence (see Clifford et al. 1984; Morauta 1987; Harris 1988). The state became Melanesianized. Led by paradoxical "double agents of law and disorder" (Gordon and Meggitt 1985: 18 n. 4), its legitimacy became *less* differentiated. The internal efficacy and credibility of Papua New Guinea withered during the 1980s. Relations of power between margin and center became increasingly equal (cf. Fitzpatrick 1980; Black 1984; A. Strathern 1974). Local-level polities were "semiautonomous," to use Sally Falk Moore's clumsy but apt term (1978b: 55). With no wider matrix of civil authority depriving them of the right to create and enforce moral order, villages, more or less, returned to being unsupervised political-moral fields endowed with the agency to devise rules, symbols, customs, and the right to attempt to enforce compliance to them. This revival in indigenous orders instanced Griffiths' (1986) strong notion of legal pluralism in the sense that the local forms of social control were independent of those of the state. When we first returned to the Murik Lakes in 1986, we learned that the remedy deployed in 1982, the *baas* feast, was not the only solution to conflict to which villagers were committed. In this chapter, I examine a second conflict in order to contrast the mode of reconciliation to which it gave rise with the one described above. In the 1980s, local forms of social control coexisted dialogically with the law, I shall argue, rather than semiautonomously.

"We know the design on your mother's vulva"

Another brawl had broken out in January 1984. Darapap fought Aramot, one of the three politically autonomous hamlets that make up Big Murik, the village about two miles to the west of Darapap (see Map 2). A taboo was again instituted, although in this second incident, police were called in to intervene. This book was very much a work-in-progress when we returned to the Murik Lakes in 1986; so upon hearing about it, I immediately started asking people from both villages about what had happened. The following summary draws upon those interviews.

A young Aramot woman named Veronica Woker had moved to Darapap to marry Levi. She returned home soon afterwards to visit her family and allegedly had an affair with her ex-husband, a youth named Philip Kangai. Levi's collateral "brothers" there heard about the tryst and sent word to him. Levi quickly fetched her back. A short time later, Aramot youths went to play soccer on the beach at Karau. Passing through Darapap on their long walk home, Levi accosted his wife's ex-husband and beat him up. Soon afterwards, Levi accompanied James Kaparo, his "elder brother," to a sago and vegetable market in one of the inland villages across the lakes. On the way, Kaparo dropped him off to visit kin in Wokumot, the hamlet adjacent to Aramot in Big Murik. Philip Kangai, the wife's ex-husband, and two friends sought him out there, knocked him down, and began to taunt him. "In your village, you are happy," they shouted. "But when you come here, you feel it: you eat dirt! You eat your mother's vulva! We know the design on your mother's vulva!"

En route back from the market, James Kaparo stopped to pick up his "younger brother." Upon hearing about the attack, Kaparo became incensed. Levi was being looked after by a "grandfather," an elder with an old score to settle with the same three assailants. This man connived with Kaparo and offered him docking privileges and magical support should he choose to lead a reprisal. But next morning when Kaparo returned to Darapap with his plan, he found the community burying its most senior man, who died the day before. Kaparo nevertheless went to the male cult house and began beating a warning on the slit-drum to signal a pending attack.

Some fifty to seventy-five youths spent the night in the cult house absorbing the powerful "heat" of the *Kakar* spears into their bodies. "To blind" the eyes of their enemies, cult elders blessed them with their spittle, which they mixed with the ashes from the firepit below the shelf of the *Kakar* "ancestors."

Word of the Darapap attack preceded their arrival: the hamlet of Aramot appeared empty when they went ashore at dawn. People were either hiding in their houses or had fled to other hamlets. Aramot informants estimated that there were about five Darapap to every Aramot. In 1988, my Darapap informants omitted to mention their superior numbers to me when they recalled the brawl. But neither did they perceive it as an even match. The Aramot "used their [magical] power," one youth told me. "But we came with the wind. We . . . chased them like pigs and dogs. We beat up the three men who had attacked Levi, knocked them out and threw them into the water." Both sides agreed that many youths were beaten unconscious, that much blood was spilled and that a great deal of Aramot property was damaged.

When the fighting subsided, Darapap youths withdrew to the cult house of their ally in the adjacent hamlet, where he made their bodies "cold." Encircling their heads with bespelled betel nuts and giving them cigarettes which he himself put into their mouths, he ended their food taboos by serving them a meal. The warriors then left for home, where they crawled through a tunnel of legs made by the *Kakar* cult elders inside the cult house, after which they were allowed to return to their girlfriends, wives and families.

Aramot elders sent a letter to the administrative center at Angoram to request police intervention. A police boat was dispatched down the Sepik. A big group of Darapap youths were ordered to appear before the magistrate in court and stand trial. The trial resulted in three-month jail terms for some, but not all, of them. The verdicts were greeted with outrage and confusion in both communities. The wrong people, it was felt, had been jailed. Men who had either not led the fighting or had not been involved at all went to jail while those who had been in the middle of the fracas were freed. The Darapap collected K200 "to compensate" the magistrate and oblige him to reverse his decisions. But he refused, claiming that he lacked the authority to do so. The Darapap wanted to sue.

In 1986, Wangi, a leading *sumon goan* in Aramot, told me that he was still so enraged about the Darapap attack that he wanted to maintain the taboo between the two villages for ten more years. Other Aramot people were simply demanding that the Darapap stage a pig feast for them. And indeed several men I knew there were vying to sponsor it, although two years later none had succeeded.

By 1991, Kiso, Sauma's senior co-wife (see Chapter 8), had done so on behalf of Levi, her firstborn grandson. She purchased a pig from Sepik River trading partners. An outrigger canoe then took her, her *sumon* insignia and grandson, together with the rest of the village youth, to Aramot, bearing ginger leaves. The Aramot set canoe-rollers out for them and pulled their vessel up into a central spot in the village. The Darapap put Kiso's two heraldic baskets (Kauk and Wankau), together with those representing her grandson's assailant, out to display on the platform of the canoe for everyone to see. Darapap youths went about tying ginger leaves on the pilings of the male cult house and the rest of the dwellings in the hamlet and spent the rest of the morning and early afternoon at a *baas* feast in the men's cult house The speeches and ambience were evidently happy. The Aramot vowed to return the pig and prepare compensation payments.

Later on in the day, some time after the Darapap had departed for home, Wangi, the furious *sumon*-holder to whom I had spoken in 1986, returned to the village from Wewak town. His rage about the Darapap attack seven years earlier unabated, he proceded to rip down all the ginger leaves.

There are many similarities between this case and the one I discussed in the preceding chapter. The perceived cause of the conflict, an Oedipal violation of a young man's conjugal rights and honor, was identical. A photograph was the contested image of desire which fomented the first conflict which I interpreted as an updating of the "Two Brothers" legend. In this latter case, however, conflict explicitly recapitulated cosmogony. The taunt that the husband's rivals knew "the design on his 'mother's' [e.g., his wife's] vulva" was quoted from the "Two Brothers" story. The step from dyadic to intervillage conflict was no less abetted by the male cult in Darapap, which again "swallowed" the moral bodies of the youths and "spat" them out in "heat" to fight, before the men "gave birth" to their moral bodies again. The same ritual mode of social control – the institution of an avoidance relationship concluded amid heraldic displays and the renewal of commensality – was implemented. Two clear differences stand out: the resort to state intervention and the suspended outcome.

Legal anthropologists (Malinowski 1922; Barnes 1961; Comaroff and Roberts 1981; Rosen 1984) have repeatedly sought to locate the legitimacy and power granted authority in any settlement process "from below," that is, as arising from the disputants. Rather than descending from the sovereignty of the law, Gluckman (1955) argued, the "reasonableness" of Barotse justice stemmed from its capacity to follow the exigencies of social relations which had to be maintained. Arguing more broadly, Abel (1973) proposed that the mode of redress disputants select will be determined by inequities in their relationship. As differences in status, intimacy, size, degree of socioeconomic interdependency and cultural homogeneity increase, so will the likelihood that appeal will be made to a more highly specialized, repressive and unilateral form of authority. As power, in a word, becomes more unequal, the weaker party may seek to neutralize his disadvantage by engaging a third party upholding an official discourse. The "location" of the intermediary in social space should be roughly equidistant between himself and the stronger adversary. The problem may be that the more differentiated remedy will likely be less satisfactory, being less flexible, and less culturally informed, not to mention less authentic.

What, then, were the differences between the two pairs of adversaries in these two cases? One was size. The Darapap attack was not mounted against the whole village of Big Murik in 1984. With the collusion of an elder living in an adjacent hamlet, they only fought one of the three hamlets of which the larger community is composed. Without accounting for absenteeism of various types, our 1982 census data show that Darapap was about twice the size of its second opponent. In the earlier conflict, by contrast, the population of Darapap was roughly the same as that of Mendam. The degree of social and economic interdependence was also different. While ceremonial relations cross-cut both pairs of communities at a more or less equal level, there were demonstrably fewer affinal relationships between the second pair of villages. Again, the 1982 census data show that nearly 13 per cent of female household heads living in Darapap (6 of 35) were Mendam women. The Mendam faction comprised an even larger percentage of the adult women (23 per cent) in the village. However, there were only two Aramot women living in Darapap in this second conflict. Both of them had just recently married and taken up virilocal residence there. The young woman over whom the two men fought was the third. In neither Mendam nor Aramot, I should add, were there any Darapap women living virilocally.

There were, in addition to demographic and affinal differences, relevant material differences. The narrow beach on which the three hamlets of Big Murik stood was no more than a few hundred yards wide in the middle 1980s. Unlike Mendam, which owned large sago stands located relatively near the village, Aramot possessed no conveniently available resources. Not settling the second conflict, in short, cost the Darapap very little in terms of lost access to resources. Reciprocally, the one thing Aramot routinely sought from Darapap, surplus coconuts, was available to them through several alternative, if somewhat less accessible, sources. From the standpoints of their relative social and economic interdependence, in other words, disparities between the second pair of antagonists were far greater than between the first pair. Lacking the support of the other two hamlets in their village and sharing few affinal ties with their attackers, the outmanned Aramot did not opt to espouse and defend the autonomy of "custom" as Darapap and Mendam had done in 1981. They offset weakness (and retaliated) by soliciting "nonpartisan authority," the police of the postcolonial state. At the same time, however, they also instigated a local remedy. As the state decentralized, the relationship between its law and local mechanisms of social control depended upon the political capital which might be made out of them in the village arena. Judicial institutions, despite being contingent upon local-level politics, were unanimously criticized in both villages for being stupid, rigid and ineffective. The Darapap felt insulted for having had to make their own way to court and angry that the official was unwilling to accept their offer of "compensation." The jail terms, both sides agreed, had been meted out to the wrong people. The state did not repair their relationship at all. But, as the image of the irate *sumon goan* ripping down the knotted ginger leaves from the houseposts indicates, neither had the pig exchange nor the *baas* feast compelled a comprehensive resolution. The rule of recognition required local consensus: thus the suspended ending.

Our leavetaking in 1988

Sauma, the reticent hero of the Mendam settlement rite, died in 1985 (amid sorcery allegations his family leveled against kin of the first husband of his second wife). Some time later, Dakuk, his second wife, left Darapap and married an old flame, an Aramot man who had recently retired to live as a pensioner in Wewak after a career of police service. In September 1988, we were preparing to return home once again, and were seeing a fair bit of Dakuk and this man, Barlow's new "father." The night before we were to leave town, they gave us a farewell party[2] which took place in the house of the husband's elder brother, Wangi, the angry *sumon goan*.[3] Dakuk, it seemed, wanted to place her still relatively new affines in her debt. At his mother's instructions, Barlow's elder brother advised us to favor her new husband, the husband's elder brother, his wife, Manog, and their children, with our departure gifts.

The small room was jammed with people. Under the sharp light of several pump lamps, Barlow and I were served an absolutely enormous sandcrab set atop a huge, shallow bowl of sago pudding. As we ate and ate, but failed to make any impact on the grand portion, I began to look around to see what was going on and noticed that Luke, one of my adopted younger brothers, was nowhere in sight, although both his co-wives and their children were in attendance. Above the din, I called out to one of my *mwara* sister's sons who was sitting nearby and asked after Luke's whereabouts. The eyes of my sister's son darted across the room to Wangi, our host. "The talk! The talk!" he whispered, holding up the palm of his hand to silence me. "Understand?!"

The night before, I quickly remembered, Luke had arrived from Darapap to pick up his second wife and children, who had been staying in town visiting Dakuk, her mother, and take her back to the village. When Luke's canoe reached the foreshore below his mother-in-law's house, he called out to his wife to help unload the smoked fish he had brought to market. Wangi heard him from the adjacent house, and refused to grant him docking privileges, citing the taboo between Darapap and Aramot. The two men argued: Luke claimed that the taboo did not apply in town and Wangi relented. Although his two wives came to our party, Luke was still so upset about the incident that he was refusing to enter, or eat in, our host's house.

The Murik ethnotheory of the process of conflict and conflict resolution advances in four, more or less, discrete phases: (1) "private" tensions between coeval youths arise from sexual jealousy; (2) they escalate into widespread violence which encompasses or is encompassed by preexisting social divisions; the conflict may be (3) suspended by the institution of avoidance relations that may (4) be "resolved" by the exchange of reciprocal pig feasts and the display of heraldry, should the parties be willing to forgo their hostilities. In the course of this process, mystically defiled, interior images of the body, such as sexual jealousy, cultic aggression, and bloody injury, become metamorphosed into pure, exterior images of a body, adorned with heraldry, bound with knots, weighted down by abundant prestations, renewed commensality and so forth. The cleansing and ridding the body of polluted fluids, I have argued, metaphorically recapitulates the process of reproduction as it is locally understood: the truce or embargo phase is therefore equivalent to the defiled, "gestative" phase of pregnancy, when the maternal body is withdrawn into interior, feminized spaces and avoids collective life outside of the domestic group. Parties may not eat together during this period, but may go on with routine exchanges outside of domestic and communal space if they so wish. In part, Luke's refusal to attend our farewell party matched the refusal of our host to allow him to dock the night before. Having been refused entry, he refused to enter our host's house. Luke's sulk was a signal of equality between antagonists, to be sure, but it was also a signal that the relationship between them, as well as between their communities, continued to "gestate," as it were, continued to be impure, continued "in utero," in a "femininized" state. Luke's absence sent a message to Wangi that their relationship remained incomplete and dangerous. As such, it also suggested the possibility of a moral conclusion, a ceremonial "birth" of a pure, canoe-body of commensal relations.

Avoidance

The theorization of informal mechanisms of social control, such as avoidance, in stateless societies has remained little more than a footnote to Durkheimian functionalism. The degree of centralization of political authority, it has been argued, determines the structure of legal forms (Durkheim 1933; see also Lukes and Scull 1983). For example, Koch et al. (1977) distinguished triadic modes of conflict management, in which external intervention is engaged, from dyadic ones, in which disputants negotiate grievances directly with each other. The latter, conducted either as negotiation or self-help, is found in small, kinship-based societies with low-production economies and no centralized authority. Legal theory has therefore had to concern itself with a broad, unlimited social field made up of kinship, ritual and politics rather than discrete judicial institutions (Kroeber 1925; Radcliffe-Brown 1933; Hoebel 1954; Gibbs 1963; Comaroff and Roberts 1981). As Malinowski clearly understood (1926), legal order in stateless societies derives from everyday life, the reciprocal expectations of others sanctioned informally through shaming, ridicule or avoidance.

Avoidance, in particular, has been further analyzed in terms of Popperian methodological individualism. In this view, moral institutions – culture – are understood to arise from the decisions and transactions of individual actors pursuing rational interests (Popper 1966: 98; Barth 1959; 1966; Gellner 1973; Leach 1961b; Giddens 1979). Avoidance may be a tolerable response to conflict so long as "the costs" incurred by the parties who ignore each other ("the costs" being equal to the loss of mutual resources) can be absorbed and endured (Felstiner 1974). These "costs" may be compounded by the difficulty of having to find replacement values and by the extent to which innocent people who were not party to the original conflict may be caused to suffer. Avoidance may alleviate trouble, but it may not solve either the original problem or any lingering hostility associated with it, a hostility which may then be redirected against someone else, or even against the self. Avoidance can be an "expensive" mode of social control. If, on theoretical grounds, it seems to be a superficial remedy, why does avoidance possess such great moral salience in Murik?

Avoidance is part of a more comprehensive metaphor advanced by those who live and espouse the quietist values of the maternal schema and their ambivalent allies, the Spirits of Wealth, but contested by those whose bodies are at once emboldened, yet beset, by passion and fertility. What issue is the subject of their dialogue? The moral reproduction of society. Avoidance stands for the defiled phase of gestation during which "uterine" withdrawal has yet to be metamorphosed into pure relationships of dependency upon hollow, maternal canoe-bodies. Both male and female possess interior toxicities, fluids, rages, or sacra that challenge their ascetic, exterior forms. The interiority of the body thus offers up imagery for conflict discourse. Where are antagonisms metaphorically stored? They are "swallowed" into the "stomach/womb" (*sar*). From where do "hot" warriors emerge? From the "stomach/womb" of the male cult house. Symbols of moral association, by contrast, appear as hollow mannequins, vehicles, or receptacles, whose perfect beauty is undefiled by internal impurities.

Thus baskets, the exemplary representation of moral identity in Murik, evoke the visible, decorated body of a woman (see Plate 32). They are spoken of as having "breasts," "skirts" and "hands," but inside, of course, they are either empty or filled with extrasomatic goods, or sometimes, with babies. Neither the heraldic standardbearers of the descent groups, the *sumon*-holders, nor the "creeper vines," the bilateral groups of commensal siblings which they lead, possess internal embodiments. No consubstantial fluid, or discrete gender, differentiates them. They are androgynous – at once masculine yet decidedly maternal – "canoe-bodies." Perhaps the dwelling, as a moral image of interior space, might seem to contradict this overall logic between inner defilement and outer purity. But the interior spaces of houses are virtually free of uterine imagery or passion. The dwellings of firstborn elders (*pot iran*) receive *sumon* insignia and personal names when consecrated. Outbursts of anger, conflict or violence are then prohibited within their confines. When fleeing a fracas in village avenues, people know to retreat inside houses for the sanctuary they offer. Birth and sexuality are also excluded from these domestic spaces. Only death is not taboo in them. Mourners' bodies then become defiled and must remain secluded inside until they have undergone a Washing Feast. The house itself has not become defiled, only the mourners. What gender is attributed to the dwelling? Although constructed by men, her "soul" is feminine. "Her" hearth and housefire are the site of a mother's cooking

Plate 32: Murik baskets represent decorated, exterior bodies of women

and nurture of children. "She" is precisely where children are found "hanging on their mother's skirts." The dwelling is a privileged site for the values of commensality and peace of the domestic group. The nurturant, ascetic and quietist elements of the maternal schema are clearly recreated inside of "her" while her procreative force and passion are cast out (see Plate 1).

The meanings of food and nutritional value possess the same form. Foodgiving is classed as a maternal act which can affect the identity of a recipient. But accepting and eating "Murik" foods – sago jelly, seafood, or *aragen* porridge – transfers no "essential component of the body" (Wagner 1972: 41). Food is understood as a token in exchanges which cause a recipient to be indebted and thereby make him or her belong to one group rather than another. Murik foods possess such visible qualities as name, size, shape and number. Women adhere to the dietary restrictions of pregnancy because of the potential of the fetus to mimic visible features of the foods. Giving food is coded as an act of duty, ambition, mystical allure or aggression, but giving food staves off no internal states of being, such as "hunger," as in Milne Bay (Kahn 1986). When a dish of sago jelly is given as a gesture of postcoital gratitude – a woman's reciprocity to her lover – it denotes an act of intercourse, not fluids exchanged or possible procreative consequences. The substances contained within Murik foods are toxic. Mystical contact with the youthful bodies of menstruating, or sexually active, women may contaminate food and cause shortness of breath, or arthritic bones, in senior men. Like the dwelling house, the meanings of food are engaged in the selfsame logic of the inner and outer body. Inner defilements have a strong dialogical association with uterine processes. This is why avoidance is taken for granted as a "gestative" response in Murik modes of conflict management. Avoidance, no less than any of the other manifestations of the grotesque body I have analyzed in this book, is part of an ongoing dialogue between the inner, uterine body and the exteriorized reproduction of the maternal schema.

This schema, I need to emphasize, privileges no specific category, or principle, of relationship, such as matriliny (see A.B. Wiener 1976). Its relationship to cultural processes and symbols is presuppositional. The terms in which order and conflict are lived recreate, in part or in whole, its distinctive inner–outer tensions. The willingness of elder siblings to "give in" and of *sumon*-holders to abdicate their authority, the grotesque guardianship rendered by the Spirits of Wealth, the chastity of the knot in defense of the ascendancy of his moiety in the *Kakar* cult and, lastly, the Oedipal meanings of conflict and the gestational imagery in social control processes, all attest to the pervasive influence of this schema upon the poetics of moral order. Its effect on the cultural construction of the material world is no less discernible. Murik society is thus cast as an "empty body" of relationships that is forever dependent, like children upon their mother, upon exogenous resources. The reproduction of society is intricately bound up with the movement of canoe-bodies and the import of intertribal surpluses. "Her" fertility, that is to say, depends upon relations which take place outside the communal "body" rather than within "her." Murik men and women, as "her" children, contrive to supplement "the hollow mother," not with substances from the soil, or with the gift of life, but with travel and exchange outside the boundaries of "her body." They become the "children" of "trade-mothers" who import their staples. They voyage "like women" in their outrigger canoes "to seduce" their island trading partners with their spells and beauty to import ceremonial wealth. Through the triumph of their external activities, an androgynous image of a heraldic body is assembled and displayed before the community. A placeless group of men and women is situated in a vehicular/body/space amid an abundance of food and aesthetic affirmations of identity, hierarchy and autonomy. All the values that are otherwise missing from, or chronically problematic in, the society – resources, gender relations and authority – are then resolved by an ephemeral image of a perfect, vulnerable image of a maternal canoe-body (see Plate 16).

But this resolution, as I have shown, is but momentary. Viewed more broadly, a metaphor of self set apart from, or opposed to, the other is locked in dialogue with a metaphor of self that remains forever part of the other in Murik culture (Meeker, Barlow and Lipset 1986). The one is advocated by immoral, single-sexed voices whose bodies are full of impure, cosmic fluids, desires and rage. They are countered by moral, androgynous voices who present a quietist and interdependent face. Although both men and women take part in this dialogue, the privileged position afforded the maternal body as the referential object to which they both respond raises imagery of womanhood to the forefront of this eternal quarrel.

Elsewhere in Melanesia, the self has been understood to be "empty" (Leenhardt 1979: 153). Or, to put it another way, the self has been understood as almost indivisibly engaged in reciprocal relations with the other (see, e.g., Malinowski 1922/1959). Indeed, one could take the view that the anthropomorphized gift-exchange Mauss so admired elsewhere in the insular Pacific (1925/1967) suggests an even wider distribution of this concept of self. A central point made in this book has been that the concept of self as part of the other, which is implicit in acts of generosity and moral intervention, comprises one side of a great, gendered dialogics. Removed and estranged as they are from its central image, Murik men have been shown to adopt tactics of artifice and, to a lesser extent, denial, which amount to an attempt to claim maternal capacities for themselves and redefine them, if not as distinctively masculine, then at least as heterosexual. In the service of such a strategy, gift-exchange, which, in everyday practice, is ethically associated with motherhood, becomes, in its grandest, ceremonial forms, associated with both men and women. At the same time as Murik men take pains to make public displays of their abilities to mother, processes and attributes ethically associated with them – e.g., collective aggression and other magico-religious forms of agency – recede into cultic secrecy or have as their purpose a transformation of women's sexual desires to male powers. By detaching themselves emotionally from women, Murik men used to try to purify themselves in order to gain invincibility. But no such achievement of autonomy was possible. The qualities with which Murik women are culturally endowed forever challenge masculinity. Like culture itself, male assertions to agency and authority continue to be unalterably compromised by women's bodies and voices. No matter how reified, authoritative and pristine men purport their bodies, beliefs and agency to be, they remain conditional and stylized dispositions in the shadow of a barely disguised maternal figure. In Murik, culturally particular conditions have given "her" a distinctive form: a moral, hollow canoe-body, on the one hand, and a grotesque body, on the other. Murik masculinity does not lie beyond dialogue; it is defined in and emerges from terms that are simultaneously inside and outside of authoritative discourse. Notions of hegemony, domination and so forth fail to capture the intense dialogicality and attendant ambivalence which suffuse masculine images of moral order. In the 1980s and early 1990s, I have shown that paradoxes of gender identity, rather than assertions of unilateral control, continued to constitute the definitive mysteries of Mangrove Man.

Glossary

A'iin: "Talk." Several types of "talk" are distinguished: the "ancestor-spirit talk" (*pot a'iin*) is of tales about the legendary exploits of Murik culture heroes and heroines. Flirtatious "play talk" (*gwaga'iin*) takes place among classificatory affines and in the bawdy comedy of the Spirits of Wealth. The commensal domestic group engages in "true talk" (*nogo'iin*) and there is also the "stand-up talk" (*dekara'iin*), which is genealogical knowledge shared or withheld by kin. The truce, or embargo, phase of the Mendam–Darapap conflict, during which an avoidance relationship was instituted, was called "there remains [unresolved] talk" (*a'iinaro*). Talk, it is also said, can be "killed" by food distributed to and shared by disputants.

Arabopera Gar: The "Washing Feast." A kind of secondary burial celebration in which the ghost of the deceased is banished from the community and the mourners are released from the impurities with which they have lived following their contact with the corpse. The mourners wash in the sea, receive haircuts and new clothes from the Spirits of Wealth (*mwara* kin). The Washing Feast honors the latter for all they did for the deceased and the mourners by offering them a lavish banquet and an opportunity to perform a folk opera. The end-of-mourning celebration is glossed as a kind of metaphoric "murder" of the Spirits of Wealth by the hosts.

Aragen: A sweet-tasting porridge made of coconut milk and meats, mixed with sago breads which is prepared by the male cult to celebrate various rites of passage and grade-takings. The porridge is a cosmic, male displacement of mother's milk fed aggressively to infant firstborn children by their alters in the male or female cults. In an esoteric tableau revealed to initiates in a long defunct phase of male initiation, *aragen* porridge is also said to be the polluted sexual excretions from the vagina of a culture heroine called Namiit.

Asamot goan or *asamot ngasen*: Literally "trading partner son" or "trading partner daughter" with whom ego maintains hereditary trading relations of varying types depending on their location in the region.

Asamot ngain: "Trade mothers." They are hereditary inland sago-suppliers residing either in small villages which dot the hinterlands north of the Murik Lakes or along the lower Sepik River. These latter are sometimes called the "Bush Murik" or the "Kakra" peoples. The riverine sago-suppliers are also known as "The Number Two Murik." Both groups view the Murik as poor, dependent "children" who depend upon them for sago and garden produce which they supply them on a year-round basis. In the past, sago was bartered in "silent trade" for Murik seafood. Today this pattern is supplemented by the sale of sago for cash, when, in particular, the river dwellers come to Murik villages in their motor-canoes and market their sago packages for K2 each.

Awo: "Goodbye" or "Farewell." The signature valediction in Murik culture.

Baas: A recreational but somewhat rivalrous banquet and house palaver staged by the male Spirits of Wealth to conclude a major rite of passage. The poetics of the banquet are aggressive. The Spirits of Wealth try "to spear" their alters to "death" by presenting each other with plates of seafood garnishing sago pudding, which they encircle about the heads of their rivals just before they deposit a plate in their laps. Following the meal, a palaver should be convened to air and resolve any disputes which have arisen during the stage-management of the just concluded ritual event, as in the case of the conflict between Darapap and Mendam, a state of war was formally ended.

Be iran: "Birth house." This outbuilding is set on the margins of Murik villages because it is held to be defiled by the mystical traces of birth fluids, which pose a hazard to male health.

Brag: Male, tutelary war spirits. The *brag* are manifest in wooden masks (*brag sebug*) and bamboo flutes. They are household spirits which are stored in dwellings. The *brag* are personified in daily life and in ritual by the Spirits of Wealth (*mwara* kin), who engage in joking relations and exchanges in their names. The "Two Brothers," Sapendo and Sendam, are *brag* spirit-men. During the initiation of male youths in the the cult of the *brag* flutes, the novices are "swallowed up" into the "stomach/womb" of the *brag* spirits, who "teach" them how to play the flutes and protect their health and cosmic agency from feminine sexual pollution.

Dag: "Skirts" worn by women. A metonymic reference to and euphemism for sexual services (*dago'mariin*) a woman may provide an older man in several ritual contexts so that her husband may gain various sorts of cosmic authority or agency in return. The rival matrimoieties in the *Kakar* cult are said to have "fought with skirts." Defunct today, the competitive exchange of "skirts" has been replaced by the "payment" of cash money. The evocation of womanhood represented by "skirts" also appears in distinctively masculine contexts such as the cult house, whose exterior walls are decorated with a bunting called its "skirt." An ornament consisting of a series of "skirts" hangs above the floor inside the hall of this building. The sides of the big slit-drums which manifest male ancestor-spirits are decorated with a "skirt" bunting. When newly consecrated, the gunwales and sides of outrigger canoes are also decorated with "skirts." Lastly, both male and female dancers wear "skirts" during the performances in which they appear in full regalia.

Dapag: "Coconuts." An important evocation of maternal attachment. Wet, feminine coconuts are taboo to young warriors as they seclude themselves to purify their bodies before going into battle. Mature coconuts are broken, their milk expectorated by kinsmen to honor the public appearance of heraldry during rites of passage. A gift of coconuts (or firewood) must be made prior to the death of a senior *sumon*-bearer in order to fulfill the protocols leading to the outbreak of the rite of reversal called *noganoga'sarii*.

Debun: Literally, a "drumbeat." There are many kinds of *debun*: e.g., to announce the death of a warrior (*pre debun*), to call an individual out fishing in the mangrove lagoons (*brag debun*), to herald the exchange of wealth (*mwara debun*), or the exchange of a boar's tusk (*sumon debun*).

Dekara: "To stand up." A metaphor for the capacities of agency and autonomy in Murik poetics. "Stand up!" is demanded of a corpse by his *mwara* sisters' sons as they try to revive their mother's brother from his asocial state. "Stand up talk" (*dekara'iin*) is an individual's genealogical knowledge through which he or she claims citizenship, resources and so forth in society. Many men engaged in it during the *baas* palaver. The Murik also refer to the exchanges which repair the unity of a *sumon* insignia which has been "broken" by conflict as "standing up" the *sumon*. The opposite of "standing up" is "falling down" (*turetikum*), the infantile, shameful state of dependency, which is sanctioned by the grotesque parody of the Spirits of Wealth.

Gai'iin: "Canoe." There are five types of canoe: the river canoe (*gai'iin*), lagoon, fishing canoe (*bor*), oceangoing outrigger (*sev gai'iin*), airplanes (*pise gai'iin*) and automobiles (*yabar gai'iin*). The ritual process by which outrigger canoes are built and consecrated, as well as their iconography, evokes an androgynous body, an image which is at once that of a beautiful woman, a mother, a newborn baby and a newly initiated male warrior. As metaphor, "canoe" is applied to the womb, the rib cage and the body as a whole. A group of five siblings is called their parents' "canoe," just as their firstborn is their "prow" (*gai'kev goan*). When a father learns that his son has withstood the pain of learning to expel sexual impurities through his urethra during his initiation, he might exclaim, "Oh! My canoe has come ashore!" The relationship of heraldry to a titleholder is also likened to that of passenger to a canoe. Elders are said "to steer" youth with their guidance. The wicker framework into which a dancer inserts himself in order not only to dance as an ancestor-spirit effigy but to be possessed by it is called its "canoe." The idea of a "canoe" is used also to describe a kind of ethnosemiotic rather than a totemic relationship between a *brag* ancestor-spirit and his or her various manifestations: e.g., a shark is the "canoe" of the *brag* named Marenor and the kookebura is the "canoe" of the *brag* called Senge. The handles of plates and slit-drums are also called "prows." The image of a hollow body and agency of which the canoe presents a crucial metaphorical prototype for an exteriorized body which combines with a maternal schema in the poetics of the moral order.

Gaingiin: A generic category of leafy, masked spirit-monsters which roam the avenues of villages and their adjacent coconut groves. The most junior grade of the male, public masking society holds effigies called *gaingiin*. The monster into which mourners cram themselves as they cross the beach to rinse their bodies of death pollution is also called a *gaingiin*.

Gai'suumon: The "canoe shed," which is a bivouac of the male cult house where outrigger canoes are built and from which women are debarred. This structure is the symbolic equivalent of the birth house, from which men are debarred.

Gar: Any ritual feast or ceremonial exchange. The ritual system of competitive status attribution instigated and managed by the *sumon*-holders who bestow their heraldry, especially upon children and the deceased as they move out of the matricentric domestic group and into the androgynous, cosmic/jural environment. The poetics of Murik feasting evokes images of nonuterine capacities of motherhood in which both sexes participate.

Goan: Son, as in *sumon goan* (the firstborn son, a ritual leader), *asamot goan* (hereditary male trading partner), *kaik goan* (Iatmul, literally the "last sons" in the upriver trade network) and *tata'goan* (elder brother in the *Kakar* cult).

Gorongol: Literally, "ginger leaves" which are used to demark sociomoral purity. They were tied about the prows of canoes and house pilings to conclude the state of war between Mendam and Darapap.

Hausboi or *haus tambaran*: Tokpisin terms for male cult house; see *taab*.

Iran: The dwelling house; symbolically a chaste, maternal space.

Jakum: An aggressive term of address (and endearment) used by Spirits of Wealth in place of personal names or kin terms.

Kakar: Paramount war spirits in Murik religion. They are manifest in carved, individually named, anthropomorphic spear images which are stored in a secret loft deep within the male cult house. Once sacralized, they are held to be very dangerous to the health of women, noninitiates and children. Their aggressive and tutelary powers are vested in a complex dual organization consisting of rival matrimoieties called alternatively "Two Brothers" or the "Big" and "Small" halves of the cult. The moieties may once have been exogamous but this is unclear today. They are subdivided into two age-grades, each of which holds rights to different cultic insignia and power. Cultic partners representing the senior grades in each moiety used to compete to wrest authority over and the cosmic agency of the *Kakar* spirits from each other through the ritual prostitution of their wives. Grade-taking in the cult required the provision by a husband's wife of sexual services to his cultic "elder brother." The *Kakar* cult has been subjected to various forms of repression during the twentieth century and Murik women no longer go to the male cult house to seduce the *Kakar*-holders. Today, grade-taking is contingent upon the payment of money instead of sexual intercourse.

Kandimboang: Male spirits whose images are carved in wood. The figures may serve as a surrogate for youths during their initiation into the descent group status. They are also engraved on the prows of canoes, the handles of paddles, plates and slit-drums.

Katres: Tokpisin term meaning "cutlass" or "cartridge" used by Spirits of Wealth as an aggressive term of address, but also of endearment.

Kiinumb: Coup sticks made of sago midribs which are given to a new initiate when he is promoted into the "younger brother" grade of the *Kakar* cult by his cultic "elder brother" (*tata'goan*). The initiate must "repay" the latter man by sending his wife to the male cult house and offer him her sexual services.

Masok: A supernatural soul or spirit possessed by persons, spirits and canoes.

Menumb: Impure spirits said to be the mystical traces of polluted exchanges of fluids that harm the recipient's health. There are *be menumb*, or "birth spirits," and *son menumb*, or "the spirits of feminine sexual blood" which enter a man's body during intercourse and defile it. There are also *brag menumb*, which are male spirits that tend to make young women infertile.

Merogo or *numero*: "Woman," as in *nog merogo*, which means "kinswoman." The image of a kinswoman is an authoritative sign of moral intervention in the midst of conflict. A *numero apo* is "big woman." Such a woman has been initiated and therefore has jural rights to trade, instigate or attend ceremonial exchange. Her body may not suffer domestic abuse by her husband.

Moan sikemo: Literally "to show a thing," but taken to mean "to set an example" (*soim pasin* in Tokpisin). This is a tactic used by elder siblings to disgrace negligent junior kin who have failed to fulfill some often unspoken expectation. It is also used in exaggeration in the aggressive nurture and sympathetic mimicry practiced by *mwara* kin to shame each other for committing some minor *faux pas* such as not having a spoon or falling down in public.

Mwak: Bodily impurities obtained from contact with a corpse which cause a mourner to withdraw her- or himself from jural society and engage in minor self-mortifications until such time as ritual ablutions are staged. See Washing Feast.

Mwara kin: The Spirits of Wealth. This is a category of joking partners and ritual courtiers who personify *brag* spirits. *Mwara* kin are matrilateral relations of men and patrilateral relations of women. Their duties are varied, and their presence is ubiquitous in each other's lives. The poetics of their discourse, both verbally and kinesically, parodies the staid, maternal conventions of the moral body. They commonly tangle through a kind of social hyperbole which amounts to a grotesque parody of motherhood. As ritual courtiers, they donate *mwaran* and provide services to their charges in exchange for sumptuary rights at banquets. In such settings as early childhood rites, initiation, succession, or death, they recreate moral elements of the maternal schema rather than mock them.

Mwaran: "Valuables" or "wealth." This is a category of objects that includes *sumon* heraldry, pigs, basket reeds, pots, wooden plates, buckets of imported almonds, fishing spears, gifts of cooked food and so forth. *Mwaran* are exchanged by the Spirits of Wealth, as a result of which they are called the *mwara* kin.

Naboag: A "flying fox," or a "fruit bat." This is an important metaphor in Murik poetics. A "flying fox" is alternatively a thief, an illegitimate child, or a husband. The "flying fox mother" is a capricious womb spirit which determines the sex, personality and facial resemblance of a child to one of its parents.

Nabran: The "soul" of a person, spirit or house. In humans, the soul is attributed no gender and it possesses no "zero-sum" relationship to the ancestors. The soul of a house, however, is distinctly feminine.

Nembren or *bren*: "Pig." Pig exchange determines attribution of heraldic identity in the ceremonial economy. Generally, although not exclusively, pigs are imported from Schouten Islanders in exchange for cash or Murik baskets.

Nemot or *mot*: Village, hamlet or locality, e.g., Aramot or Boimot (Kaup).

Ngain and the maternal schema: "Mother," as in "trade mothers" (*asamot ngain*), one's "maternal name" (*ngain ya'ut*), which is assumed to be a "true name," the phrase "your elder brother should be like a mother" (*tatan ngain enambo*), or the term of address of a man's "wife" (*ngain*). In the poetics of moral order, the maternal schema consists of the elements of abundant nurture, acts of generosity, moral intervention, instruction and the maintenance of bodily purity, but largely excludes processes of uterine fertility and the passions except insofar as they can be metamorphosed into "social" acts of mothering.

Ngasen: Daughter, as in *sumon ngasen* (the firstborn daughter, a ritual leader) or *asamot ngasen* (hereditary female trading partner).

Nimbon or *bon*: "Sago pudding." Symbolically, a feminine fluid, this food is also known as the "bones of the ancestors." It is imported from the Murik "trade mothers" who reside in the Marienberg Hills or along the lower Sepik River. Sago pudding is also a sign of post-coital gratitude from a woman to a man as well as a "social" food which supplements mother's milk in infancy. Sago pudding is a "wet," feminine food young warriors forgo during the ritual ablutions they perform before going into battle.

Ningeg: "Spear." This image is used metaphorically during a *baas* palaver to refer to the presentation of a dish of food. It refers to sexual intercourse between a reigning *Kakar* spear-holder and the wife of his junior partner.

Nog or *nogo*: A single-ply "creeper vine" used to bind outrigger canoes, ornaments, women's skirts. *Mwara* brothers tie vines about the foreheads of their sisters (or vice versa) in the midst of their initial grief following the death of a loved one to protect them from soul loss. The *nog* is a bilateral network of commensal kin, or more specifically, a sibling group. Members of the *nog* are differentiated from the ceremonial, classificatory *mwara* kin. *Nogo'iin* means "truth" as opposed to *kakaowi*, fanciful nonsense or falsehood, or as opposed to *gwaga'iin*, which is flirtatious joking. The *nogo ya'ut* is the person's "true name" given to him by his or her mother and the *nogo nogam* are the Murik ancestor-refugees who are held to have floated downriver to the Murik coast where they established Murik society. Lastly, *nog merogo* are "kinswomen" whose appearance in the midst of a fight causes it to cease.

Nogam: Non-Austronesian-speaking Murik ancestor-refugees who are held to have migrated from the middle Sepik River in historical times and then to have become affiliated through marriage with multiple sibling pairs belonging to coastal and lower Sepik descent groups.

Noganoga'sarii: "The fight of sibling groups." This rite of reversal broke out in 1982 in the aftermath of the death of a senior *sumon*-holding woman. Men and women (who were otherwise affinal joking partners in daily life) wrestled each other to the muddy ground and tried to force-feed their alters with a mixture of mud, ashes and animal feces. The rite appears to be a grotesque "weaning" of society from dependency upon a lost ceremonial elder.

Nor: "Man," as in *Bar Nor* which means "Mangrove Man" or *nor apo* which means "big man."

Orub: "Age-grade." Members of a subunit of the *Kakar* cult or the *gaingiin* society. Membership is defined by a common chronological age and/or by fulfilling ceremonial requirements, such as giving a feast or squaring sex-debts or paying a fee or all three.

Pilai kandere: Tokpisin term for the Spirits of Wealth (see *mwara* kin).

Poang: In a basket, the points at which the vines meet and criss-cross to form one of the former corners (called the "breasts") at the base of the container. In a village, the cognatic descent groups which are politically equal and autonomous as they each possess their own heraldry with which to decorate and claim their membership. The descent groups are made up of age-ranked sibling sets (*nog*). They are alternatively referred to as "hearths" or "platforms" (*maig*).

Pokanog: A "knot" tied in a rope or a creeper vine or the hard knot in a tree. The knot image is a major metaphor of sociomoral and temporal boundedness. The office called the knot in the *Kakar* cult is filled by one of the men in the senior age-grade of the reigning moiety. His duty is to abstain and remain chaste in the midst of sexual license among his peers in order to protect the cosmic authority of his and their moiety. Once he is seduced by the wife of his cult "younger brother," that authority must be transferred to the rival moiety and he and his grade are forced into retirement.

Pot: Of, or relating to, ancestor spirits or ghosts. The Murik speak of the *pot nor*, the ancestor-spirit men such as the "Two Brothers," Sendam or

Sapendo, and the *pot merogo*, such as Areke, Namiit and Jari (see Meeker, Barlow and Lipset 1986). The corpus of tales about the exploits and tragedies of these figures are called *pot a'iin*, or "ancestor-spirit talk." They also refer to a numinous community possessed by each descent group called a *pot kaban*, an "ancestor-spirit space" to which ghosts are meant to repair following death or at least the Washing Feast. Ranking titleholders, such as Bate hoped to become, may seek to build a named dwelling, with some sort of signature detail like a little gable or a miniature house on its roof, as part of their incumbency. Such a house is one of the *pot iran*, which is part of corporation of each descent group. Finally, there are also "hereditary baskets," the *pot suun*. Rights to weave and carry them are either inherited or gained through marriage.

Samban merogo: Erotic, female spirit.

Sangait: "Stolen or taken." The Murik speak of "taken" or "stolen children" (*sangait najen*) or a "taken mother" (*sangait ngain*). A *sangait goan* is a "taken son." Full jural adoption used to be legitimated in an exchange called "cutting the breast." An important nonuterine system of reproduction through which masculine dependency upon biological motherhood is displaced by a heterosexual and transactional system of reproduction. Custody disputes over partial adoptions and following divorce generate, along with sexual jealousy, a constant source of conflict discourse in the society.

Sansam: The "ghost," "shade" or "shadow" of a deceased person. The Spirits of Wealth may be asked to drive away the "ghost" of the deceased from his next of kin as they mourn by "shooting his or her reflection" in a cup of water. The *sansam* is also said to be the "shadow" of a living person, which is one reason why dusk, as a time when shadows separate from persons, is thought to be a somewhat dangerous time of day. *Sansam* is also used to refer to photographic "reflections."

Sar: "Stomach/womb"; but also "fight." The seat of the emotions for both sexes. Used metaphorically during the *brag* initiation to refer to the interior of the male cult house within which the youths are secluded and morally transformed.

Schouten Islanders: Austronesian-speaking, politically centralized, off-shore gardening and pig-raising peoples (Kairiru, Wogeo, Bem, Blupblup, Kaduar and Manam), the Murik call *save gata* and from whom they import the wherewithal to fund their heraldic system of status attribution.

Succession: A pre-mortem rite sponsored by an heir for an incumbent either during the latter's end-of-mourning ceremony or separately. The rite is also called "giving paint" (*waikur mariabo*) or the "elder brother feast" (*tata'gar*), should an elder brother be the retiring incumbent. The rule of succession is determined by seniority rather than by gender. Therefore the heir might be a younger sibling, firstborn son or daughter, or firstborn sister's son or daughter. The rite culminates in a *tableau vivant* in which the incumbent and the heir's wife appear seated together on a canoe-like platform surrounded by their kinsmen and women.

Sumon: Sacred, heraldic ornaments that represent the corporation, authority, moral order and jural identity of a descent group. The *sumon* consist of outfits of individually named paraphernalia with which persons (as well as constructions, such as outrigger canoes, dwellings and male cult houses) of descent groups are adorned as they change or receive status during the ritual cycle. The most important emblems are the heraldic baskets (*sumon suun*) and boars' tusks (*usiig*). These images should be held and used in transactions only by the firstborn member, male or female, of the senior sibling group of a descent group resident in a particular village. Metaphorically, the *sumon* insignia are considered to be "hot." They evoke a vulnerable, brittle moral order whose unity can be "broken" by the outbreak of conflict and "repaired" through ceremonial exchange and acts of commensality.

Sumon goan/ngasen: The firstborn heraldic son or daughter of any sibling group. The senior ritual leader of the local section of a descent group who comes to hold title to his or her office by sponsoring a pre-mortem succession rite for an incumbent insignia-holder. Succession is organized by age and ceremonial ambition not gender. The normative role of the *sumon*-holder recreates the maternal schema. He or she should be a man or woman of hospitality and so forth. The *sumon*-holder should also be intimately involved in the fishery, outrigger manufacture and regional exchange. The type is neither a bigman nor a great man but represents some sort of intermediate category.

Suun: Any type of plaited "basket." Several categories of basket are differentiated, such as "trade baskets" (*save suun*), "heraldic baskets" (*sumon suun*), hereditary baskets (*pot suun*) as well as an assortment of differently shaped baskets. Metaphorically, a basket is a female body whose corners are "her breasts," whose handles are "her hands" and whose decorative fringe is "her skirt." The basket is a distinctive metonym of Murik identity both within the community and throughout the region. "A basket," say the Murik, "stands for the person."

Taab: The male spirit house, or cult house, in which sacra, such as *Kakar* spears or *brag* flutes, are stored that are dangerous to the health of women and uninitiated children. Also called a *hausboi* or a *haus tambaran* in Tokpisin.

Tanget: Tokpisin term referring to a sociomoral boundary. Used as a device by pairs of trading partners to synchronize a future meeting.

Tatan/dam relations: "Elder sibling/younger sibling" terms of address and reference. This is a pivotal cultural distinction. It may be applied irrespective of gender. In the kinship system, it describes a primogenitural pattern of modesty, asceticism and indirect nurturant authority by means of which elder presides over younger by maintaining the latter's dependency. It is also employed metaphorically to discriminate the rival matrimoieties in the *Kakar* war cult.

Timiit: "To do or make." The mystical agency, or sorcery, which can be operated by persons, ghosts and spirits for either moral or immoral purposes. There is the tutelary sorcery of a ghost (*nabran timiit*) or an ancestor spirit (*brag timiit*). There is also pig-trading sorcery (*nembren timiit*).

Usiig: A "boar's tusk." The equivalent of a pig in the precontact era. One of the most important valuables (*mwaran*) in the material culture. An important motif in *sumon* regalia, e.g., Makembo, the mother of Malai, one of the youths originally involved in the conflict which escalated into the intervillage brawling, wore a net bag vest *sumon* studded with six boars' tusks when she greeted the rival community as they came ashore to end the state of war. Boars' tusks were also exchanged in the compensation payments made during the subsequent reconciliation rite. Rights to make transactions with boars' tusks are exclusively vested in initiated firstborn men or women.

Waikur: Red ochre "paint" applied to the bodies of initiates, titlebearers or dancers who appear arrayed in their regalia. "To give paint" is a metonym for staging an initiation or a succession rite in honor of a youth or the holder of a *sumon*ship.

Wajak: A ceremonial "double." The bilaterally inherited heraldic feasting/joking rival. His or her primary duties serve to facilitate the ritual transaction of *sumon* emblems. In return for services provided, the *wajak* is honored with the best, most prestigious portions of the feast. He or she is particularly charged with safeguarding the sacred unity of heraldry. Thus there was a failed attempt to activate *wajak* relations during the Darapap–Mendam negotiations. These partners are classed as Spirits of Wealth (*mwara yakabor*). They personify war spirits (*brag*). In daily life, a strict property avoidance, sanctioned by an expensive penalty, separates *wajak* doubles.

Wandiik: Respectful term of address used in senior–junior affinal avoidance relations, e.g., between elder sister and younger sister's husband.

Its emotionally repressed, modest ethos contrasts with bawdy, flirtatious joking relations that obtain between noncommensal, classificatory affines, e.g., a younger brother and his elder brother's wife.

Yakabor: Literally, "path," but used to refer to relationships or categories of relationships, such as the *mwara yakabor*. It also refers metaphorically to an outrigger canoe or an outboard motor.

Yanga nabiin: Literally, "our eye." This phrase is used to mean "shame," especially that which affines are expected to display in each other's presence. During the speechmaking to conclude the state of war between the two feuding villages, men professed to feeling the sentiment as a result of the fighting, but had different reasons. Some said they were ashamed because of affinal modesty which had been breached, others admitted to feeling ashamed because of common genealogical and heraldic ties which had been breached. *Mwara* kin are said to "share the shame" of their alters when they "fall down" or commit some of any number of minor *faux pas*.

Yarok: The moieties which "fight with skirts" for ascendancy in the *Kakar* cult. While, *vis-à-vis* each other, the senior members of each moiety are "elder brother" to "younger brother," internally, each *yarok* consists of a "father" and a "son" age-grade which holds (and exchanges) insignia of rank in the cult. The *Kakar* spears are held by the "father" grade in the ascendant moiety while daggers are held by the "son" grade in their moiety; lances are held by the "father" grade in the subordinate grade while war magic leaves are held by the "son" grade in this moiety. Each moiety also holds rights to sit around one of two firepits on the floor of the cult house, which they do on a daily basis and whenever the cult formally assembles. See *Kakar*, *orub*, *kiinumb* and *pokanog*.

Notes

Introduction

1 On *naven* rites, see Forge 1971; Keesing 1982; Korn 1971; Gewertz 1983, Handelman 1979; D'Amato 1979; Marcus 1985; Herdt 1984; Clay 1977; Lindenbaum 1987; Morgenthaler, Weiss and Morgenthaler 1987; Stanek 1990; Weiss 1990; Wassman 1990; Silverman 1993; and Houseman and Severi 1994.

2 My original research plan had focused on the relationship of local forms of "Melanesian leadership" to political development of the nation state (see Lipset 1989).

3 On schemata, see Rumelhardt 1978; see also Lakoff and Johnson 1980; Casson 1983; and Johnson 1987: 19ff. Barlow first used "maternal schema" in 1985.

4 This is the first full-scale application of a Bakhtinian concept of culture in a non-Western ethnography (cf. Karp 1987). But a steady trickle of anthropologists have made limited use of particular ideas; see Battaglia 1990; Briggs 1993; Crapanzano 1990; Da Matta 1991; Feldman 1991; Gottlieb 1989; Hereniko 1994; Limon 1989; Mumford 1989; Sahlins 1985; Trawick 1988. For an excellent, more general discussion of the influence of Bakhtin upon ritual studies see Kelly and Kaplan 1990. See also Bauman and Briggs 1990 on the relationship of Bakhtin to linguistic anthropology. For a valuable exposition of Bakhtin's view of psychoanalysis, see Daelemans and Maranhao 1990. See also Tedlock and Mannheim 1995.

5 For two recent surveys of a vast literature, see Lock 1993 and B. Turner 1991.

6 For evaluation of Rosaldo's argument, see Comaroff 1987; Josephides 1985; Nash and Leacock 1977; Quinn 1977; Rapp 1979; Reiter 1975; Rogers 1975; Rosaldo 1980; Sacks 1975; M. Strathern 1980; 1981; 1988a; Yanagisako 1979.

7 For discussion of Collier and Rosaldo's typology in Papua New Guinea, see Feil 1987; Kelly 1993; and Wood 1987.

8 In Papua New Guinea, this kind of aquatic adaptation is comparable to that of the Manus of the Admiralty Islands (Fortune 1935; Mead 1956; Schwartz 1963; see also Carrier and Carrier 1989) and the Siassi of the Vitiaz Strait (Freedman 1979; Harding 1968; Pomponio 1990; 1992).

9 By contrast, in pastoral regions, work and warfare do provide extrasocial and antisocial opportunities which are culturally honored and elaborated.

Representations of self and other in the Middle East and East Africa, for example, stress opposition and boundedness. The capacities of individual men who thieve or defend themselves from attackers are glorified (Meeker 1989).

2 A predicament in space

1 There are also three other subgroups of Lower Sepik speakers (Karawari, Yimas and Chambri) which dispersed from the lower river groups about 2,500 years ago (Foley 1986).

2 Tiesler developed this argument in the German Democratic Republic, during the late 1960s when non-Marxist analyses were considered politically subversive by the state, which retaliated by banning him from doing research or going to professional meetings outside its borders.

3 Compare this image of the elder brother spirit with the people the explorer Finsch described around the mouth of the Sepik River in 1885:

> Their hair was done up in a thick pigtail standing away from the head and put in a neatly woven little basket . . . Their richly decorated loin-cloths made from tapa was something I saw nowhere else . . . Body decorations were plentiful. Strings of little shells and dogs' teeth were worn over the forehead, neck and hips . . . Bracelets . . . were woven from ferns . . . Some were engraved . . . Rosettes of dogs' teeth used as breast decorations and body-strings made of peculiar shells pleated onto strips of raffia that was probably used as money, too, all were new to me . . . Most of the men wore pretty knitted bags around their necks, beautified with splinters of cymbim shell. They carried special baskets as well, woven from palm leaves adorned with shells and colourful dyed ferns that looked like velvet.
> *(quoted in Whittaker et al. 1975)*

4 Among all their island trading partners between Muschu and Bem, the Karau are known as the "Ut." And indeed, the Karau continue to call their village "Ut'amb" or "Uta'mot." I should add that both Murik and Sub men translate the word *utim'kara* to mean "big prick."

5 I, however, shall distinguish between these communities on the basis of their location in space. The subdivisions of the communities which compose the constituency of a single local government councilor, such as Big Murik and Kaup, but are adjacent, I shall call hamlets. The communities which are separated by water, but identify themselves as having common ethnohistorical origins, such as Karau, I shall call villages.

6 Rights to fish in the ocean are not divided.

7 Store-bought, fiberglass dinghies were also coming into use during the early 1990s.

8 Even in premodern times, initiated Murik women traveled on trade expeditions.

9 This was the case all along the river: the waterfront groups viewed themselves as militarily and culturally superior to their inland neighbors although they depended upon them for their staple carbohydrate resource. See Barlow 1985a; Bateson 1936/1958; Gewertz 1983: 27; Harrison 1982; Hauser-Schaublin 1977; Lipset 1985; Schindlbeck 1980; Tiesler 1969/70.

10 In Darapap in 1981–2, each household made three trips to Wewak during the dry season, and returned to the village with an average of K23 (US$30) per household net profit.

3 The maternal schema and the uterine body

1 For beliefs about reproduction in particular Sepik cultures, see Bohm 1983; Errington and Gewertz 1987: 146 n. 8; Forge 1971; Hogbin 1970; Kulick 1992: 92–9; Lewis 1980: 175–6; Williamson 1983. For comparative discussions of this subject in Papua New Guinea at large, see Bulmer 1971; Herdt and Poole 1982; and Jorgensen 1983.

2 Barlow and I have gone so far as to argue that women's conduct during the construction, launching ritual and voyaging of the outrigger is the symbolic equivalent of the couvade behavior which Murik men are expected to enact during birthing (see Barlow and Lipset 1997).

3 Magic to speed a difficult birth and make the child strong is also known to certain senior men.

4 In 1981–2, 20–25 per cent of all children living in domestic groups in Darapap and Karau were adopted and there were no domestic groups lacking in adoptees or children who had been given up for adoption. On Murik adoption, see Barlow 1985b; Lipset and Stritecky 1994; on adoption elsewhere in the insular Pacific, see also Carroll 1970 and Brady 1976.

5 Actually, custody rights remain very tense. Child custody is a delicate and intensely disputed issue, especially since first marriages so frequently end in divorce. There is competition to recruit children into the sibling group of each parent. Domestic groups very often consist of adopted children and children of several different parents. Individuals commonly maintain multiple ties to natal, step- and adoptive parents, and an individual's loyalty, affection and work may be competitively claimed by several sets of kin at once. Most frequently first-born or last-born children are adopted. The incidence of divorce among the newly married seriously weakens the otherwise privileged position of the first-born offspring. When adopted into one of the families of divorced parents, his or her status as firstborn heir may be challenged by the firstborn child of his parents' second marriage. See Chapters 4 and 5 for further discussion of this issue.

6 On siblingship in Pacific societies, see Burridge 1959; Brady 1976; Kelly 1977; J.F. Wiener 1982; and Marshall 1983.

7 Alternatively, a sibling group of five offspring of the same sex may be called a "canoe" (*gai'iin*) by its parents.

8 This pattern also obtains between younger sister's husband and wife's elder sister.

9 One of Barlow's adoptive mothers, for example, insisted on calling me "Kathy."

10 As may classificatory younger sisters and their elder sisters' husbands.

4 The heraldic body

1 But Harrison does note that Manambu agnatic names are also referred to as "bone names" (1990a: 60).

2 The Kaian apparently believe that they came to the North Coast in outrigger canoes. The "platform" is possibly a metonym for the vessel on which male (but not female) passengers and crew were permitted to sit while at sea.

3 In this sense, it might be apt to call these groups "lineages," but I am hesitant

to go ahead and do so since their membership is horizontal and bilateral rather than lineal. I am aware, however, that the Polynesianists/Micronesianists have not been so nervous about using the term to describe just these kinds of groups (see Howard and Kirkpatrick 1989).

4 Kulick (1992) mentions that in Gapun village, located a few miles east of the Murik Lakes, a firstborn daughter received the "*siman*" insignia of her matri-clan during a rite of passage staged in the 1970s. But he too does not translate the term directly. Elsewhere in this quite peculiar book, Kulick seems to trans-late the word "*suman*" as meaning "big" (1992: 287–8, n. 2).

5 Neither were they tokens of affinal exchange. Rights in the disposal of women's sexuality are vested in the commensal sibling group not the *poang*. According to Father Schmidt, however, during 1911–40, the *poang* were associated with marriage prohibitions.

> Through initiation, the children receive *sumon* from the father's *poang* and the mother's [*sic*]. Therefore the children ought not marry members of the mother's *poang*, at least as long as the fact (of marriage and descent) is still fresh in their thinking. If there is quite a long time in between, the children's children may marry into the mother's *poang*. The father can also give special dispensation. An example: Philip Osega is of the Unaroman *poang*, his wife is from the Kania *poang*, but was adopted into the Unaroman *poang* and was considered a member of this new clan. Osega could not marry his present wife without hesitation. But his father gave permission and said, "You may marry her quietly."
>
> *(1933: 670)*

I heard about these restrictions in 1981–2. But I never heard any reference to them with regard to the marriages or divorces that actually occurred then. Schmidt's ambiguous example (and genealogies I collected) confirm that the *sumon*-holding groups were not rigorously exogamous even during his time on the Murik coast.

6 Except for Koisop, a spirit-man who underwent a sex change because a female woodsprite was taking revenge against the jealousy of his two co-wives. Yet even Koisap's body turned entirely to that of a woman.

7 There is one esoteric institution called "turning over all the floorboards of the male cult house" (*tibur tokobun iarara*), through which a single insignia-holder may achieve a paramount status in and discursive authority over an entire com-munity for the duration of his lifetime. But the achievement of this singular title, which is unnamed but is signified by the privilege of carrying a distinctive basket (*ya'suun*), is not hereditary and does not derive from membership in a particular descent group. It is open to any *sumon goan* who fulfills its ritual prerequisites, namely mounting a feast (*ya'gar*) and completing other highly secret protocols.

8 Murik kinship terminology does not differentiate cross-cousins from other same-generation collaterals. With respect to *sumon* competition, however, fathers' sisters' children are called "skirts" (*dag*) who are opposed by "loin-cloths" (*nimberon*), the fathers' brothers' children. These two groups are expected to compete with each other to sponsor *sumon* displays and should monitor each other's ritual activity (cf. M. Marshall 1983).

9 This interpretation also fits with the idea of "cutting the breast" discussed in Chapter 3.

10 By contrast, on Wogeo Island ornaments were brought out to "frighten the

dying individual into prolonging the struggle" because the corpse will be decorated in them after death (Hogbin 1970: 157).

11 But not in the two western ones of Big Murik or Kaup.

12 Mourners are said to "sit with ashes" or lime powder on their faces.

13 The literature on bigman leadership in Melanesia is extensive. For discussions of specific cases see P. Brown 1963; Burridge 1975; Clay 1993; Epstein 1968; Hawkes 1978; Langness 1973; Keesing 1978; Meggitt 1973; Oliver 1955; Read 1959; Salisbury 1964; Sillitoe 1978; 1979; A.J. Strathern 1966; 1979; Watson 1973. For comparative discussions see M. Allen 1981; 1984; Berndt and Lawrence 1972; Chowning 1979; B. Douglas 1979; Flanagan 1989; Forge 1972; Godelier and M. Strathern 1991; Lindstrom 1981; 1984; McDowell 1990; Modjeska 1982; and Sahlins 1963.

14 The literature on Melanesian chieftainships is less extensive: see Hau'ofa 1971; 1981; Hogbin 1978; Lutkehaus 1990; Malinowski 1922; Mosko 1985; Powell 1960; Wedgwood 1934. See also Brunton 1975.

15 Which is the case in Wogeo as cited above.

16 Except in the case of an heir who has staged an installation rite (*ya'gar*) in the course of which his wife seduces *every* insignia-holder in his generation. The spouse of such an ambitious man, it is said, has "turned over all the floorboards in the male cult house," the last of which is his elder brother, father, or mother's brother, e.g., the incumbent he wants to succeed. He then assumes a paramount position in his community. He is said to have achieved a discursive authority no man may challenge. He may give any order to his rival *sumon goan*, who must obey him because he can shame them by saying, "Oh! Haven't you put your leg on my 'canoe' [i.e., had sexual intercourse with my wife]?!" before going to seal a secret pact with the rest of the insignia-holders to conspire to ensorcell such an insubordinate man. Such a rarely won success is not transferable to an heir.

17 For the classic examples of ceremonial foodgiving in Melanesia see Feil 1978; Glass and Meggitt 1969; Malinowski 1935; Meggitt 1974; 1977; Oliver 1955; Rappaport 1968; Serpenti 1965; Sillitoe 1979; A. Strathern 1971; Tuzin 1976; A. Wiener 1976; Young 1971; 1984.

18 In Murik, heraldic titleholders believe that mystical powers over the material world – particularly over the sea and weather – belong to the Schouten Island leaders. Agency in ceremonial exchange, for which multiple insignia-bearers compete, benefits from all kinds of magical knowledge they do possess. But none of these powers are cosmogonic. None of them establish and protect the preconditions of ceremonial exchange; none of them account for the origin of the world. They promote production, industry, protect travel, ensure bountiful gift exchanges and so forth, but they make no more primordial claims. Of course, it would be a distortion to omit the point that they believe that they have power over the life and death of individuals who commit "crimes" against their insignia or kinswomen. But knowledge of sorcery is ubiquitous in the region.

5 Who succeeded Ginau?

1 Matthew Tamoane, the firstborn son of Ginau's rival, was also planning to stand in the 1987 election. Tamoane did so but was defeated (see May 1989).

2 At some point between 1988 and 1993, Marabo did manage to retrieve the walking-stick *sumon*. "I sent a pig to [my feasting partner] together with a stick," Marabo recalled, when I asked him about the outcome of the conflict. "He attached boars' tusks to it and returned it to us."

3 An heir, it is also said, should repay a fosterage debt; the tributary feast is meant to cleanse his *sumon goan* for having soiled himself with excreta during his or her infancy.

6 A body more carnal

1 This story was told to me by Peter Kanari of Big Murik in 1982.

2 Andena and Arena, the "Two Brothers", are also classed as *brag* spirits. Recall the similar expression of rage by the elder brother as he tried to kill his younger brother by crushing him with the centerpost of the cult house he was building, "He raped my wife! He raped my wife! He is no good! Kill him!"

3 As a premodern, or precapitalist, category of ceremonial goods, *mwaran* include *sumon* insignia, which are inalienable, as well as such alienable ornaments as pigs, almonds, pots, plates and basket reeds.

4 On the female spirits, see Barlow 1992 and 1995.

5 The image conveyed by the Tokpisin term – *pilai kandere* – which is used widely, stresses familiarity among matrilateral kin rather than the exchange of goods.

6 In Chapter 5, Ginau repeatedly sought to oblige his classificatory sisters' sons to perform ritual services for him.

7 Since both identical services and goods are exchanged, here again we see Collier and Rosaldo's distinction between the body in brideservice and bridewealth systems confounded.

8 *Moan apo!*, which literally means "Big thing!", being one of the most common euphemisms; another is *Delik apo!* meaning "Big stick!".

9 Mothers, that is to say, spoon-feed young children sago pudding.

10 I have not observed this initiation: my discussion is drawn from a composite of informants' accounts, Father Schmidt (1933) and the autobiography of Sir Michael Somare (1975).

11 The bones of the father or an important elder used to be saved and then displayed during the *brag* initiation. If the *mwara* mother's brother noticed insects chewing on the bones, he would assume that the novice, or his father, had seduced his wife and would kill him by deliberately inserting a rotten reed into the youth's urethra.

12 While Mead reported that the Mountain Arapesh men viewed penis bleeding as a symbolic form of male menstruation, as did Hogbin in Wogeo (see also Bettelheim 1954), Lewis could not elicit this association among the Gnau (1980; see also Roscoe 1990: 413 n. 18). Tuzin, who actually observed the practice among Ilahita Arapesh, has added that it also has an autoerotic meaning. I did not see it performed and did not collect any evidence that Murik men construe it either as pleasurable or as a form of male menstruation.

13 This threat has been a subsidiary theme in their joking relationship.

14 The soul is thought to escape through the fontanel, which feels as if it is

splitting open, throbbing in pain. The vine binds it closed and protects the mourner from soul-loss.

15 And vice versa, brothers will attend their grieving *mwara* sisters.

16 Eating this pork is sometimes referred to as "eating the skin" of the deceased.

17 Dignitaries may also be given more major presents by their hosts, such as a spirit mask, a slit-drum or an outrigger canoe name, in appreciation of their attendance: in Tokpisin such an important gift is referred to as "buying the head" of the deceased.

18 Sleep is also impure: the soul may come into contact with ghosts of the dead and the body wakens with the supernatural residues of this nocturnal interaction. More fastidious Murik therefore "wash" every day at dawn. Each new day is marked by "birth."

19 The only "endocannibalism" found in Murik repays the *mwara* kin who "eat the skin" of the deceased in the form of cooked pork fed them during the Washing Feast. Rather than introjecting the potency of the deceased, as among the Gimi, the meal resolves the debt owed by the deceased for their provision of burial services. Endocannibalism, in other words, is removed from the realm of substance and reformulated in terms of indebtedness and the exchange of services.

7 The sexuality and aggression of the cosmic body of man

1 There are two types of cult building (*taab* and *kamasan*). I discuss the former one in this chapter. The *kamasan* store no sacra, or the sacra they contain are not toxic to noninitiates. Senior women occasionally approach them and will put their heads inside to ask their husbands or sons about something. In 1982, there were three *taab* in the five Murik villages and six *kamasan*. In 1988, the same number of *taab* and *kamasan* still stood, but two more *taab* were being built and the construction of a third was being debated.

2 See Bateson 1936, Plates I and VII; Swadling et al. 1988: 217; Forge 1966; or Tuzin 1980.

3 The bunting is also meant to block mosquitoes coming through side-vents.

4 I should mention that one usually finds the bedding of a few men – bachelors, or husbands wanting to avoid noisy children or menstruating women – hung on the walls. Formerly, only initiated men had rights to sleep in the cult house and apparently many regularly took this option.

5 The story was told to me on separate occasions in Tokpisin by Nangumwa of Mendam and Saub Sana of Karau.

6 In another version of this myth published by Beier and Somare (1973: 4), the rape and birth occur in the middle of a lake which was not a real lake but the blood and fluids which Arake emits during labor.

7 The commensal mother's brother makes his claim by splashing white lime powder on the mother's pregnant stomach and then, after the birth itself, the claim should be confirmed by having food brought to the mother while she is still in the birth house.

8 Did this moiety system organize marriage? Murik women were supposed to belong to the same moiety as their husbands. Whenever I asked about the relationship of marriage to the *Kakar* cult (and I asked many different people

repeatedly in many different ways), I was put off with the same perplexing answer: "Before, we used to exchange sisters. Today, people marry according to their own wishes." But I was also given to understand that, should a man marry a woman who belonged to the moiety opposite his own, he could then cross over and sit with the other moiety at his own discretion. This was presented as an anomalous case.

If a woman married a man and *shared his moiety status*, then the recruitment of her firstborn son by her brother would have placed the father and her son in rival groups and united mother's brother and sister's son (see Figure 16). During the childhood of his protégé, the cultic role of the commensal mother's brother is one of nurture and instruction. The mother's brother should teach his nephew which food taboos are required by their common *brag* and *Kakar* spirits. He should teach him which designs (*sigia*) to engrave, or at least identify, on the material culture of their *nog*, namely, canoe prows, spears, paddles, house posts, plates, and so forth. He should teach him to beat the slit-drum calls of their spirits (*brag debun*). And most importantly, the mother's brother should teach his charge the name of and the legends associated with their mutual spirit-ancestor. Before the boy is initiated, and is otherwise barred from entry into the hall, the mother's brother may invite his nephew to join him during minor feasts. The youth should eat quickly during such illicit meals and then get out. Long before initiation, he has thus become familiar with the interior of the cult house and some of its sacra.

9 But the meaning given by Beier and Somare for *pokanog* is "tail" (1973: 10), presumably the opposite number of the "head."

10 Although the gender is different, tying collective status to an image of chastity safeguarded from sexual license suggests an interesting comparison with the sacred maid complex in old Samoa (see Shore 1981).

11 In Gapun village, located about 30 km from the Watam Lakes, which are the eastern estuary of the Sepik, Kulick was told about the belief that men should not have sexual intercourse with women during the day because the sight of their genitals would excite them too much and cause them to become unable to concentrate on the goings on in the male cult house as they would be "worrying about who was having intercourse with it" in their absence (1992: 287 n. 8).

12 Such alliances shifted and did not permanently confederate the Murik with other cults. Indeed, they did not even unite the five Murik villages into a single tribal sodality, much less a single village. Intrigues involving "half-moon men" who double-crossed their own communities by secretly relaying maneuvers to enemy villages were commonplace. Such betrayals was always a retaliation for the alleged rape or seduction of married women. Harrison rightly concludes that Sepik River military alliances were between "actor focused" segments of descent groups rather than between whole villages (1990b: 593).

13 After a male cult house becomes too dilapidated for further use, the junior moiety will also "attack" and "kill" the building (tearing down its remnants) as if it were a dying warrior.

14 *Aragen*, the sweet-tasting, milky porridge, made of fresh coconut milk and meats mixed with shredded sago breads, that is prepared by the cult heirs in honor of the "retiring" senior grade, does have a relevant esoteric meaning.

During a scene in a long defunct phase of male initiation, a grade of married youths are taken by their "elder brothers" to a secluded spot on the edge of the lakes. They are there shown a figure of a mythic woman, called Namiit, whose face and breasts have been carved onto the trunk of a big mangrove tree so that two aerial roots appeared spread out like her "legs" (cf. Mead 1935/1963: 271). On the ground, beneath Namiit's "legs," they have placed a big bowl of *aragen* porridge, which the novices are meant to eat. While they do so, their initiators reveal a secret meaning of the porridge: it is "really" sexual excreta which has dripped from Namiit's vagina. The "elder brothers" begin to chant – Vulvas! Vulvas! (*Puniim! Puniim!*) – after which they are allowed to have intercourse with the novices' wives. The "wife-givers" then receive rights to exchange the image and reveal the secret meaning of Namiit to the next group of novices during a subsequent initiation. Thus, the milky substance explicitly condenses the values of maternal nurture, feminine sexuality and cultic reproduction but, once again, the meanings of the symbol do not concern uterine fertility.

15 In 1993, I did learn about an esoteric sense in which the defiling power of the *Kakar* was said to depend upon masculine blood but not upon the exchange of semen. In order to sacralize a newly carved *Kakar* spear it is necessary "to wash" it with blood expelled from the carver's penis. The carver should mix his blood with red ochre and ginger leaves and rub the spear image all over with it. Only then, and precisely because of the presence of male blood in this mix, does the *Kakar* become lethal to women and children. Then, should a woman accidentally tread or make love upon the isolated spot where a carver has "washed" a new spear, or where the spears have been taken for some initiatory purpose, a woman's body will gradually become dehydrated. She will bleed vaginally and lose the rest of her fluids, before dying. The cause of this disease is understood to be her having had contact with the blood with which the carver has rubbed and bespelled the spear image. The cure involves two steps: first a sister of the patient should be brought to the cult house by the woman's husband to have sexual intercourse with the knot of the reigning moiety. They make love in the presence of a single spear which the knot has taken down from the loft and stood up next to them. Having been obliged to do so, he will then mix some of the ashes/blood taken from the firepit located below their loft with some water and a few ginger leaves. He will then "wash" and "cleanse" the sick woman's body with this concoction and give her to drink what remains of the liquid: the blood of the *Kakar*, with the help of the cosmic spirit of the knot, is said to cure her and "retrieve" her defiled soul (*masok bomunge*). The cosmic spirit of the knot will enter her body and rid her of the spirit of the *Kakar*. Thus here again, we see that the meaning given to cosmic intercourse is not procreative but indebting.

16 Thurnwald (1916) reported a complicated instance of wife-lending among the Banaro of the Keram River, which is a tributary of the lower Sepik. But in this thoroughly cognate case, the sexual exchanges are expressly associated, if not with fertility beliefs, then certainly with the priority of the cult in the bodily reproduction of society. In the Banaro male cult, which was a moiety system that regulated marriage selection, sister-exchange was said to have occurred simultaneously between the moieties. Inside the cult house in the presence of

the flute spirits, the cult partners of the grooms' fathers would then take the privilege of initiating the two brides into "the mysteries of married life" (Thurnwald 1916: 261). The two husbands indeed were not to touch their wives until each of these women had given birth to a cult child, who was conceived during this cosmic intercourse.

17 In the ritual heterosexual exchanges which have been reported, many cults do indeed show concern with fertility. Asmat (Eyde 1967) ritual intercourse was largely associated with promoting cosmic and social fertility in feasting and warfare. The Kiwai (Landtman 1927) mingled fluids during ritual intercourse which were collected to ensure fertility and strength (wives were also offered to guests as an act of hospitality). Marind Anim (Van Baal 1966; 1984) men and women also made use of fluids mixed during sexual intercourse to increase their strength and fertility. Among the Kolopom (Serpenti 1965; 1984), fluids created by ritual intercourse between men and women were rubbed on initiates to stimulate their growth and were rubbed on men's bodies before they went to war (see also Kehoe 1970 for a brief comparative discussion of the role of sexual exchanges in Plains Indians religious societies which extends Lowie 1916).

18 Today, images of *Kakar* spears appear on the state's K100 notes.

8 Conflict and the reproduction of society

1 Four democratically elected councilors (Tokpisin: *konsil*) represent the five Murik villages. In 1982, the four were middle-aged, firstborn men, who were all heirs to heraldic titles. Each man was assisted by a second (Tokpisin: *komiti*) and these ancillary officers were also firstborn. Their main duties were to conduct regular Monday morning village-wide meetings, during which they mediated conflict and organized public works projects, and to attend quarterly assemblies in the district center. In Darapap, during the 1980s, the weekly councilors' meetings were held on a basketball court adjacent to the male cult houses.

2 They were, in addition, forbidden to smoke. Tobacco, although a dried substance, is prepared by all ages of women. The exchange of tobacco, moreover, is a moral gesture.

3 Note that mourners are also said "to sit with ashes."

4 The sons had honored the cult; the necklaces were grade-taking insignia.

5 The one cost K75 (US$95) and one hundred baskets; while the other had been purchased for K45 (US$65) and the promise of twenty more baskets.

6 I was presumably still trying to establish myself on better moral footing there.

7 The frond is stuck in the ground. A knot is tied in a cluster of leaves which then droop forward, forming a V-shape in the frond which represents a symbolic "vulva"; the leaves knotted together are called a "penis," which is forbidden entry (see Glass 1988).

8 On knot imagery see also van Gennep 1914: 28–40; on knot imagery elsewhere in the insular Pacific see Shore 1989; Lessa 1966: 45; White 1990: 61; Boggs and Chun 1990; Ito 1985; and Shook 1985.

9 In other words, Kaparo was reminding his audience that not he, but others, had originally caused the conflict which had led to the brawling.

9 Social control and law

1 This policy had been initiated with the creation of an independent system of village courts in 1972 (Scaglion 1979; 1987; Scaglion and Whittingham 1985; Zorn 1990).

2 Not only does conflict recapitulate cosmogony in Murik, so does hospitality; leave-taking reenacts the departure of Andena, the elder brother spirit, as he ventured through the region in search of his lost brother. A host should stage a farewell party the night before the departure of his trading partner, during which the guest should offer some small departure-gifts (see Chapter 2). Of course, however embedded in legend, these final acts of generosity are motivated.

3 Wangi was staying in town to receive treatment for tuberculosis.

References

Abbott, Stan. 1978. Murik verb morphology, unpublished manuscript, Ukarumpa PNG, Summer Institute of Linguistics

Abbott, Stan and Jo Deanne Abbott. 1978. Murik grammar: from clause to word, unpublished manuscript, Ukarumpa PNG, Summer Institute of Linguistics

Abel, Richard L. 1973. A comparative theory of dispute institutions in society, *Law and Society* 8: 217–347

Abu-Lughod, Lila. 1986. *Veiled Sentiments: Honor and Poetry in Bedouin Society*, Berkeley, University of California Press

1990. The romance of resistance: tracing transformations of power through Beduoin women, *American Ethnologist* 17: 41–55

1993. *Writing Women's Worlds: Bedouin Stories*, Berkeley, University of California Press

Alder, William F. 1922. *The Isle of Vanishing Men: A Narrative of Adventure in Cannibal Land*, New York, Century Company

Allen, Michael. 1967. *Male Cults and Secret Initiations in Melanesia*, Melbourne, Melbourne University Press

1981. *Vanuatu: Politics, Economics and Ritual in Island Melanesia*, San Francisco, Academic Press

1984. Elders, chiefs and big men: authority legitimization and political evolution in Melanesia, *American Ethnologist* 11: 20–44

1988. The 'hidden power' of male ritual: the North Vanuatu evidence. In D.B. Gewertz (ed.), *Myths of Matriarchy*, Sydney, University of Sydney Press, pp. 74–96

Allott, A.N., A.L. Epstein and M. Gluckman. 1969. Introduction. In M. Gluckman (ed.), *Ideas and Procedures in African Customary Law*, Oxford, Oxford University Press, pp. 1–96

Arno, Andrew. 1990. Disentangling indirectly: the joking debate in Fijian social control. In K. Watson-Gegeo and G. White (eds.), *Disentangling*, Stanford, Stanford University Press, pp. 241–89

Avruch, Kevin and Peter Black. 1993. Conflict resolution in intercultural settings: problems and prospects. In D. Sanhole and H. Van der Merwe (eds.), *Conflict*

Resolution Theory and Practice: Integration and Application, Manchester, Manchester University Press, pp. 131–45

Avruch, Kevin, Peter Black and Joseph Scimecca. 1991. *Conflict Resolution: Cross-Cultural Perspectives*, New York, Greenwood Press

Baal, J. Van. 1966. *Dema: Description and Analysis of Marind-Anim Culture (South New Guinea)*, The Hague, Martinus Nijhoff

1984. The dialectics of sex in Marind-Anim culture. In Gilbert H. Herdt (ed.), *Ritualized Homosexuality in Melanesia*, Berkeley, University of California Press, pp. 167–210

Bachofen, J.J. 1861 (1967). *Myth, Religion and Mother Right: Selected Writings of J.J. Bachofen*, trans. and ed. Ralph Manheim, Princeton, Princeton University Press

Bailey, Frederick. 1969. *Strategems and Spoils: A Social Anthropology of Politics*, New York, Schocken

Bakhtin, Mikhail. 1965 (1984a). *Rabelais and His World*, trans. Helene Iswolsky, Bloomington, Indiana University Press

1981 (1985). *The Dialogic Imagination: Four Essays by M.M. Bakhtin*, ed. M. Holquist, trans. C. Emerson and M. Holquist, Austin, University of Texas Press

1984b. *Problems of Dostoevsky's Poetics*, trans. Caryl Emerson, Minneapolis, University of Minnesota Press

Bakhtin, Mikhail and P.N. Medvedev. 1978 (1991). *The Formal Method in Literary Scholarship*, trans. Albert J. Wehrle, Baltimore, Johns Hopkins University Press

Bandlamudi, Lakshmi. 1994. Dialogics of understanding self/culture, *Ethos* 22: 460–93

Barlow, Kathleen. 1985a. The role of women in Murik trade, *Annual Review of Research in Economic Anthropology* 7: 95–122

1985b. The social context of infant feeding in the Murik Lakes of Papua New Guinea. In L.B. Marshall (ed.), *Infant Care and Feeding in the South Pacific*, New York, Gordon and Breach, pp. 137–54

1990. The dynamics of siblingship: nurturance and authority in Murik society. In N.C. Lutkehaus et al. (eds.), *Sepik Heritage*, Durham, Carolina Academic Press, pp. 325–36

1991. Murik. In T.E. Hays (ed.), *Encyclopaedia of World Cultures*, vol. II: *Oceania*, Boston, G.K. Hall, pp. 220–3

1992. Dance when I die!: context and role in the clowning of Murik women. In W. Mitchell (ed.), *Clowning as Critical Practice*, Pittsburgh, University of Pittsburgh Press, pp. 58–87

1995. Achieving womanhood and the achievements of women in Murik: cult initiation, gender complementarity and the prestige of women. In Nancy Lutkehaus and Paul B. Roscoe (eds.), *Gender Rituals: Female Initiation in Melanesia*, London, Routledge and Kegan Paul, pp. 121–42

Barlow, Kathleen, Lissant Bolton and David Lipset. 1988. Trade and society in transition along the Sepik coast: Technical Report on regional research in the East Sepik and Sundaun Provinces, Papua New Guinea for the Sepik Documentation Project. Sponsored by the Australian Museum, Sydney, Australian Museum

Barlow, Kathleen and David Lipset. 1997. Dialogics of material culture: male and female in Murik outrigger canoes, *American Ethnologist* 24(1): 4–36

Barnes, J.A. 1961. Law as politically active: an anthropological view. In Geoffrey Sawer (ed.), *Studies in the Sociology of Law*, Canberra, Australian National University Press, pp. 167–96

1962. African models in the New Guinea highlands, *Man* (n.s.) 62: 5–9

Barth, Frederic. 1959. *Political Leadership among Swat Pathans*, London School of Economics 19, New Haven, Yale University Press

1966. *Models of Social Organization*, Occasional Paper 23, London, Royal Anthropological Institute

Barton, R.F. 1919 (1969). *Ifugao Law*, Berkeley, University of California Press

Bascom, William R. 1948 (1970). Ponapean prestige economy. Reprinted in Thomas Harding and Ben Wallace (eds.), *Cultures of the Pacific*, New York, The Free Press, pp. 85–93

Bateson, Gregory. 1932. Social structure of the Iatmul people of the Sepik River, *Oceania* 2: 245–91, 401–53.

1935. Music in New Guinea, *The Eagle, St. John's College* 47: 158–70

1936 (1958). *Naven: A Survey of the Problems Suggested by a Composite Picture of the Culture of a New Guinea Tribe Drawn from Three Points of View*, Stanford, Stanford University Press

1946. Arts of the South Seas, *The Arts Bulletin* 2: 119–23

1955 (1973). A theory of plan and fantasy. Reprinted in *Steps to an Ecology of Mind*, New York, Ballantine, pp. 177–93

1956 (1973). Toward a theory of schizophrenia. Reprinted in *Steps to an Ecology of Mind*, New York, Ballantine, pp. 201–27

Bateson, Gregory and Margaret Mead. 1942. *Balinese Character*, New York, New York Academy of Sciences

Battaglia, Deborah. 1985. 'We feed the father': paternal nurture among the Sabarl of Papua New Guinea, *American Ethnologist* 12 (3): 427–41

1990. *On the Bones of a Serpent: Person, Memory and Mortality in Sabarl Island Society*, Chicago, University of Chicago Press

Bauman, Richard and Charles L. Briggs. 1990. Poetics and performance as critical perspectives on language and social life, *Annual Review of Anthropology* 19: 59–88

Behar, Ruth. 1990. Rage and redemption: reading the life story of a Mexican marketing woman, *Feminist Studies* 6: 223–58.

1993. *Translated Woman: Crossing the Border with Esperanza's Story*, Boston, Beacon Press

Behrmann, W. 1922. *Im Stromgebeit des Sepik*, Berlin: August Scherl

Beier, Uli and Michael Somare. 1973. The Kakar images of Darpoap, *Record of the PNG Museum and Art Gallery* 3: 1–16

Benedict, Ruth. 1934 (1959). *Patterns of Culture*, Boston, Houghton Mifflin

1946 (1989). *Chrysanthemum and the Sword*, Boston, Houghton Mifflin

Berndt, R.M. and R. Lawrence (eds.). 1973. *Politics in New Guinea*, Seattle, University of Washington Press

Bettelheim, Bruno. 1954. *Symbolic Wounds: Puberty Rites and the Envious Male*, New York, The Free Press

Bhabha, Homi K. 1994. *The Location of Culture*, London, Routledge

Biersack, Alletta. 1982. Ginger gardens for the ginger woman: rites and passages in a Melanesian society, *Man* (n.s.) 17: 239–58

Black, Donald. 1984. *Toward a Theory of Social Control* (vol. I), Orlando, Academic Press

Blake, William. 1800? (1975). *The Marriage of Heaven and Hell*, Oxford, Oxford University Press

Bloch, Maurice (ed.). 1975. *Political Language and Oratory in Traditional Society*, New York, Academic Press

1982. Death, women and regeneration. In M. Bloch and J. Parry (eds.), *Death and the Regeneration of Life*, Cambridge, Cambridge University Press, pp. 211–30

Bloch, Maurice and Jonathan Parry. 1982. Introduction: Death and regeneration of life. In M. Bloch and J. Parry (eds.), *Death and the Regeneration of Life*, Cambridge, Cambridge University Press, pp. 1–44

Boggs, S.T. and M.N. Chunn. 1990. Ho'oponopono: a Hawaiian method of solving interpersonal problems. In K. Watson-Gegeo and G. White (eds.), *Disentangling*, Stanford, Stanford University Press, pp. 122–60

Bohannon, Paul. 1963. *Social Anthropology*, New York, Holt, Rinehart and Winston

Bohm, K. 1983. *The Life of Some Island People of New Guinea*, ed. N. Lutkehaus, Berlin, Dietrich Reimer.

Bourdieu, Pierre. 1977. *Outline of a Theory of Practice*, Cambridge, Cambridge University Press

1990. *The Logic of Practice*, trans. R. Nice, Stanford, Stanford University Press

Bowden, Ross. 1983. *Yena: Art and Ceremony in a Sepik Society*, Oxford, Pitt-Rivers Museum

Brady, Ivan (ed.). 1976. *Transactions in Kinship: Adoption and Fosterage in Oceania*, ASAO Monograph 4, Honolulu, University of Hawaii Press

Brenneis, Don and Frederick Myers (eds.). 1984 (1991). *Dangerous Words: Language and Politics in the Pacific*, Prospect Heights, Ill., Waveland Press

Briggs, Charles. 1993. Personal sentiments and polyphonic voices in Warau women's ritual wailing: music and poetics in a critical and collective discourse, *American Ethnologist* 95: 929–57

Brison, Karen. 1989. All talk and no action? Saying and doing in Kwanga meetings, *Ethnology* 28 (2): 97–115

Brown, B.J. 1969. *Fashion of Law in New Guinea: Being an Account of the Past, Present and Developing System of Laws in Papua and New Guinea*, Sydney, Butterworths

Brown, Paula. 1963. From anarchy to satrapy, *American Anthropologist* 65: 1–15

1964. Enemies and affines, *Ethnology* 3: 335–56

Brunton, Ron. 1975. Why do the Trobriands have chiefs?, *Man* (n.s.) 10: 544–58

1980. Miscontrued order in Melanesian religion, *Man* (n.s.) 15: 11–28

Bulmer, R.N.H. 1971. Traditional forms of family interaction in New Guinea. In *Population Growth and Socio-economic Change*. New Guinea Research Unit 42: 137–62

Burman, Sandra B. and Barbara E. Harrell-Bond (eds.). 1979. *The Imposition of Law*, New York, Academic Press

Burridge, K.O.L. 1957a. The *Gagai* in Tangu, *Oceania* 28: 56–72
  1957b. Disputing in Tangu, *American Anthropologist* 59: 763–80
  1959. Siblings in Tangu, *Oceania* 30: 128–54
  1975. The Melanesian manager. In J.H.M. Beattie and R.G. Lienhardt (eds.), *Studies in Social Anthropology, Essays in Memory of E.E. Evans-Pritchard*, Oxford, Clarendon Press, pp. 86–104
Carrier, James and Achsah H. Carrier. 1989. *Wage, Trade and Exchange in Melanesia: A Manus Society in the Modern State*, Berkeley: University of California Press
Carroll, Vern (ed.) 1970. *Adoption in Eastern Oceania*, Honolulu, University of Hawaii Press
Casson, Ronald W. 1983. Schemata in cognitive anthropology, *Annual Review of Anthropology* 12: 429–62
Chodorow, Nancy. 1978. *The Reproduction of Mothering: Psychoanalysis and the Sociology of Gender*, Berkeley: University of California Press
Chowning, Ann. 1979. Leadership in Oceania, *Journal of Pacific History* 14: 66–84
Clay, Brenda J. 1977. *Pinikindu: Maternal Nurture, Paternal Substance*, Chicago, University of Chicago Press
  1993. Other times, other places: agency and the big man in Central New Ireland, *Man* (n.s.) 27: 719–33
Clifford, William, Louise Morauta and Barry Stuart. 1984. *Law and Order in Papua New Guinea*, vol. I: *Report and Recommendations*, Port Moresby, Institute of National Affairs
Codere, Helen. 1951. *Fighting with Property: A Study of Kwakiutl Potlatching and Warfare 1792–1930*, New York, American Ethnological Society
Collier, Jane F. 1975. Legal processes, *Annual Review of Anthropology* 4: 131–63
  1984. Two models of social control in simple societies. In Donald Black (ed.), *Toward a General Theory of Social Control*, Vol. II: *Selected Problems*, New York, Academic Press, pp. 105–40
  1988. *Marriage and Inequality in Classless Societies*, Stanford, Stanford University Press
Collier, Jane and Michelle Rosaldo. 1981. Politics and gender in simple societies. In Sherry Ortner and Harriet Whitehead (eds.), *Sexual Meanings*, Cambridge, Cambridge University Press, pp. 231–54
Collier, Jane Fishburne and Sylvia Junko Yanagisako (eds.). 1987. *Gender and Kinship: Essays toward a Unified Analysis*, Stanford, Stanford University Press
Colson, Elizabeth. 1953. Social control and vengeance in Plateau Tonga society, *Africa* 23 (3): 199–212
Comaroff, Jean. 1985. *Body of Power, Spirit of Resistance: The Culture and History of a South African People*, Chicago, University of Chicago Press
Comaroff, John L. 1987. Sui genderis: feminism, kinship theory and structural domains. In Jane Fishburne Collier and Sylvia Junko Yanagisako (eds.), *Gender and Kinship: Essays Toward a Unified Analysis*, Stanford, Stanford University Press, pp. 53–86
Comaroff, John and Jean Comaroff. 1992. *Ethnography and the Historical Imagination*, Boulder, Colo., Westview Press
Comaroff, John and Simon Roberts. 1981. *Rules and Processes: The Cultural Logic of Dispute in an African Context*, Chicago, University of Chicago Press

Coser, Lewis A. 1956. *The Functions of Social Conflict*, Glencoe, Ill., The Free Press

Counts, Dorothy (ed.). 1990. *Domestic Violence in Oceania*, Special Issue of *Pacific Studies*

Crapanzano, Vincent. 1980. *Tuhami: Portrait of a Moroccan*, Chicago, University of Chicago Press

1990. On dialogue. In Tullio Maranhao (ed.), *The Interpretation of Dialogue*, Chicago, University of Chicago Press, pp. 269–92

Curtin, Philip D. 1985. *Cross-Cultural Trade in World History*, Cambridge, Cambridge University Press

Daelemans, Sven and Tullio Maranhao. 1990. Psychoanalytic dialogue and the dialogical principle. In Tullio Maranhao (ed.), *The Interpretation of Dialogue*, Chicago, University of Chicago Press, pp. 219–41

D'Amato, John. 1979. The wind and the amber: notes on headhunting and the interpretation of accounts, *Journal of Anthropological Research* 35: 61–84

Da Matta, Roberto. 1991. *Carnivals, Rogues and Heroes: an Interpretation of the Brazilian Dilemma*, trans. J. Drury, South Bend, Ind., University of Notre Dame Press

Damon, Frederick H. 1989. Introduction. In F.H. Damon and R. Wagner (eds.), *Death Rituals and Life in the Societies of the Kula Ring*, DeKalb, Ill., N. Illinois University Press, pp. 3–22

D'Andrade, Roy and Claudia Strauss (eds.). 1992. *Human Motives and Cultural Models*, New York, Cambridge University Press

Derrida, Jacques. 1970 (1972). Structure, sign and play in the discourse of the human sciences. In Richard Macksey and Eugenio Donato (eds.), *The Structuralist Controversy: The Languages of Criticism and the Sciences of Man*, Baltimore, Johns Hopkins University Press, pp. 247–65

Devereux, George. 1967. *From Anxiety to Method in the Behavioral Sciences*, New York, Humanities Press

Douglas, Bronwyn. 1979. Rank, power, authority: a reassessment of traditional leadership in Island Melanesia, *Journal of Pacific History* 14: 2–27

Douglas, Mary. 1966. *Purity and Danger: An Analysis of the Concepts of Pollution and Taboo*, London, Routledge and Kegan Paul

1970. *Natural Symbols: Explorations in Cosmology*, London, Cresset Press

1975. *Implicit Meanings*, London, Routledge and Kegan Paul

Dundes, Alan. 1976. A psychoanalytic study of the bullroarer, *Man* (n.s.) 11: 220–38

1994. *The Cockfight: A Casebook*, Madison, University of Wisconsin Press

Dureau, C.M. 1991. Death, gender and regeneration: a critique of Maurice Bloch, *Canberra Anthropology* 14 (1): 24–44

Durkheim, Emile. 1915 (1965). *The Elementary Forms of the Religious Life*, trans. Joseph W. Swain, New York, The Free Press

1933 (1964). *The Division of Labor in Society*, trans. George Simpson, New York: The Free Press

1938 (1966). *The Rules of Sociological Method*, trans. Sarah Solovay and John H. Mueller, New York, The Free Press

Eco, Umberto. 1984. The frames of comic freedom. In Thomas A. Sebeok (ed.), *Carnival*, Berlin, Mouton, pp. 1–10

Eggan, Fred. 1954. Social anthropology and the method of controlled comparison, *American Anthropologist* 56: 743–63

Epstein, A.L. 1968. Power, politics and leadership: some central African and Melanesian contrasts. In M.J. Swartz (ed.), *Local-Level Politics*, Chicago, Aldine, pp. 53–68

1974. *Contention and Dispute: Aspects of Law and Social Control in Melanesia*, Canberra, Australian National University Press

1979. Tambu: the shell-money of the Tolai. In R.H. Hook (ed.), *Fantasy and Symbol: Studies in Anthropological Interpretation*, London, Academic Press, pp. 149–91

1984. *The Experience of Shame in Melanesia: An Essay on the Anthropology of Affect*, Occasional Paper 40, London, Royal Anthropological Institute of Great Britain and Ireland

Errington, Frederick. 1974. *Karavar: Masks and Ritual in a Melanesian Society*, Ithaca, Cornell University Press

Errington, Frederick and Deborah Gewertz. 1986. The confluence of powers: entropy and importation among the Chambri, *Oceania* 57: 99–113

1987. *Cultural Alternatives and a Feminist Anthropology: An Analysis of Culturally Constructed Gender Interests in Papua New Guinea*, Cambridge, Cambridge University Press

Evans-Pritchard, E.E. 1940 (1979). *The Nuer: A Description of the Modes of Livelihood and Political Institutions of a Nilotic People*, New York, Oxford University Press

1963. The comparative method in social anthropology. In *The Comparative Method in Social Anthropology*, London, Athlone Press, pp. 2–30

1965. The position of women in primitive societies and in our own. In *The Position of Women in Primitive Societies and Other Essays in Social Anthropology*, London, Faber and Faber, pp. 37–59

Eyde, David B. 1967. Cultural correlates of warfare among the Asmat of southwest New Guinea, Ph.D. dissertation, Department of Anthropology, Yale University

Feil, D.K. 1978. Enga women in the Tee exchange, *Mankind* 11: 220–30

1984. *Ways of Exchange: The Enga Tee of Papua New Guinea*, St. Lucia, University of Queensland Press

1987. *The Evolution of Highland Papua New Guinea Societies*, Cambridge, Cambridge University Press

Feldman, Allen. 1991. *Formations of Violence: The Narrative of Violence and Political Terror in Northern Ireland*, Chicago, University of Chicago Press

Felstiner, William L.F. 1974. Influences of social organization on dispute processing, *Law and Society Review* 9 (1): 63–94

Firth, Raymond. 1940. The analysis of *Mana*: an empirical approach, *Journal of the Polynesian Society* 40: 483–510

1970. Sibling terms in Polynesia, *Journal of the Polynesian Society* 79: 272–81

Fitzpatrick, Peter. 1980. *Law and State in Papua New Guinea*, New York, Academic Press

Flanagan, James G. 1989. Hierarchy in simple "egalitarian" societies, *Annual Review of Anthropology* 18: 245–66

Foley, William. 1986. *The Papuan Languages of New Guinea*, Cambridge, Cambridge University Press

Forge, Anthony. 1962. Paint: a magical substance, *Palette* 9: 9–16

1966 (1971). Art and environment in the Sepik. In Carol F. Jopling (ed.), *Art and Aesthetics in Primitive Society*, New York, Dutton, pp. 290–314

1972. The Golden Fleece, *Man* (n.s.) 7: 527–40

n.d. Sepik cultural history. Unpublished manuscript

Fortes, Meyer. 1958. Introduction. In Jack Goody (ed.), *The Developmental Cycle in Domestic Groups*, Cambridge, Cambridge University Press, pp. 1–12

1962. Ritual and office in tribal society. In Max Gluckman (ed.), *Essays on the Ritual of Social Relations*, Manchester, Manchester University Press, pp. 53–89

1969. *Kinship and the Social Order*, Chicago, Aldine

1973. On the concept of the person among the Tallensi. In *La Notion de personne en Afrique noire*, No. 544, Paris, Colloques Internationaux du Centre National de la Recherche Scientifique, pp. 289–319

Fortune, Reo F. 1935 (1969). *Manus Religion: An Ethnological Study of the Manus Natives of the Admiralty Islands*, Lincoln, University of Nebraska Press

1939. Arapesh warfare, *American Anthropologist* 41: 22–41

Foucault, Michel. 1977 (1979). *Discipline and Punish*, trans. Alan Sheridan, New York, Vintage Books

Frazer, J.G. 1922 (1963). *The Golden Bough*, New York, Macmillan

Freedman, Michael P. 1970. Social organization of a Siassi island community. In T. Harding and Ben J. Wallace (eds.), *Cultures of the Pacific*, New York, The Free Press, pp. 159–80

Freud, Sigmund. 1916 (1946). *Totem and Taboo: Resemblances between the Pyschic Lives of Savages and Neurotics*, trans. A.A. Brill, New York, Vintage Books

1923 (1960). *The Ego and the Id*, trans. Joan Riviere, New York, Norton

Frye, Northrop. 1957 (1973). *Anatomy of Criticism*, Princeton, Princeton University Press

Gal, Susan. 1991. Between speech and silence: the problematics of research in language and gender. In Micaela di Leonardo (ed.), *Gender at the Crossroads of Knowledge: Feminist Anthropology in the Postmodern Era*, Berkeley, University of California Press, pp. 175–200

Gallimore, Ronald, Joan Briggs and Cathie Jordon. 1974. *Culture, Behavior and Education: A Study of Hawaiian Americans*, Beverly Hills, Calif., Sage

Geertz, Clifford. 1973. *The Interpretation of Cultures*, New York, Basic Books

1980. *Negara: The Theatre State in Nineteenth Century Bali*, Princeton, Princeton University Press

1983. *Local Knowledge*, New York, Basic Books

Geiger, Susan N.G. 1986. Women's life histories: method and content, *Signs* 11: 334–51

Gell, Alfred. 1975. *Metamorphosis of the Cassowaries*, London, Athlone Press.

1992. Intertribal commodity barter and reproductive gift-exchange in old Melanesia. In Caroline Humphries and Stephen Hugh-Jones (eds.), *Barter, Exchange and Values: An Anthropological Approach*, Cambridge, Cambridge University Press, pp. 142–68

Gellner, E. 1973. *Cause and Meaning in the Social Sciences*, London, Routledge and Kegan Paul

Gennep, A. van. 1914 (1960). *The Rites of Passage*, trans. M.B. Vizedom and G.L. Caffee, Chicago, University of Chicago Press

Gewertz, Deborah. 1977. The politics of affinal exchange: Chambri as a client market, *Ethnology* 16: 285–98

1982. The father who bore me: the role of the *tsambunwuro* during Chambri initiation ceremonies. In Gilbert H. Herdt (ed.), *Rituals of Manhood*, Berkeley, University of California Press, pp. 286–320

1983. *Sepik River Societies*, New Haven, Yale University Press

1988. *Myths of Matriarchy Reconsidered*, Oceania Monograph 33, Sydney, University of Sydney

Gewertz, Deborah B. and Frederick Errington. 1992. *Twisted Histories, Altered Contexts: Representing the Chambri in a World System*, Cambridge, Cambridge University Press

Gibbs, James L. 1963. The Kpelle moot, *Africa* 33 (1): 1–10

Giddens, Anthony. 1979. *Central Problems in Social Theory: Action, Structure and Contradiction in Social Analysis*, Berkeley, University of California Press

1984. *The Constitution of Society: Outline of a Theory of Structuration*, Berkeley, University of California Press

1987. *Social Theory and Modern Sociology*, Berkeley, University of California Press

Gillison, Gillian. 1983. Cannibalism among women in the eastern highlands of Papua New Guinea. In Paula Brown and D.F. Tuzin (eds.), *The Ethnography of Cannibalism*, Washington, D.C., Special Publication of the Society for Psychological Anthropology, pp. 33–51.

1993. *Between Culture and Fantasy: A New Guinea Highlands Mythology*, Chicago, University of Chicago Press

Glass, Patrick. 1988. Trobriand symbolic geography, *Man* (n.s.) 23: 56–76

Glasse, R.M. and M.J. Meggitt (eds.). 1969. *Pigs, Pearlshells and Women*, Englewood Cliffs, Prentice-Hall

Gluckman, Max. 1955. *The Judicial Process among the Barotse of Northern Rhodesia*, Manchester, Manchester University Press

1956. *Custom and Conflict in Africa*, Oxford, Basil Blackwell

1963. Rituals of rebellion in south-east Africa. Reprinted in *Order and Rebellion in Tribal Africa: Collected Essays with an Autobiographical Introduction*, New York, The Free Press, pp. 110–36

1965. *The Ideas in Barotse Jurisprudence*, New Haven, Yale University Press

1969. *Ideas and Procedures in African Customary Law*, Oxford, Oxford University Press

Godelier, Maurice. 1986. *The Making of Great Men: Male Domination and Power among the New Guinea Baruya*, trans. Rupert Swyer, Cambridge, Cambridge University Press

Godelier, Maurice and Marilyn Strathern (eds.). 1991. *Big Men and Great Men: Personifications of Power in Melanesia*, Cambridge, Cambridge University Press

Goldman, Irving. 1970. *Ancient Polynesian Society*, Chicago, University of Chicago Press

Goldman, Lawrence R. 1979. Speech categories and the study of disputes: a New Guinea example, *Oceania* 50 (3): 209–27.

1983. *The Talk Never Dies: The Language of Huli Disputes*, London, Tavistock

Goodenough, Ward. 1970. Epilogue: Transactions in parenthood. In Vern Carroll (ed.), *Adoption in Eastern Oceania*, Honolulu, University of Hawaii Press, pp. 391–410

Goody, Jack. 1962. *Death, Property and the Ancestors*, Stanford, Stanford University Press

Gordon, Robert and Mervyn Meggitt. 1985. *Law and Order in the New Guinea Highlands: Encounters with Enga*, Hanover, University Press of New England

Gottlieb, Alma. 1989. Hyenas and heteroglossia: myth and ritual among the Beng of Côte d'Ivoire, *American Ethnologist* 16: 487–501

Gregory, C. 1980. Gifts to men and gifts to gods: gift exchange and capital accumulation in contemporary Papua, *Man* (n.s.) 15: 626–52

1982. *Gifts and Commodities*, London, Academic Press

Grierson, P.J. Hamilton. 1980. The silent trade. Reprinted in *Research in Economic Anthropology* 3: 1–75

Griffiths, John. 1986. What is legal pluralism?, *Journal of Legal Pluralism and Unofficial Law* 24: 1–55

Gulliver, P.H. 1969. Introduction to case studies in law in non-Western societies. In L. Nader (ed.), *Law in Culture and Society*, Chicago, Aldine, pp. 11–23

1979. *Disputes and Negotiations: A Cross Cultural Perspective*, New York, Academic Press

Haddon, A.C. and J. Hornell. 1936. *Canoes of Oceania*, Honolulu, Bishop Museum

Hallowell, A.E. 1955 (1971). The self in its behavioral environment. Reprinted in *Culture and Experience*, New York, Schocken Books, pp. 75–110

Hallpike, C. 1969. Social hair, *Man* (n.s.) 4: 256–64

Hammond, P.B. 1964. Mossi joking, *Ethnology* 3: 259–67

Handelman, Don. 1979. Is Naven ludic? Paradox and communication of identity, *Social Analysis* 1: 177–91

Handy, E.S.C. and M.K. Pukui. 1958 (1988). *The Polynesian Family System in Ku'a, Hawaii*, Rutland, Tuttle

Harding, Thomas. 1968. *Voyagers of the Vitiaz Strait*, American Ethnological Society 44, Seattle, University of Washington Press

Harris, Bruce. 1988. The rise of rascalism: action and reaction in the evolution of rascal gangs. Discussion Paper 54, Boroko, PNG: Institute for Applied Social and Economic Research

Harrison, Simon J. 1982. Yams and the symbolic representation of time in a Sepik River village, *Oceania*, 53: 141–61

1985. Ritual hierarchy and secular equality in a Sepik River village, *American Ethnologist* 12: 413–26

1989. Magical and material polities in Melanesia, *Man* (n.s.) 24 (1): 1–20

1990a. *Stealing People's Names: Cosmology and Politics in a Sepik River Cosmology*, Cambridge, Cambridge University Press

1990b. The symbolic construction of aggression and war in a Sepik River society, *Man* (n.s.) 24: 583–99

1993a. Review of *Song to the Flying Fox* by J. Wassman, *Man* (n.s.) 28 (2): 405–6

1993b. The commerce of cultures in Melanesia, *Man* (n.s.) 28 (1): 139–58

1993c. *The Mask of War: Violence, Ritual and the Self in Melanesia*, Manchester, Manchester University Press

Hau'ofa, Epeli. 1971. Mekeo chieftainship, *Journal of the Polynesian Society* 80: 152–69

1981. *Mekeo*, Hong Kong, Australian National University Press

Hauser-Schaublin, Brigitta. 1977. *Frauen in Kararau: zur Rolle der Frau bei den Iatmul am Mittelsepik, Papua New Guinea*, Basle, Basler Beitrage zur Ethnologie 18

1993. Review of *Twisted Histories, Altered Contexts* by D.B. Gewertz and F.R. Errington, *Pacific Studies* 16 (1): 106–11

Hawkes, K. 1978. Big men in Binumerian, *Oceania* 48: 161–87

Hays, Terence E. 1988. "Myths of matriarchy" and the sacred flute complex of the Papua New Guinea highlands. In Deborah Gewertz (ed.), *Myths of Matriarchy Reconsidered*, Sydney, University of Sydney Press, pp. 98–119

1993. 'The New Guinea highlands': region, culture area or fuzzy set?, *Current Anthropology* 34: 141–64

Heider, Karl. 1969. Visiting trade institutions, *American Anthropologist* 71: 462–71

1970. *The Dugam Dani: A Papuan Culture in the Highlands of West New Guinea*, Chicago, Aldine

1979. *Grand Valley Dani: Peaceful Warriors*, New York, Holt, Rinehart and Winston

Helms, Mary W. 1988. *Ulysses' Sail: An Ethnographic Odyssey of Power, Knowledge, and Geographical Distance*, Princeton, Princeton University Press

Herdt, Gilbert H. 1981. *Guardians of the Flutes* (vol. I), New York, McGraw-Hill

1984. *Ritualized Homosexuality in Melanesia*, Berkeley, University of California Press

1990. Secret societies and secret collectives, *Oceania* 60: 360–81

Herdt, Gilbert H. and Fitz John P. Poole. 1982. 'Sexual antagonism': the intellectual history of a concept in New Guinea anthropology, *Social Analysis* 12: 3–29

Hereniko, Vilsoni. 1992. When she reigns supreme: clowning and culture in Rotuman weddings. In William Mitchell (ed.), *Clowning as Critical Practice: Performance Humor in the South Pacific*, ASAO Monograph 13, Pittsburgh, University of Pittsburgh Press, pp. 167–92

1994. *Woven Gods: Female Clowns and Power in Rotuma*, Honolulu, University of Hawaii Press

Herskovits, Melville and Frances Herskovits. 1958. Sibling rivalry, the Oedipus Complex and myth, *Journal of American Folklore* 7: 1–15

Hertz, Robert. 1909 (1960). A contribution to the study of the collective representation of death. Reprinted in *Death and the Right Hand*, trans. Rodney and Claudia Needham, Glencoe, The Free Press, pp. 27–86

Hiatt, L.R. 1971. Secret pseudo-procreative rites among the Australian Aborigines. In L.R. Hiatt and C. Jayawardena (eds.), *Anthropology in Oceania: Essays Presented to Ian Hogbin*, Sydney, Angus and Robertson, pp. 77–88

Hoebel, E. Adamson. 1954. *The Law of Primitive Man: A Study in Comparative Legal Dynamics*, Cambridge, Mass., Harvard University Press

Hogbin, H. Ian. 1935. Trading expeditions in northern New Guinea, *Oceania* 5: 375–407

1940. The father chooses his heir: a family dispute over succession in Wogeo, New Guinea, *Oceania* 11: 1–39

1970. *The Island of the Menstruating Men: Religion in Wogeo, New Guinea*, Melbourne, Melbourne University Press

1978. *The Leaders and the Led: Social Control in Wogeo, New Guinea*, Melbourne, Melbourne University Press

Holy, Ladislav. 1987. Introduction: description, generalization and comparison, two paradigms. In *Comparative Anthropology*, Oxford, Basil Blackwell, pp. 1–21

Houseman, Michael and Carlo Severi. 1994. *Naven ou le donner à voir: essai d'interpretation de l'action rituelle*, Paris, CNRS-Editions

Howard, Alan, Robert H. Heighton, Jr., Cathie E. Jordon and Ronald G. Gallimore. 1970. Traditional and modern adoption patterns in Hawaii. In Vern Carroll (ed.), *Adoption in Eastern Oceania*, Honolulu, University of Hawaii Press, pp 21–51

Howard, Allan and Robert Borofsky (eds.). 1989. *Developments in Polynesian Ethnology*, Honolulu, University of Hawaii Press

Howard, Allan and John Kirkpatrick. 1989. Social organization. In A. Howard and R. Borofsky (eds.), *Developments in Polynesian Ethnology*, Honolulu, University of Hawaii Press, pp. 47–94

Huber, Mary F. n.d. Trade and the articulation of an interethnic community in Wewak. Presented at the 1983 meeting of the American Anthropological Association, Chicago, Ill.

1988. *The Bishop's Progress: A Historical Ethnography of Catholic Missionary Experience on the Sepik Frontier*, Washington, D.C., Smithsonian Institution Press

Huntington, Richard and Peter Metcalf. 1979. *Celebrations of Death: The Anthropology of Mortuary Ritual*, Cambridge, Cambridge University Press

Ito, Karen. 1985. Ho'oponopono "To make right": Hawaiian conflict resolution and metaphor in the construction of a family therapy, *Culture, Medicine and Psychiatry* 9: 201–17

Jakobson, Roman. 1971. The dominant. In L. Matajka and K. Pomorska (eds.), *Readings in Russian Formalism*, Cambridge, Mass., MIT Press

Johnson, Mark. 1987. *The Body in the Mind*, Chicago, University of Chicago Press

Jones, Ernest. 1925. Mother-right and the sexual ignorance of savages, *International Journal of Psychoanalysis* 6: 109–30

Jorgensen, Dan. 1983. Introduction: the facts of life, Papua New Guinea style, *Mankind* 14: 1–12

Josephides, Lisette. 1985. *The Production of Inequality*, London, Tavistock

Josephides, Sasha. 1982. The perception of the past and the notion of "business" in a Seventh-Day Adventist village in Madang, New Guinea. Ph.D. dissertation, University of London

Kaberry, Phyllis M. 1940–1. The Abelam tribe, Sepik District, New Guinea: a preliminary, *Oceania* 11: 283–58, 345–67

1941–2. Law and political organization in the Abelam tribe, New Guinea, *Oceania* 12: 79–95, 209–25, 331–63

Kahn, Miriam. 1986. *Always Hungry, Never Greedy: Food and the Expression of Gender in a Melanesian Society*, Cambridge, Cambridge University Press

Karp, Ivan. 1987. Laughter at marriage: subversion in performance. In David Parkin and David Nyamwaya (eds.), *Transformations of African Marriage*, Manchester, Manchester University Press, pp. 137–53

Keesing, Roger. 1978. *Elota's Story: The Life and Times of a Soloman Islands Big Man*, St. Lucia, University of Queensland Press

1982. Introduction. In Gilbert H. Herdt (ed.), *Rituals of Manhood: Male Initiation in Papua New Guinea*, Berkeley, University of California Press, pp. 1–44

1990. The power of talk. In K. Watson-Gegeo and G. White (eds.), *Disentangling*, Stanford, Stanford University Press, pp. 493–500

Kehoe, Alice B. 1970. The function of ceremonial intercourse among the Northern Plains Indians, *Plains Anthropologist* 15 (48): 99–103

Kelly, John and Martha Kaplan. 1990. History, structure and ritual, *Annual Review of Research in Anthropology* 19: 119–50

Kelly, R.C. 1976. Witchcraft and sexual relations: an exploration in the social and semantic implications of the structure of belief. In P. Brown and G. Buchbinder (eds.), *Man and Woman in the New Guinea Highlands*, A.A.A. Special Publication 8, Washington, D.C., American Anthropological Association, pp. 36–53

1977. *Etero Social Structure: A Study in Structural Contradiction*, Ann Arbor, University of Michigan Press

1993. *Constructing Inequality*, Ann Arbor, University of Michigan Press

Kirch, Patrick V. 1984. *The Evolution of Polynesian Chiefdoms*, Cambridge, Cambridge University Press

Knauft, Bruce. 1985. *Good Company and Violence: Sorcery and Social Action in a Lowland New Guinea Society*, Berkeley, University of California Press

1987. Homosexuality in Melanesia, *Journal of Psychoanalytic Anthropology* 10: 155–91

1989. Bodily images in Melanesia: cultural substances and natural metaphors. In Michel Feher, Ramona Nadaff and Nadin Tuzi (eds.), *Fragments for a History of the Human Body, Part 3*, New York, Urzone, pp. 198–279

1990. Melanesian warfare: a theoretical history, *Oceania* 60: 250–311

Koch, Klaus-Friedrich. 1979. Liability and social structure, *Ethnology* 15: 34–46

Koch, Klaus-Friedrich, Soraya Altorki, Andrew Arno and Letitia Hickson. 1977. Ritual reconciliation and the obviation of grievances: a comparative study in the ethnography of law, *Ethnology* 16 (3): 269–83

Korn, Francis. 1971. A question of preferences: the Iatmul case. In R. Needham (ed.), *Rethinking Kinship and Marriage*, London, Tavistock, pp. 99–132

Kristeva, Julia. 1967 (1986). Word, dialogue and novel. Reprinted in Torl Mori (ed.), *The Kristeva Reader*, New York, Columbia University Press, pp. 34–61

Kroeber, Alfred. 1925. *Handbook of the Indians of California*, Washington, D.C., Government Printing Office

Kulick, Don. 1992. *Language Shift and Cultural Reproduction: Socialization, Self and Syncretism in a Papua New Guinean Village*, Cambridge, Cambridge University Press

Kuper, Adam. 1988. *The Invention of Primitive Society: Transformations of an Illusion*, London, Routledge

La Fontaine, J.S. 1985. Person and individual: some anthropological reflections. In Michael Carrithers, Steven Collins and Steven Lukes (eds.), *The Category of the Person*, Cambridge, Cambridge University Press, pp. 123–40

Lakoff, George and Mark Johnson. 1980. *Metaphors We Live By*, Chicago, University of Chicago Press

Lamphere, Louise. 1974. Strategies, cooperation and conflict among women in domestic groups. In Michelle Rosaldo and Louise Lamphere (eds.), *Women, Culture and Society*, Stanford, Stanford University Press, pp. 97–112

Landtman, Gunnar. 1927 (1970). *The Kiwai Papuans of British New Guinea: A Nature-born Instance of Rousseau's Ideal Community*, New York, Johnson Reprint Co.

Langness, Louis L. 1967. Sexual antagonism in the New Guinea highlands: a Bena Bena example, *Oceania* 37 (3): 161–77

1973. Bena Bena political organization. In R.M. Berndt and R. Lawrence (eds.), *Politics in New Guinea*, Seattle, University of Washington Press, pp. 298–316

1974. Ritual power and male domination in the New Guinea highlands, *Ethos* 2 (3): 189–212

Lawrence, Peter. 1964. *Road Belong Cargo*, Manchester, Manchester University Press

1969. The state versus stateless societies in Papua and New Guinea. In B.J. Brown (ed.), *Fashion of Law in New Guinea, Being an Account of the Past, Present and Developing System of Laws in Papua and New Guinea*, Sydney, Butterworths, pp. 15–39

1988. Twenty years after: a reconsideration of Papua New Guinea Seaboard and Highlands religions, *Oceania* 59 (1): 7–28

Lawrence, Peter and Mervyn Meggitt. 1965 (1972). Introduction. In P. Lawrence and M. Meggitt (eds.), *Gods, Ghosts and Men in Melanesia: Some Religions of Australian New Guinea and the New Hebrides*, Melbourne, Oxford University Press, pp. 1–27

Laycock, Donald C. 1973. Sepik languages – checklist and preliminary classification, *Pacific Linguistics* series C, no. 1, Canberra, Australian National University Press

Leach, Edmund. 1954. A Trobriand Medusa? *Man* no. 158

1958. Magical hair, *Journal of the Royal Anthropological Institute* 88: 147–64

1961a. Two essays concerning the symbolic representation of time. In *Rethinking Anthropology*, London, Athlone Press, pp. 124–36

1961b. *Pul Eliya, a Village in Ceylon*, Cambridge, Cambridge University Press

1966. Virgin birth, *Proceedings of the Royal Anthropological Institute*: 39–49

Ledoux, Louis Pierre 1936. Murik fieldnotes, unpublished

Leenhardt, Maurice. 1979. *Do Kamo: Person and Myth in the Melanesian World*, trans. Basia Miller Celati, Chicago, University of Chicago Press

Lepowski, Maria. 1994. *Fruit of the Motherland*, New York, Columbia University Press

Lessa, William A. 1966 (1986). *Ulithi: A Micronesian Design for Living*, New York, Holt, Rinehart and Winston

314   *References*

Lévi-Strauss, Claude. 1969. *The Elementary Structures of Kinship*, trans. James H. Bell, John R. von Sturmer and Rodney Needham, Boston, Beacon Press

Levy, Robert. 1973. *Tahitians: Mind and Experience in the Society Islands*, Chicago, University of Chicago Press

Lewis, Gilbert. 1980. *The Day of Shining Red: An Essay on Understanding Ritual*, Cambridge, Cambridge University Press

Liep, John. 1991. Great man, big man, chief: a triangulation of the Massim. In M. Godelier and M. Strathern (eds.), *Big Men and Great Men*, Cambridge, Cambridge University Press, pp. 236–47

Limon, Jose. 1989. Carne, carnales and the carnevalesque: Bakhtinian bathos, disorder and narrative discourse, *American Ethnologist* 16: 471–86

Lindenbaum, Shirley. 1984. Variations on a sociosexual theme in Melanesia. In Gilbert H. Herdt (ed.), *Ritualized Homosexuality in Melanesia*, Berkeley, University of California Press, pp. 337–61

1987. The mystification of female labors. In Jane F. Collier and Sylvia J. Yanagisako (eds.), *Gender and Kinship: Essays Toward a Unified Analysis*, Stanford, Stanford University Press, pp. 221–43

Lindstrom, Lamont. 1981. "Big man": a short terminological history, *American Anthropologist* 83: 900–5

1984. Doctor, lawyer, wise man, priest: big-men and knowledge in Melanesia, *Man* (n.s.) 19: 291–301

1990. *Knowledge and Power in a South Pacific Society*, Washington, D.C., Smithsonian Institution Press

Linnekin, Jocelyn. 1985. *Children of the Land: Exchange and Status in a Hawaiian Community*, Honolulu, University of Hawaii Press

Lipset, David M. 1980 (1982). *Gregory Bateson: The Legacy of a Scientist*, Boston, Beacon Press

1984. *Authority and the Maternal Presence: An Interpretive Ethnography of Murik Lakes Society (East Sepik Province, Papua New Guinea)*, Ann Arbor, University Microfilms International

1985. Seafaring Sepiks: ecology, warfare and prestige in Murik trade, *Annual Review of Research in Economic Anthropology* 7: 67–94

1989. Papua New Guinea: the Melanesian ethic and the spirit of capitalism, 1975–86. In L. Diamond, Juan J. Linz and S.M. Lipset (eds.), *Democracy in Developing Countries* (vol. III), Boulder, Colo., Lynne Rienner Publishers, pp. 383–423

1990. Boars' tusks and flying foxes: symbolism and ritual of office in the Murik Lakes. In Nancy Lutkehaus et al. (eds.), *Sepik Heritage: Tradition and Change in Papua New Guinea*, Durham, Carolina Academic Press, pp. 286–98

n.d.a. "That obscure object of desire": culture and development in Murik (1910–1985). Presented to the annual meeting of the American Anthropological Association, Chicago, 1983

n.d.b. "Two Brothers": geography, male agency and Austronesian hegemony in North Coast cosmology. Presented to the annual meeting of the American Anthropological Association, Chicago, 1991

n.d.c. Murik lagoon prows as ethnosemiotic. Presented to the annual meeting of the American Anthropological Association, San Francisco, California, 1992

Lipset, David and Kathleen Barlow. 1987. The value of culture: regional exchange in the Lower Sepik, *Australian Natural History* 23: 156–68

Lipset, David and Jolene M. Stritecky. 1994. The metaphor of muteness: gender and kinship in Seaboard Melanesia, *Ethnology* 33: 1–20

Llewellyn, Karl N. and E. Adamson Hoebel. 1941. *The Cheyenne Way*, Norman, University of Oklahoma Press

Lock, Margaret. 1993. Cultivating the body: anthropology and epistemologies of the bodily practice and knowledge, *Annual Review of Anthropology* 22: 133–55

Lowie, R.H. 1916. *Plains Indian Age Societies: Historical and Comparative Summary*, New York, American Museum of Natural History Papers 11 (13)

Lukes, Steven and Andrew Scull. 1983. *Durkheim and the Law*, Oxford, Basil Blackwell

Lutkehaus, Nancy C. 1982. Manipulating myth and history: how the Manam maintain themselves, *Bikmaus* 3: 81–90

   1990. Hierarchy and "heroic society": Manam variations in Sepik social structure, *Oceania* 60: 179–97

McDowell, Nancy. 1978. The struggle to be human: exchange and politics in Bun, *Anthropological Quarterly* 51 (1): 16–25

   1980. It's not who you are but how you give that counts: the role of exchange in a Melanesian society, *American Ethnologist* 7: 58–70

   1990. Competitive equality in Melanesia: an exploratory essay, *Journal of the Polynesian Society* 99 (2): 179–204

   1991. *The Mundugumor: From the Field Notes of Margaret Mead and Reo Fortune*, Washington, D.C., Smithsonian Institution Press

Mahler, M., F. Pine and A. Bergman. 1975. *The Psychological Birth of the Human Infant*, New York, Basic Books

Malinowski, B. 1922. *Argonauts of the Western Pacific*, London, Routledge and Kegan Paul

   1926 (1959). *Crime and Custom in Savage Society*, Paterson, N.J., Littlefield, Adams and Company

   1927. *Sex and Repression in Savage Society*, London, Routledge and Kegan Paul

   1929 (1987). *Sexual Life of Savages in North-Western Melanesia*, Boston, Beacon Press

   1935. *Coral Gardens and Their Magic*, London, Allen and Unwin

   1948 (1954). *Magic, Science and Religion and Other Essays*, Garden City, N.Y., Doubleday Anchor

Marcus, George E. 1985. A timely rereading of *Naven*: Gregory Bateson as oracular essayist, *Representations* 12: 66–82

Marshall, Leslie B. (ed.). 1985. *Infant Care and Feeding in the South Pacific*, New York, Gordon and Breach

Marshall, Mac. 1983. Introduction: approaches in siblingship in Oceania. In Mac Marshall (ed.), *Siblingship in Oceania: Studies in the Meaning of Kin Relations*, ASAO Monograph 8, Lanham, Md., University Press of America, pp. 1–17

Mauss, Marcel. 1925 (1967). *The Gift: Forms and Functions of Exchange in Archaic Societies*, trans. I. Cunnison, New York, Norton Library

   1935 (1979). The notion of body techniques. In *Sociology and Psychology*, trans. Ben Brewster, London, Routledge and Kegan Paul, pp. 97–123

1938 (1985). A category of the human mind: the notion of person; the notion of self. Reprinted in Michael Carrithers, Steven Collins and Steven Lukes (eds.), *The Category of the Person*, Cambridge, Cambridge University Press, pp. 1–25

May, R.J. 1989. Angoram open. In Michael Oliver (ed.), *Eleksin: The 1987 National Election in Papua New Guinea*, Hong Kong, Colorcraft, pp. 111–22

Mead, Margaret. 1935 (1963). *Sex and Temperament in Three Primitive Societies*, New York, Morrow

1938. The Mt. Arapesh, Pt. I: An importing culture, *Anthropological Papers of the American Museum of Natural History* 36: 139–349

1940. The Mt. Arapesh, Pt. II: Arts and Supernaturalism, *Anthropological Papers of theAmerican Museum of Natural History* 37 (3): 319–451

1956. *New Lives for Old: Cultural Transformation – Manus, 1928–1953*, New York, Morrow

Meeker, Michael. 1979. The twilight of a South Asian heroic age: a rereading of Barth's study of Swat, *Man* (n.s.) 15: 682–70

1989. *The Pastoral Son and the Spirit of Patriarchy: Religion, Society and Person among East African Stock Keepers*, Madison, University of Wisconsin Press

Meeker, Michael, Kathleen Barlow and David M. Lipset. 1986. Culture, exchange and gender: lessons from the Murik, *Cultural Anthropology* 1: 6–73

Meggitt, Mervyn. 1964. Male–female relationships in the highlands of Australian New Guinea, *American Anthropologist* 66: 204–24

1965. *The Lineage System of the Mae-Enga of New Guinea*, London, Oliver and Boyd

1973. The pattern of leadership among the Mae-Enga of New Guinea. In R.M. Berndt and P. Lawrence (eds.), *Politics in New Guinea*, Seattle, University of Washington Press, pp. 191–206

1974. Pigs are our hearts! *Oceania* 54: 165–203

1977. *Blood Is Their Argument*, Palo Alto, Mayfield

Meigs, Anna. 1976. Male pregnancy and the reduction of sexual opposition in the New Guinea highlands, *Ethnology* 9: 393–407

1984. *Food, Sex and Pollution: A New Guinea Religion*, New Brunswick, N.J., Rutgers University Press

Meillassoux, Claude. 1972. From reproduction to production, *Economy and Society* 1: 83–105

1981. *Maidens, Meal and Money: Capitalism and the Domestic Community*, Cambridge, Cambridge University Press

Meiser, Leo. 1955. The platform phenomenon along the northern coast of New Guinea, *Anthropos* 50: 265–72

Merry, Sally E. 1988. Legal pluralism, *Law and Society Review* 22: 869–96

1992. Anthropology, law and transnational processes, *Annual Review of Anthropology* 21: 357–79

Metraux, Rhoda. 1978. Aristocracy and meritocracy: leadership among the eastern Iatmul, *Anthropological Quarterly* 51 (1): 47–60

Mitchell, William. 1992. *Clowning as Critical Practice: Performance Humor in the South Pacific*, Pittsburgh, University of Pittsburgh Press

Modjeska, Nicholas. 1982. Production and inequality: perspectives from central New Guinea. In A. Strathern (ed.), *Inequality in New Guinea Highlands Societies*, Cambridge, Cambridge University Press, pp. 50–108

Moore, Sally Falk. 1978a. Comparative studies. In Sally F. Moore, *Law as Process: An Anthropological Approach*, London, Routledge and Kegan Paul, pp. 135–80

1978b. Law and social change: the semi-autonomous social field as an appropriate subject of study. In Sally F. Moore, *Law as Process: An Anthropological Approach*, London, Routledge and Kegan Paul, pp. 54–81

Morauta, Louise. 1987. *Law and Order in a Changing Society*, Political and Social Change Monograph 6, Canberra, Australian National University Press

Morgan, Lewis H. 1877 (1964). *Ancient Society*, ed. Leslie A. White, Cambridge, Mass., Belnap Press of Harvard University

Morgenthaler, F., Florence Weiss and M. Morgenthaler. 1987. *Conversations au bord d'un rive mourant: ethnopsychoanalyze chez les Iatmouls de Papousie/Nouvelle-Guinée*, Geneva, Editions Zoë

Morson, Gary Saul and Caryl Emerson. 1990. *Mikhail Bakhtin: Creation of a Prosaics*, Stanford, Stanford University Press

Mosko, Mark. 1985. *Quadripartite Structures: Categories, Relations and Homologies in Bush Mekeo Culture*, Cambridge, Cambridge University Press

1989. The developmental cycle in public groups, *Man* (n.s.) 24: 470–84

1992. Other messages, other missions; or Sahlins among the Melanesians, *Oceania* 63 (2): 97–113

Mumford, Stan. 1989. *Himalayan Dialogue: Tibetan Lamas and Gurung Shamans in Nepal*, Madison, University of Wisconsin Press

Munn, Nancy. 1973. The spatiotemporal transformation of Gawan canoes, *Journal de la Société des Océanistes* 33: 39–52

1986. *The Fame of Gawa*, Cambridge, Cambridge University Press

Murphy, Robert F. 1959. Social structure and sex antagonism, *Southwestern Journal of Anthropology* 15 (2): 89–98

Myers, Frederick and Donald Brenneis. 1984 (1991). Introduction: language and politics in the Pacific. In Frederick Myers and Donald Brenneis (eds.), *Dangerous Words: Language and Politics in the Pacific*, Prospect Heights, Ill., Waveland, pp. 1–29

Nader, Laura. 1980. *The Disputing Process*, New York, Columbia University Press

Nash, June and Eleanor Leacock. 1977. Ideologies of sex: archetypes and stereotypes, *Annals of the New York Academy of Science* 285: 618–45

Newman, P. 1965. *Knowing the Gururumba*, New York, Holt, Rinehart and Winston

Newton, Douglas. 1987. Shields of the Manambu (East Sepik Province, Papua New Guinea), *Baessler-Archiv* (n.s.): 35: 249–59

Obeyesekere, Gananath. 1981. *Medusa's Hair: An Essay on Personal Symbols and Religious Experience*, Chicago, University of Chicago Press

1990. *The Work of Culture: Symbolic Transformation in Psychoanalysis and Anthropology*, Chicago, University of Chicago Press

1992. *The Apotheosis of Captain Cook: European Mythmaking in the Pacific*, Princeton, Princeton University Press and the Bishop Museum

Ogan, Eugene. 1966. Drinking behavior and race relations, *American Anthropologist* 68: 181–8

O'Hanlon, Michael. 1989. *Reading the Skin: Adornment, Display and Society among the Waghi*, London, British Museum

Oliver, Douglas. 1955. *A Soloman Island Society: Kinship and Leadership among the Siuai of Bougainville*, Cambridge, Mass., Harvard University Press

Ong, Aihwa. 1987. *Spirits of Resistance and Capitalist Discipline: Factory Women in Malaysia*, Albany, State University of New York Press

Oosterwal, G. 1961. *People of the Tor*, Assen, The Netherlands, van Gorcum

Ortner, Sherry. 1974. Is female to male as nature is to culture? In Michelle Z. Rosaldo and Louise Lamphere (eds.), *Woman, Culture and Society*, Stanford, Stanford University Press, pp. 67–88

    1981. Gender and sexuality in hierarchical societies: the case of Polynesia and some comparative implications. In Sherry Ortner and Harriet Whitehead (eds.), *Sexual Meanings*, Cambridge, Cambridge University Press, pp. 359–410

    1984. Theory in anthropology since the sixties, *Comparative Studies in Society and History* 26 (1): 126–66

    1990. Gender hegemonies, *Cultural Critique* 14: 35–80

Ortner, Sherry and Harriet Whitehead. 1981. *Sexual Meanings: The Cultural Construction of Gender and Identity*, Cambridge, Cambridge University Press

Ottenberg, Simon. 1988. Oedipus, gender and social solidarity, *Ethos* 16 (3): 326–52

Paige, Karen E. and Jeffery M. Paige. 1981. *The Politics of Reproductive Ritual*, Berkeley, University of California Press

Parmentier, Richard J. 1987. *The Sacred Remains: Myth, History and Polity in Belau*, Chicago, University of Chicago Press

Paul, Robert. 1989. Psychoanalytic anthropology, *Annual Review of Anthropology* 18: 177–202

Penn, Mischa and David Lipset. 1991. Chambri mysticism and Sepik regional exchange: a critique of Errington and Gewertz, *Oceania* 62: 59–65

Petersen, Glenn. 1982. Ponapean matriliny: production, exchange and the ties that bind, *American Ethnologist* 9: 129–44

    1984. *One Man Cannot Rule a Thousand: Fission in a Ponapean Chiefdom*, Ann Arbor, University of Michigan Press

Pomponio, Alice. 1990. Seagulls don't fly into the bush: cultural identity and the negotiation of development on Mandok Island. In Jocelyn Linnekin and Lin Poyer (eds.), *Cultural Identity and Ethnicity in the Pacific*, Honolulu, University of Hawaii Press, pp. 43–71

    1992. *Seagulls Don't Fly into the Bush: Cultural Identity and Development in Melanesia*, Belmont, Calif., Wadsworth

Poole, Fitz J.P. 1982. The ritual forging of identity: aspects of person and self in Bimin-Kuskusmin male initiation. In Gilbert H. Herdt (ed.), *Rituals of Manhood*, Berkeley, University of California Press, pp. 99–155

Popper, Karl. 1966. *The Open Society and Its Enemies* (vol. II), London, Routledge and Kegan Paul

Potter, Michelle. 1973. *Traditional Law in Papua New Guinea*, Canberra, Australian National University Press

Powell, H. 1960. Competitive leadership in Trobriand political organization, *Journal of the Royal Anthropological Institute* 90: 118–48

Quinn, Naomi. 1977. Anthropological studies of women's status, *Annual Review of Anthropology* 6: 181–225

Radcliffe-Brown, A.R. 1922. *The Andaman Islanders*, Cambridge, Cambridge University Press

1933. Primitive law. In *Encyclopaedia of the Social Sciences* (vol. 9), New York, Macmillan and the Free Press, pp. 531–4

1951. The comparative method in social anthropology, *Journal of the Royal Anthropological Institute* 81: 15–22

1965a. *Structure and Function in Primitive Society*, New York, The Free Press

1965b. On joking relationships. Reprinted in *Structure and Function in Primitive Society*, New York, The Free Press, pp. 90–104

1965c. A further note on joking relationships. Reprinted in *Structure and Function in Primitive Society*, New York, The Free Press, pp. 105–16

Rapp, Rayna. 1979. Review essay: anthropology, *Signs* 4: 497–513

Rappaport, Roy. 1968. *Pigs for the Ancestors: Ritual in the Ecology of a New Guinea Tribe*, New Haven, Yale University Press

Read, K.E. 1952. Nama cult of the central highlands, *Oceania* 23 (1): 1–25

1955. Morality and the concept of the person among the Gahuku-Gama, *Oceania* 25: 233–82

1959. Leadership and consensus in a New Guinea society, *American Anthropologist* 61: 425–36

Reay, Marie. 1959. Two kinds of ritual conflict, *Oceania* 29: 290–6

Recher, Harry and P. Hutchings. 1980. The waterlogged forest, *Australian Natural History* 20: 87–96

Reik, Theodor. 1964. The puberty rites of savages. In *Ritual: Four Psychoanalytic Studies*, New York, Grove Press, pp. 96–116

Reiter, Rayna R. 1975. Men and women in the south of France: public and private domains. In Rayna R. Reiter (ed.), *Toward an Anthropology of Women*, New York, Monthly Review Press, pp. 252–82

Riesman, Paul. 1986. The person and the life cycle in African social life and thought, *African Studies Review* 29(2): 71–138

Roberts, Simon. 1979. *Order and Dispute: An Introduction to Legal Anthropology*, New York, St. Martin's Press

Rodman M. and M. Cooper. 1979. *The Pacification of the Pacific*, Ann Arbor, University of Michigan Press

Rogers, Susan Carol. 1975. Female forms of power and the myth of male dominance: a model of female/male interaction in peasant society, *American Ethnologist* 2: 727–56

Roheim, Geza. 1942. Transition rites, *Psychoanalytic Quarterly* 11: 336–74

1943 (1971). *The Origin and Function of Culture*, Garden City, N.J., Doubleday Anchor

Rosaldo, Michelle. 1974. Women, culture and society: a theoretical overview. In Michelle Z. Rosaldo and Louise Lamphere (eds.), *Woman, Culture and Society*, Stanford, Stanford University Press, pp. 17–42

1980. *Knowledge and Passion: Ilongot Notions of Self and Social Life*, Cambridge, Cambridge University Press

Roscoe, Paul B. 1980. Male initiation among the Yangaru Boiken. In N. Lutkehaus et al. (eds.), *Sepik Heritage*, Durham, N.C., Carolina Academic Press, pp. 402–17

1994a. Who are the Ndu? Ecology, migration and linguistic and cultural change in the Sepik Basin. In A.J. Strathern and G. Sturzenhofecker (eds.), *Migrations and Transformations: Regional Perspectives in New Guinea*, Pittsburgh, University of Pittsburgh Press, pp. 49–84

1994b. Amity and aggression: a symbolic theory of incest, *Man* (n.s.) 29(1): 49–76

Roseberry, William. 1982. Balinese cockfights and the seduction of anthropology, *Social Research* 49: 1013–28

Rosen, Lawrence. 1984. *Bargaining for Reality: The Construction of Social Relations in a Muslim Community*, Chicago, University of Chicago Press

Rubel, Paula G. and Abraham Rosman. 1978. *Your Own Pigs You May Not Eat*, Chicago, University of Chicago Press.

Rumelhardt, David R. 1978. Schemata: the building blocks of cognition. In R. Spiro, B. Bruce and W. Brewer (eds.), *Theoretical Issues in Reading Comprehension*, Hillsdale, N.J., Erlbaum Association, pp. 46–67

Sacks, Karen. 1975. Engels revisited: women, the organization of production and private property. In Rayna Rapp Reiter (ed.), *Toward an Anthropology of Women*, New York, Monthly Review Press, pp. 211–34

Sahlins, Marshall. 1960 (1970). Production, distribution and power in a primitive society. Reprinted in Thomas Harding and Ben Wallace (eds.), *Cultures of the Pacific*, New York, The Free Press, pp. 78–84

1963. Rich man, poor man, big man, chief: political types in Melanesia and Polynesia, *Comparative Studies in Society and History* 5: 285–303

1968. *Tribesmen*, Englewood Cliffs, N.J., Prentice-Hall

1972. *Stone Age Economics*, Chicago, Aldine-Atherton

1981. *Historical Metaphors and Mythical Realities*, Ann Arbor, University of Michigan Press

1985. *Islands of History*, Chicago, University of Chicago Press

1995. *How Natives Think: About Captain Cook, for Example*, Chicago, University of Chicago Press

Salisbury, R.F. 1964. Despotism and Australian administration in New Guinea highlands, *American Anthropologist* 66: 225–38

1965. The Siane of the eastern highlands. In P. Lawrence and M. Meggitt (eds.), *Gods, Ghosts and Men in Melanesia*, Melbourne, Oxford University Press, pp. 50–77

Scaglion, Richard. 1979. Formal and informal operations in a village court in Maprik, *Melanesian Law Journal* 7: 116–29

1981. Samakundi Abelam conflict management: implications for legal planning in Papua New Guinea, *Oceania* 52: 463–86

1985. Kiaps as kings: Abelam legal change in historical perspective. In D. Gewertz and E.L. Schieffelin (eds.), *History and Ethnohistory in Papua New Guinea. Oceania* Monograph 28, Sydney, Oceania Publications, pp. 77–99

1987. Customary law and legal development in Papua New Guinea, *The Journal of Anthropology* 6 (1)

1990. Legal adaptation in a Papua New Guinea village court, *Ethnology* 29: 17–33

Scaglion, Richard and R. Whittingham. 1985. Female plaintifs and sex-related disputes in rural Papua New Guinea. In S. Toft (ed.), *Domestic Violence in Papua*

New Guinea, Port Moresby, Law Reform Commission of Papua New Guinea, pp. 120–33

Schieffin, Edward L. 1976. *The Sorrow of the Lonely and the Burning of the Dancers*, New York, St. Martin's Press

1982. The *Bau A* ceremonial hunting lodge: an alternative to initiation. In G. Herdt (ed.), *Rituals of Manhood*, Berkeley, University of California Press, pp. 155–200

Schindlbeck, Marcus. 1980. *Sago bei den Sawos (Mittelsepik, Papua New Guinea): Untersuchungen über die Bedeutung von Sago in Wirtschaft, Sozialordnung und Religion*, Basle: Basler Beitrage zur Ethnologie 19

Schmidt, Joseph. 1922–3. Die Ethnographie der Nor-Papua (Murik–Kaup–Karau) bei Dallmanhafen, Neu-Guinea, trans. K. Barlow, *Anthropos* 18–19: 700–32

1926. Die Ethnographie der Nor-Papua (Murik–Kaup–Karau) bei Dallmanhafen, Neu-Guinea, trans. K. Barlow, *Anthropos* 21: 38–71

1933. Neue Beitrage zur Ethnographie der Nor-Papua (Neuguinea), trans. K. Barlow, *Anthropos* 28: 321–54, 663–82

Schneider, David M. 1968. *American Kinship: A Cultural Account*, Englewood Cliffs, N.J., Prentice-Hall

1972. What is kinship all about? In Priscilla Reining (ed.), *Kinship Studies in the Morgan Centennial Year*, Washington, D.C., Anthropological Society of Washington, pp. 32–63

1984. *A Critique of the Study of Kinship*, Ann Arbor, University of Michigan Press

Schwartz, Ted. 1963. Systems of areal integration, *Anthropological Forum* 2: 56–97

1973. Cult and context: the paranoid ethos in Melanesia, *Ethos* 1: 153–74

Schwimmer, Eric. 1973. *Exchange in the Social Structure of the Orokaiva*, London, Hurst

Serematakis, C. Nadia. 1991. *The Last Word: Women, Death and Divination in Inner Mani*, Chicago, University of Chicago Press

Serpenti, L.M. 1965. *Cultivators in the Swamps: Social Structure and Horticulture in a New Guinea Society*, Assen, Van Gorkum

1984. The ritual meaning of homosexuality and pedophilia among the Kimam-Papuans of South Irian Jaya. In Gilbert H. Herdt (ed.), *Ritualized Homosexuality in Melanesia*, Berkeley, University of California Press, pp. 292–336

Service, Elman. 1962. *Primitive Social Organization: An Evolutionary Perspective*, New York, Random House

Sharman, A. 1969. "Joking" in Padhola: categorical relationships, choice and social control, *Man* (n.s.) 4: 103–17

Shils, Edward. 1981. *Tradition*, Chicago, University of Chicago Press

Shook, Victoria E. 1985 (1989). *Ho'oponopono: Contemporary Uses of a Hawaiian Problem Solving Process*, Honolulu, East–West Center, Institute of Culture and Communication

Shore, Brad. 1981. Sexuality and gender in Samoa: conceptions and missed conceptions, In Sherry Ortner and Harriet Whitehead (eds.), *Sexual Meanings*, Cambridge, Cambridge University Press, pp. 192–215

1989. *Manu* and *tapu*. In A. Howard and R. Borofsky (eds.), *Developments in Polynesian Ethnography*, Honolulu, University of Hawaii Press, pp. 137–73

Sillitoe, Paul. 1978. Big men and war in New Guinea, *Man* (n.s.) 13: 252–71

1979. *Give and Take: Exchange in Wola Society*, New York, St. Martin's Press

Silverman, Eric K. 1993. *Tambunam: New Perspectives on Eastern Iatmul (Sepik River, Papua New Guinea) Kinship, Marriage and Society*, Ann Arbor, University Microfilms

Simmel, Georg. 1955. *Conflict and the Web of Group Affiliation*, trans. R. Bendix, New York, The Free Press

1964. Adornment. In K.H. Wolf (ed.), *The Sociology of Georg Simmel*, New York, The Free Press, pp. 338–44

Simmons, Dave. 1983. Moko. In S.M. Mead and B. Kernot (eds.), *Art and Artists of Oceania*, Palmerston, NZ, Dumore Press, pp. 226–43

Singer, P. and Daniel E. DeSole. 1967. The Australian subincision ceremony reconsidered: vaginal envy or kangaroo bifid penis envy? *American Anthropologist* 69: 355–8

Smith, Michael French. 1985. White man, rich man, bureaucrat, priest: hierarchy, inequality and legitimacy in a changing Papua New Guinea village, *South Pacific Forum* 2 (1): 1–24.

1994. *Hard Times on Kairiru Island: Poverty, Development and Morality in a Papua New Guinea Village*, Honolulu, University of Hawaii Press

Somare, Michael. 1975. *Sana: An Autobiography of Michael Somare*, Hong Kong, Niugini Press

Spencer, Baldwin and F.J. Gillen. 1899. *The Native Tribes of Central Australia*, London, Macmillan

Spiro, Melford E. 1965. Religious systems as culturally constituted defense mechanisms. In *Context and Meaning in Cultural Anthropology*, New York, The Free Press, pp. 100–13

1982. *Oedipus in the Trobriands*, Chicago, University of Chicago Press

Spradley, James P. 1969 (1978). *Guests Never Leave Hungry: The Autobiography of James Sewid, a Kwakiutl Indian*, Montreal, McGill-Queen's University Press

Stanek, Milan. 1990. Social structure of the Iatmul. In N. Lutkehaus et al. (eds.), *Sepik Heritage: Tradition and Change in Papua New Guinea*, Durham, Carolina Academic Press, pp. 266–73

Starr, June and Jane F. Collier. 1989. *History and Power in the Study of Law: New Directions in Legal Anthropology*, Ithaca, Cornell University Press

Steiner, Christopher. 1990. Body personal and body politic: adornment and leadership in cross-cultural perspective, *Anthropos* 85: 431–45

Stephens, William. 1962. *The Oedipus Complex: Cross-Cultural Evidence*, New York, The Free Press

Strathern, Andrew J. 1966. Despots and directors in the New Guinea highlands, *Man* (n.s.) 1: 356–67

1970. Male initiation in the New Guinea highlands societies, *Ethnology* 9 (4): 373–9

1971 (1979). *The Rope of Moka: Big-Men and Ceremonial Exchange in Mount Hagen*, Cambridge, Cambridge University Press

1972. *One Father, One Blood: Descent and Group Structure among the Melpa People*, London, Tavistock

1974. When dispute procedures fail. In A.L. Epstein (ed.), *Contention and Dispute*, Canberra, Australian National University Press, pp. 240–70

1979. *Ongka*, London, Gerald Duckworth

Strathern, A. and M. Strathern. 1971. *Self-Decoration in Mount-Hagen*, London, Duckworth

Strathern, Marilyn. 1980. No nature, no culture: the Hagen case. In C. MacCormack and M. Strathern (eds.), *Nature, Culture and Gender*, Cambridge, Cambridge University Press, pp. 174–222

1981. Culture in a netbag: the manufacture of a subdiscipline in anthropology, *Man* (n.s.) 4: 665–88

1985a. Kinship and economy: constitutive orders of a provisional kind, *AmericanEthnologist*: 12 (2): 191–209

1985b. Discovering "social control," *Journal of Law and Society* 12 (2): 111–34

1987. Conclusion. In M. Strathern (ed.), *Dealing with Inequality*, Cambridge, Cambridge University Press, pp. 278–302

1988a *The Gender of the Gift*, Berkeley, University of California Press

1988b. Self-regulation: an interpretation of Peter Lawrence's writing on social control in Papua New Guinea, *Oceania* 59 (1): 3–6

Swadling, Pamela. 1977. Introduction. In *Myths of Samap by J. Gehberger*, Port Moresby, Institute for Papua New Guinea Studies, pp. 1–13

1989. Research report: a late Quaternary inland sea and early pottery in Papua New Guinea, *Archaeology of Oceania* 24: 106–9

1990. Sepik prehistory. In Nancy Lutkehaus et al. (eds.), *Sepik Heritage*, Durham, Carolina Academic Press, pp. 71–86

Swadling, Pamela, Brigitta Hauser Schaublin, Paul Gorecki and Frank Tiesler. 1988. *The Sepik-Ramu: An Introduction*, Boroko, National Museum of Papua New Guinea

Swadling, Pamela and Geoff Hope. 1992. Environmental change in New Guinea since human settlement. In John Dodson (ed.), *The Native Lands: Prehistory and Change in Australia and the Southwest Pacific*, Melbourne, Longman Cheshire, pp. 13–42

Swartz, Marc J., Victor Turner and Arthur Tuden. 1966. Introduction. In Marc J. Swartz, Victor W. Turner and Arthur Tuden (eds.), *Political Anthropology*, Chicago, Aldine, pp. 1–42

Swingewood, Alan. 1987. *Sociological Poetics and Aesthetic Theory*, New York, St. Martin's Press

Tamoane, Matthew. 1977. Kamoai of Darapap and the legend of Jari. In G. Trompf (ed.), *Prophets of Melanesia*, Port Moresby, Institute of Papua New Guinea Studies, pp. 174–211

Taussig, Michael. 1992. *The Nervous System*, New York, Routledge

Tedlock, Dennis. 1983. The analogical tradition and the emergence of a dialogical anthropology. In *The Spoken Word and the Work of Interpretation*, Philadelphia, University of Pennsylvania Press, pp. 302–11

Tedlock, Dennis and Bruce Mannheim (eds.). 1995. *The Dialogic Emergence of Culture*, Urbana, University of Illinois Press

Thomas, Nicholas. 1989. The force of ethnology: origins and significance of the Melanesia/Polynesia division, *Current Anthropology* 30 (1): 27–42

Thurnwald, Richard. 1916. Banaro society: social organization and the kinship system of a tribe in the interior of New Guinea, *American Anthropological Association Memoirs* 3: 253–391

1936 (1970). Adventures of a tribe in New Guinea (the Tjimundo). In Raymond Firth, Bronislaw Malinowski and Isaac Shapera (eds.), *Essays Presented to C.G. Seligman*, Westport, Conn., Negro Universities Press, pp. 345–60

Tiesler, Frank. 1969/70. *Die intertribalen Beziehungen an der Nordkust Neuguineas in Gebiet der kleiner Schouten-Inseln*, trans. K. Barlow, Abhandlungen und Berichtet des Statlichen Museums fer Volkerkunde, Dresden, Akademie Verlag

Tiffany, Sharon W. and Kathleen J. Adams. 1985. *The Wild Woman: An Inquiry into the Anthropology of an Idea*, Cambridge, Mass., Schenkman

Todorov, Tzvetan. 1984 (1989). *Mikhail Bakhtin: The Dialogical Principle*, trans. Wlad Godzich, Minneapolis, University of Minnesota Press

Trautmann, Thomas R. 1987. *Louis Henry Morgan and the Invention of Kinship*, Berkeley, University of California Press

Trawick, Margaret. 1988. Spirits and voices in Tamil song, *American Ethnologist* 15: 193–215

Tsing, Anna L. 1993. *In the Realm of the Diamond Queen: Marginality in an Out of the Way Place*, Princeton, Princeton University Press

Turner, Bryan S. 1991. Recent developments in the theory of the body. In Mike Featherstone, Mike Hepworth and Bryan S. Turner (eds.), *The Body: Social Process and Cultural Theory*, London, Sage, pp. 1–35

Turner, Terrance. 1980. The social skin. In Jeremy Cherfas and Roger Lewis (eds.), *Not by Work Alone*, London, Temple Smith, pp. 1–35

Turner, Victor W. 1957. *Schism and Continuity in an African Society: A Study of Ndembu Village Life*, Manchester, Manchester University Press

1967a. Betwixt and between: the liminal period in *rites de passage*. Reprinted in *The Forest of Symbols*, Ithaca, Cornell University Press, pp. 93–111

1967b. *The Forest of Symbols*, Ithaca, Cornell University Press

1968. *The Drums of Affliction*. London, Oxford University Press

Tuzin, Donald F. 1972. Yam symbolism in the Sepik: an interpretive account, *Southwestern Journal of Anthropology* 28 (3): 230–53

1974. Social control and the *tambaran* in the Sepik. In A.L. Epstein (ed.), *Contention and Dispute*, Canberra, Australian National University Press, pp. 317–44

1976. *The Ilahita Arapesh: Dimensions of Unity*, Berkeley, University of California Press

1977. Reflections of being in Arapesh water symbolism, *Ethos* 5 (2): 195–223

1980. *The Voice of the Tambaran: Truth and Illusion in Ilahita Arapesh Religion*, Berkeley, University of California Press

1984. Miraculous voices: the auditory experience of numinous objects, *Current Anthropology* 25: 579–96

Valeri, Valerio. 1985. *Kingship and Sacrifice: Ritual and Society in Ancient Hawaii*, Chicago, University of Chicago Press

Wagner, Roy. 1967. *The Curse of Souw*, Chicago, University of Chicago Press

1972. *Habu: The Innovation of Meaning in Daribi Religion*, Chicago, University of Chicago Press

Wassman, Jurg. 1990. Nyaura concepts of space and time. In Nancy Lutkehaus et al. (eds.), *Sepik Heritage*, Durham, Carolina Academic Press, pp. 23–35

1991. *Song to the Flying Fox: The Public and Esoteric Knowledge of the Important Men in Kandingei about Totemic Songs, Names and Knotted Chords (Middle Sepik)*, Boroko, Papua New Guinea, National Research Institute

Watson, J.B. 1973. Tairora: the politics of despotism in a small society. In R.M. Berndt and R. Lawrence (eds.), *Politics in New Guinea*, Seattle, University of Washington Press, pp. 224–75

Watson-Gegeo, Karen A. and Geoffrey White. 1990. *Disentangling: Conflict Discourse in Pacific Societies*, Stanford, Stanford University Press

Weber, Max. 1949. *The Methodology of the Social Sciences*, trans. Edward A. Shils and Henry A. Finch, New York, The Free Press

Wedgwood, Camilla. 1934. Report on research in Manam Island, Mandated Territory of New Guinea, *Oceania* 4: 373–403

1959. Manam kinship, *Oceania* 29: 239–56

Weiss, Florence. 1990. The child's role in the economy of Palimbei. In N. Lutkehaus et al. (eds.), *Sepik Heritage*, Durham, Carolina Academic Press, pp. 337–42

White, Geoffrey and Karen Watson-Gegeo. 1990. Disentangling discourse. In *Disentangling: Conflict Discourse in Pacific Societies*, Stanford, Stanford University Press, pp. 3–52

Whitehead, Harriet. 1986. The varieties of fertility cultism in New Guinea, *American Ethnologist* 13: 80–99, 271–89

1987. Fertility and exchange in New Guinea. In Jane F. Collier and Sylvia J. Yanagisako (eds.), *Gender and Kinship: Essays toward a Unified Analysis*, Stanford, Stanford University Press, pp. 244–67

Whiting, John W.M. 1941. *Becoming a Kwoma*, New Haven, Yale University Press

Whiting, John W.M., R. Kluckhohn and A. Anthony. 1958. The function of male initiation ceremonies at puberty. In E.E. Maccoby, T.M. Newcomb and E.L. Hartley (eds.), *Readings in Social Psychology*, New York, Holt, Rinehart and Winston, pp. 359–70

Whiting, John W.M. and Stephen Reed. 1939. Kwoma culture: report on field work in the Mandated Territory of New Guinea, *Oceania* 9: 170–216

Whittaker, J.L., N.G. Nash, J.F. Hookey and R.J. Lacey. 1975. *Documents and Readings in New Guinea History: Prehistory to 1889*, Hong Kong, Jacaranda Press

Wiener, Annette B. 1976. *Women of Value, Men of Renown: New Perspectives on Trobriand Exchange*, Austin, University of Texas Press

1978. The reproductive model in Trobriand society, *Mankind* 11: 175–86

1979. Trobriand kinship from another point of view: the reproductive power of women and men, *Man* (n.s.) 14: 328–48

1980. Reproduction: a replacement for reciprocity, *American Ethnologist* 7: 71–85

1983. "A world of made is not a world of born": doing kula on Kiriwina. In *The Kula: New Perspectives on Massim Exchange*, Cambridge, Cambridge University Press, pp. 147–71

Wiener, James F. 1982. Substance, siblingship and exchange: aspects of social structure in New Guinea, *Social Analysis* 11: 3–35

1988a. *The Heart of the Pearlshell: The Mythological Dimension of Foi Sociality*, Berkeley, University of California Press

1988b. Durkheim and the Papuan male cult: Whitehead's views on social structure and ritual in New Guinea, *American Ethnologist* 15: 567–73

Williams, F.E. 1936. *Papuans of the Trans-Fly*, Oxford, Clarendon Press

Williamson, Margaret Holmes. 1983. Sex relations and gender relations: understanding Kwoma conception, *Mankind* 14: 13–23

Wood, Mike. 1987. Brideservice societies and the Kamula, *Canberra Anthropology* 10: 1–23

Yanagisako, Sylvia J. 1979. Family and household: the analysis of domestic groups, *Annual Review of Anthropology* 8: 161–205

Young, Michael. 1971. *Fighting with Food: Leadership, Values and Social Control in a Massim Society*, Cambridge, Cambridge University Press

1985. Abutu in Kalauna: a retrospect, *Mankind* 15 (2): 198–202

Zorn, Jean G. 1990. Customary law in Papua New Guinea village courts, *The Contemporary Pacific, A Journal of Island Affairs* 2 (2): 279–312

Index

Tangu, 76
taxes, 50
Tchambuli, 31
Tedlock, Dennis, 5
Tiesler, Frank, 22–3, 27, 47, 290 n. 2
Tiffany, Sharon, 144
Todorov, Tzvetan, 6
tourism
  artifacts for, 41, 242
  World Discoverer, 241–2
trade, 22–3, 42, 53, 82, 278
  gender and, 44–9, 290 n. 8
  hosting visits, 160
  impact of capitalism on, 52
  inland sago partners ("Bush Murik"), 1,
    29–31, 32, 44, 45–6, 48–9, 111, 267
  military alliances and, 29
  legends involving, 26–8
  overseas partners, 10, 31, 41, 46–9, 90,
    111, 118, 127–31, 156, 158, 222–3, 228,
    248–9, 253, 274
  participation in rites, 111–15
  vulnerability to feminine sexuality, 40
Trobriand Islands, 10, 169
Tsing, Anna, 9
Turner, Victor, 260
Turubu village, 26
Tuzin, Donald, 39, 56, 168
Two Brothers, 25–8, 31–2, 73–4, 81, 182–7,
    192–4, 200, 204, 231–2, 267, 284–5

urban settlements
  markets, 44
  participation in village life, 52
  remittances from, 51, 84
  in Wewak, 50, 109, 243, 246
uterine fertility, 39, 80, 96
  agency of, 42, 107, 178, 263
  dependence upon, 88, 98, 169
  dialogue with maternal schema, 109–11,
    274
  exclusion from succession rite, 85
  external form of, 54–6

feminine sexuality and, 29–32, 46, 41,
    76
  masculine fear of, 61
  relation to maternal body, 59, 92, 274
  vulnerability to, 37
  see also feminine sexuality; maternal
    body; maternal schema

Valeri, Valerio, 79, 100
Van Baal, J., 96

wage labor, 44, 49–50, 123, 127
Wagner, Roy, 274
wajak, see heraldic double; insignia-holder
Walis Island, 31
warfare, 28–31, 177, 234
  metaphor of, 198–200
  postcolonial state and, 264
  trade and, 46
  war spirits, *see* spirit-men
Washing Feast (*Arabopera Gar*),109, 119,
    162, 258, 260, 277
  as birth, 168
  preparation for, 111, 114–16, 125–6
  see also death; mortuary rites
Wassman, Jurg, 61
Watson-Gegeo, Karen, 223
Wau, 50
Weber, Max, 96
Wedgewood, Camilla, 86, 174
We Island, 111, 116
Wewak, 50, 127, 226, 243, 246, 269
White, Geoffrey, 223
Whitehead, Harriet, 8–9, 209
Wiener, Annette, 10, 24
Wiener, James, 209
Williams, F.E., 96
Wogeo Island, 7, 80
world system, 75
World War II, 50

Yanagisako, Sylvia, 8
Young, Michael, 150

Cambridge Studies in Social and Cultural Anthropology

*available in paperback